AF885429

SOTHIS PRESS

Sothis Press aims to completely reimagine the public domain book with an emphasis on universal access to captivatingly designed print editions. We believe that technology will enable every single book ever published to be perpetually accessible in beautiful print form. Sothis is beauty in ubiquity.

www.sothis.press
1 Waterhouse Square, London, ECIN 2ST

Applied Mechanics for Beginners

J. Duncan

SOTHIS PRESS

ISBN 978-1-0365-1462-4

Applied Mechanics for Beginners
is in the public domain.

Mechanics' Laboratory, West Ham Technical Institute.

APPLIED MECHANICS FOR BEGINNERS

BY

J. DUNCAN, WH.EX., A.M.I.M.E.

HEAD OF THE DEPARTMENT OF CIVIL AND MECHANICAL ENGINEERING AT THE MUNICIPAL TECHNICAL INSTITUTE, WEST HAM

London
MACMILLAN AND CO., LIMITED
NEW YORK: THE MACMILLAN COMPANY
1902

GLASGOW: PRINTED AT THE UNIVERSITY PRESS
BY ROBERT MACLEHOSE AND CO.

PREFACE.

In the preparation of this little book, the object has been to provide students of engineering and allied constructive arts with a practical statement of the principles of Mechanics essential to an intelligent interest in their occupations. While the book will serve as a sufficient preparation for the elementary examination in Applied Mechanics of the Board of Education, its contents are not limited by the syllabus of this or any other examining body.

Before the student can profitably take up the study of Mechanics, he should be acquainted with the elementary portions of Practical Geometry, Machine or Building Construction, and Practical Mathematics. This preliminary knowledge has therefore been taken for granted.

A constant endeavour has been made to avoid producing a mere collection of rules and formulae, sufficient explanations being given to ensure for a careful reader a systematic knowledge of the principles discussed. To preserve a constant connection between theory and practice, numerous worked out examples of problems which present themselves in everyday work are scattered throughout the chapters. Additional exercises, suitable for home or class work, will be found at the end of each chapter; those with a date are from examination papers of the Board of Education, South Kensington.

Importance should be attached to the performance by the student of typical experiments. Descriptions of suitable forms of apparatus are given, and practical exercises to be worked out with them are suggested. These experiments have been arranged in the form of a Laboratory Course, the subjects of which are brought together at the end of the book.

The opportunity is here taken to make several grateful acknowledgments: to Prof. Rowden, of Glasgow, from whom I received my first lessons in Mechanics, and to whom many of the methods, used in developing the principles elucidated, are due; to Messrs. A. Walker and R. Macmillan, Demonstrators in Engineering in West Ham Technical Institute, for kindly reading the proofs; and to Prof. R. A. Gregory and Mr. A. T. Simmons for useful hints and guidance in preparing the book.

Thanks must also be given to the following firms for permission to reproduce certain illustrations from their copyright lists: Messrs. Tangyes, Ltd., for Figs. 258, 310, 311, 312, 316, 318; Messrs. Buck & Hickman, for Fig. 9; Messrs. Charles Churchill & Co., for Fig. 6; Messrs. Hoffman Mfg. Co. Ltd., for Figs. 193, 228; Messrs. Ludw. Loewe & Co., for Figs. 2, 3, 4, 5, 10; and Messrs. The Empire Roller Bearing Co. Ltd., for Fig. 194.

The Tables of Logarithms and Trigonometrical Ratios are reprinted from Mr. F. Castle's *Practical Mathematics for Beginners* (Macmillan).

It is almost too much to hope that a book containing so many exercises will be found entirely free from numerical errors, and I shall be very grateful if any such are pointed out to me.

J. DUNCAN.

WEST HAM,
July, 1902.

CONTENTS

CHAPTER XVI.

CHAPTER XVII.

CHAPTER XVIII.

CHAPTER XIX.

APPLIED MECHANICS FOR BEGINNERS.

CHAPTER I.

INTRODUCTORY.

MEASURING INSTRUMENTS.

Straight Edge.—The engineer's *straight edge* consists of a long strip of metal with one edge bevelled, this edge being such that a straight line drawn from two points situated near the ends of the edge will lie wholly in the edge. If a straight edge has to be originated, it is necessary to make three at the same time, then by a continual process of comparing one with the others and removing the faulty parts by scraping, the edges of all three may be brought nearly true.

Surface Plate.—The *surface plate* consists of a rigid plate of cast iron, having three feet on its under side, in order always to distribute the supporting forces in the same manner and so prevent the plate warping. The upper surface of the plate is brought, as far as possible, all to lie in one plane. This is done by constructing three plates at a time. The upper surfaces of the plates having been carefully planed, are compared one with the others. A little oil and colouring matter rubbed

FIG. 1.—Surface plate.

on the surface show those spots which are in error, and these are removed by scraping. The operation is finished when any two of the three, on being brought together, show contact over a large number of bearing spots evenly distributed. These spots will then all lie in one plane. Any one of the plates may now be used for the reproduction of other plane surfaces.

External and Internal Gauges.—Sir Joseph Whitworth, by constructing bars having plane ends perpendicular to their axes, subdivided the standard yard into inches, etc. Engineers' steel rules, subdivided with considerable accuracy into inches, tenths, etc., are one of the results of his work, and can be used for producing objects having required dimensions. For standards of reference in the shops, *cylindrical external and internal gauges* are used. These are shown in Fig. 2, and consist of a collar with a hole, and a plug. Both hole and plug are brought nearly to size by machining, then hardened, ground and lapped down to size by hand. Gauges such as this are turned out true to $\frac{1}{10000}$th inch.

FIG. 2.—Standard cylindrical gauges.

Standard gauges are not suitable for the reproduction of dimensions to a given degree of accuracy, as **calipers** have to be used in transferring the dimension from the gauge to the work. The calipers are adjusted as nearly as the workman can tell by touch to the standard gauge and then are applied to the work. In this process too much is left to the skill and discretion of the workman, and no limits of accuracy can be stated and worked to. To secure good results, **limit gauges** are necessary. Two of these are shown in Figs. 3 and 4. The **internal** limit gauge is for measuring inside cylindrical holes. One end is made slightly larger in diameter than the other, the difference being determined by the limits of accuracy

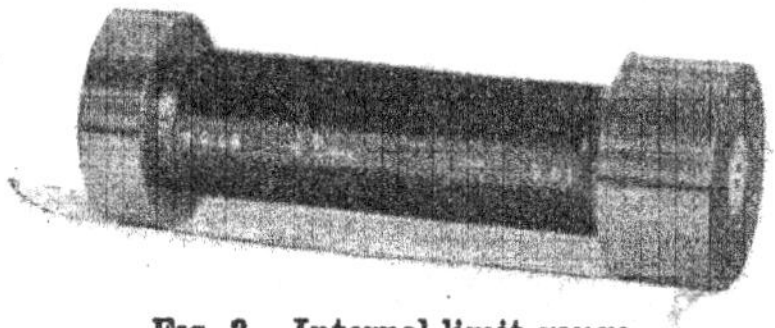

FIG. 3.—Internal limit gauge.

required in the work under execution. The smaller end must "go in" to the finished hole, and the larger end must "not go in." Consequently, the finished hole is larger in diameter than the small end of the gauge and smaller in diameter than the large end, and so the work is kept within the desired limits of accuracy. For example, a hole to be 1 inch approximately in diameter would be bored, using an internal limit gauge having diameters 1·006″ and 0·994″ respectively. The total variation in the diameter of the finished hole cannot exceed 0·0012″.

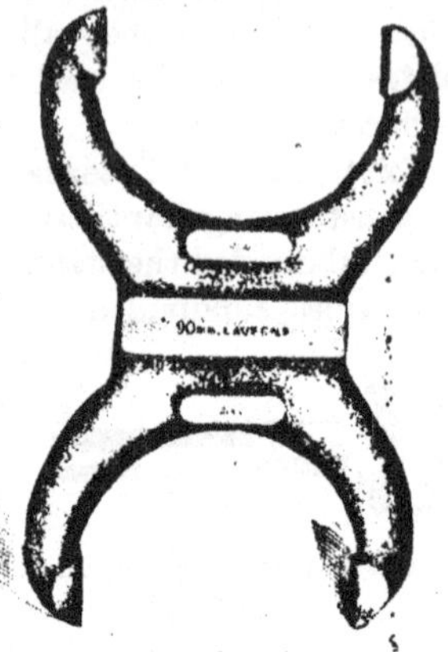

FIG. 4.—External limit gauge.

The **external** limit gauge is used for turning cylindrical pieces down to size, and is used in a similar manner. As nothing is left to the discretion of the workman, interchangeable parts can easily be produced by the use of these gauges.

Standard screw-gauges are also useful for reference, and help to produce accurate work. Two of these, one external and one internal, are shown in Fig. 5.

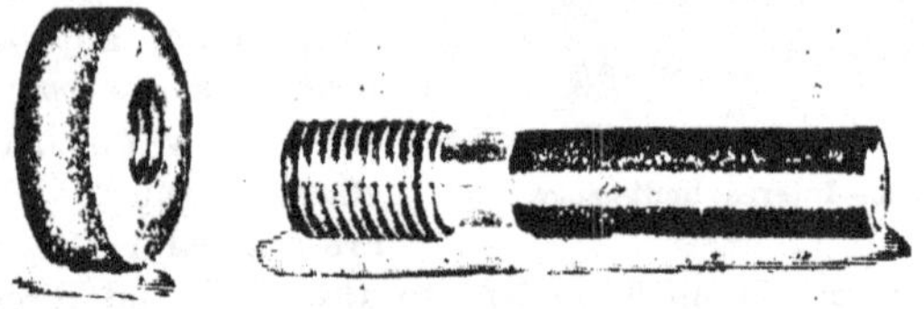

FIG. 5.—Standard screw-gauges.

For the more accurate measurement of dimensions than can be secured by the use of calipers, **micrometers** are used. In Fig. 6 a micrometer is shown having its outer parts shown transparent. The instrument consists of a very finely cut screw, which may be rotated by turning the outer milled thimble. This screw works in a split nut, fitted with an adjusting nut to take up wear, and terminates, as shown, between the jaws of the instrument, in a cylindrical portion having its end brought plane and square to the axis of the screw. The screw is protected from dust and grit by the outer thimble casing. The object to be measured is placed in the jaw of the instrument, and the thimble turned until the object is gently nipped. The dimension is then read from two scales, one engraved along the barrel longitudinally, and the other circularly round the edge of the thimble. In the instrument shown, the screw has 40 threads to an inch, and one inch on the longitudinal scale is divided into tenths of an inch, each tenth being subdivided into four parts; each part will therefore be $\frac{1}{40}''$ or 0·025″ long. One revolution of the thimble will consequently advance the screw one part on the longitudinal scale, or a distance of 0·025″ The circular scale on the thimble has 25 divisions round the complete circumference, consequently revolving the thimble through one division will advance the screw

$$\tfrac{1}{25} \times 0{\cdot}025'' = 0{\cdot}001''.$$

The instrument can therefore be used for taking dimensions to $\frac{1}{1000}$ inch. To read the scales, suppose, as in Fig. 6, that the longitudinal scale shows three parts beyond 0·1″. This will be $0{\cdot}1 + (3 \times 0{\cdot}025) = 0{\cdot}175''$. The circular scale is set at 0 or 25, so that in this position nothing need be allowed for it. The dimension, as set, is therefore 0·175″.

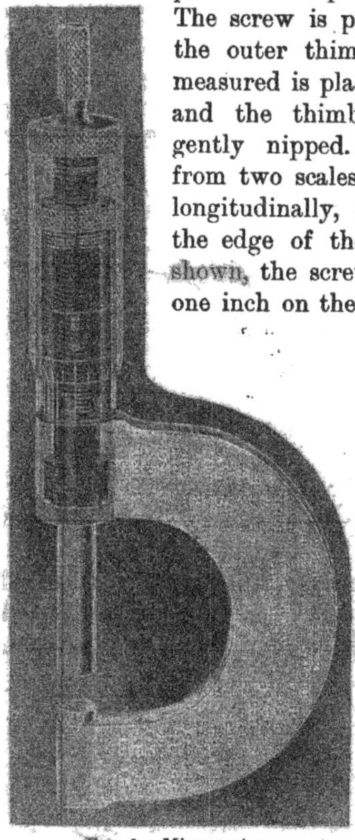

FIG. 6.—Micrometer.

If the circular scale had been beyond the 0·175″ mark on the longitudinal scale, by, say, 14 divisions, then we should have added 0·014″ to the above reading, giving $0{\cdot}175 + 0{\cdot}014 = 0{\cdot}189''$ as the required dimension.

Micrometers should be tested occasionally for zero error by running the screw right home until the points touch, and ascertaining if the 0 or 25 mark on the circular scale comes opposite to the line on the longitudinal scale. If this is not the case, the instrument can be adjusted by the screw point on the opposite jaw, or a *zero error* may be allowed for in subsequent readings.

Verniers.—The *vernier* is a device for subdividing the parts of a scale into divisions that would be too fine to be read by the eye. It consists of a sliding piece fitted to a main scale and having a suitable scale engraved on it. In the case shown in Fig. 7 the vernier scale has 10 divisions of total length equal to 9 divisions on the main scale. Each division on the vernier is therefore $\frac{1}{10}$th shorter than a division on the main scale, so that if set with the arrow opposite a division on the main scale, the next two divisions will be $\frac{1}{10}$th of a division apart, the next pair of divisions $\frac{2}{10}$ths and so on. To read the instrument, note the division on the main scale to the left of the vernier arrow, in the case shown in Fig. 8 this is 0·7; then look along the vernier to find a division on it exactly opposite a division on the main scale and note the vernier division, in the example this is 4; so that the vernier arrow is $\frac{4}{10}$ths of a main scale division beyond the 0·7 mark; the reading is therefore 0·74.

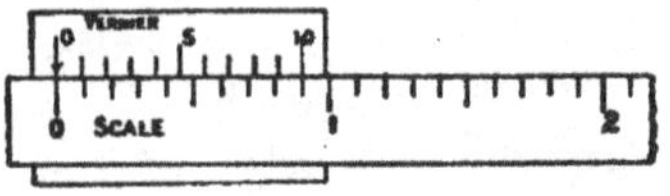

FIG. 7.—Vernier, set at zero.

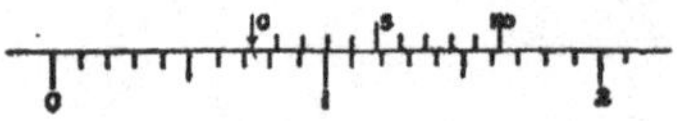
FIG. 8.—Vernier, set at 0·74.

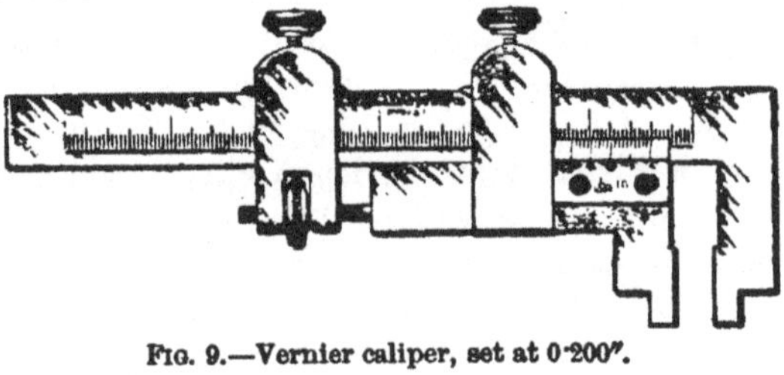
FIG. 9.—Vernier caliper, set at 0·200″.

Calipers for use in the workshop are often fitted with verniers. The one shown (Fig. 9) can read up to about $1\frac{3}{4}''$ by $\frac{1}{1000}$th of an inch. The main scale has inches divided into tenths, and each tenth is subdivided into five parts, each part being therefore $\frac{1}{50}$ inch. The vernier has 20 divisions of total length equal to 19 divisions on the main scale, so that each vernier division is $\frac{1}{20}$th shorter than a main scale division. The vernier therefore reads to $\frac{1}{20} \times \frac{1}{50} = \frac{1}{1000}$ inch. As set in Fig. 9, the instrument reads 0·200″ on the main scale and 0 on the vernier, the dimension is therefore 0·200″. Had the vernier been set say at 11, the reading would be 0·211″. Readings of this instrument should be corrected for zero error in the same manner as for micrometers.

Other Devices.—**End measuring rods** (Fig. 10) are very useful for calipering holes or distances between parallel faces, when the dimensions are large. The ends of these rods are made spherical, so that they cannot be jammed when in position, no matter how they are turned. The rod is held by the vulcanite handle at its middle. **Flexible steel tapes** may be used for taking the diameter of large cylindrical pieces. By passing these tapes round such a cylinder its circumference can be obtained, from which measurement the diameter of the cylinder may be calculated.

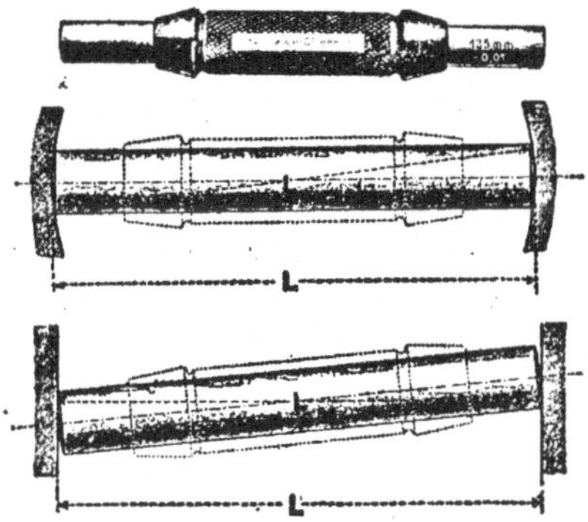

FIG. 10.—End measuring rods, with spherical ends.

MENSURATION.

Determination of Areas.—Some of the ordinary rules of mensuration are given here for future reference.

Square, side s ; area $= s^2$.

Rectangle, adjacent sides a and b ; area $= a \times b$.

Triangle, base b, perpendicular height h ; area $= \frac{1}{2} b \times h$.

Parallelogram, area = one side × perpendicular distance from that side to the opposite one.

Any irregular figure bounded by straight lines; to find its area split it up into triangles, find the area of each triangle separately, and take the sum of these areas for the area of the figure.

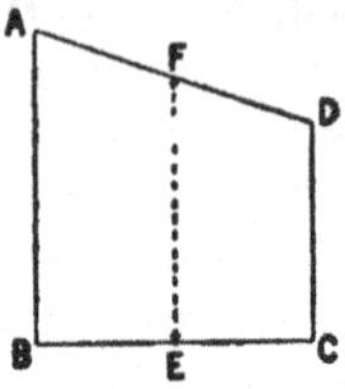

FIG. 11.

Trapezoid, such as $ABCD$ (Fig. 11), area $= BC \times$ average height.

The average height will be EF, drawn from the centre of BC, and will be equal to $\frac{1}{2}(AB+CD)$.

Area of trapezoid $=\frac{BC}{2}(AB+CD)$.

Suppose we have a figure consisting of a number of trapezoids (Fig. 12), all of the same breadth a, of which the area is required. We may proceed to find the area of each separately and to sum for the total area.

$$\text{Thus, area} = \left(\frac{h_1+h_2}{2}\right)a+\left(\frac{h_2+h_3}{2}\right)a+\left(\frac{h_3+h_4}{2}\right)a+\left(\frac{h_4+h_5}{2}\right)a,$$

$$=a\left\{\frac{1}{2}h_1+\frac{1}{2}h_2+\frac{1}{2}h_2+\frac{1}{2}h_3+\frac{1}{2}h_3+\frac{1}{2}h_4+\frac{1}{2}h_4+\frac{1}{2}h_5\right\},$$

$$=a\left\{\frac{h_1+h_5}{2}+h_2+h_3+h_4\right\}.$$

This gives us the **trapezoidal rule** for such areas, viz.—take half the sum of the first and last ordinate, add to this the sum of all the intermediate ordinates, and multiply the result by the common distance between the ordinates.

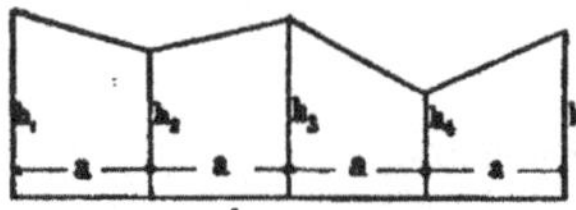
FIG. 12.

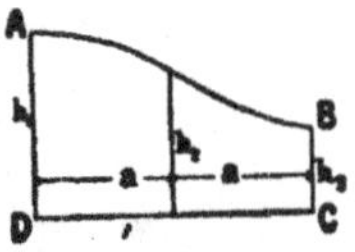

FIG. 13.

Simpson's Rule for finding the area of a plane figure bounded by a curve and end ordinates perpendicular to the base is founded on the assumption that the curve is parabolic. The rule only is given here. For a curve such as $ABCD$ (Fig. 13) divide it by an ordinate h_2 so as to bisect the base DC.

Let $a=$ the distance between the ordinates and h_1, h_2, $h_3=$ the heights of the ordinates.

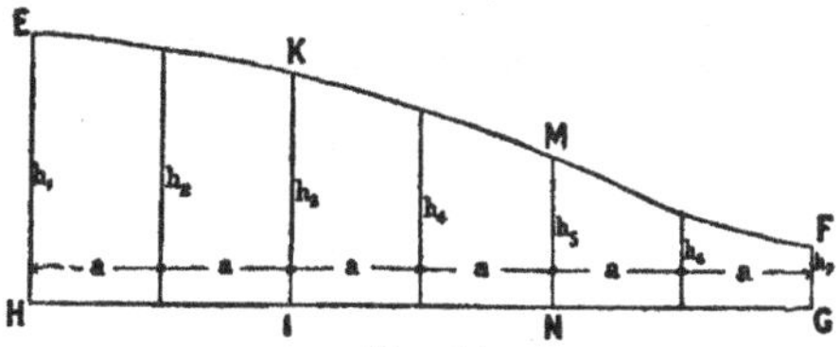

FIG. 14.

Then,

$$\text{area}=\frac{a}{3}(h_1+4h_2+h_3).$$

It is sometimes convenient to take more ordinates as is shown for the area $EFGH$ (Fig. 14). The number of ordinates must always be odd and they must be equidistant.

In this case,
$$\text{area } EKLH=\frac{a}{3}(h_1+4h_2+h_3),$$
$$\text{area } KMNL=\frac{a}{3}(h_3+4h_4+h_5),$$
$$\text{area } MFGN=\frac{a}{3}(h_5+4h_6+h_7),$$

and total area $=\frac{a}{3}(h_1+4h_2+2h_3+4h_4+2h_5+4h_6+h_7)$.

This rule may be stated thus:—Add the first and the last ordinates; to this sum add four times the sum of the even intermediate ordinates and twice the sum of the odd intermediate ordinates; multiply this total sum by one third the common distance between the ordinates.

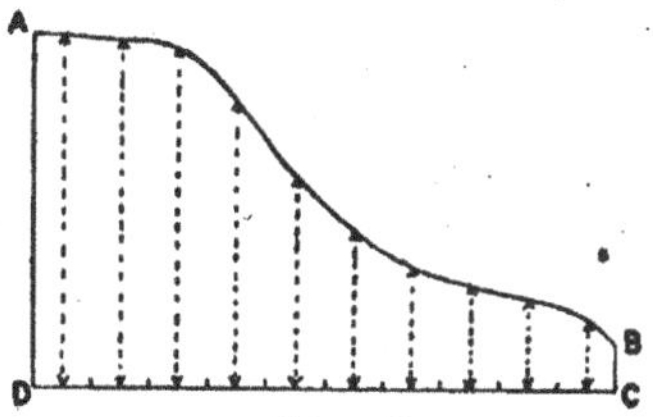

FIG. 15.

These and other convenient rules for finding the areas of figures bounded by curves are more used by naval architects than by engineers. The ordinary process used by engineers for finding the area of a figure such as $ABCD$ (Fig. 15) is to divide DC into 10 equal parts and to measure the height of the diagram at the centre of each part. Sum these heights and divide by 10. The result gives the average height of the diagram and if this be multiplied by DC the final result will give the area.

Circle, diameter, d or radius, r; circumference $=\pi d=2\pi r$.

π denotes the ratio of the circumference of a circle to its diameter and is represented by the number 3·1416. For many engineering calculations the value $\frac{22}{7}$ is correct enough.

$$\text{Area of circle}=\frac{\pi d^2}{4}=\pi r^2=0{\cdot}7854\,.\,d^2.$$

Determination of Volumes.—*Cube*, edge, s; volume $=s^3$.

Prism, having its ends perpendicular to its axis; volume $=$ area of one end $\times$ length of prism.

Sphere, radius, r; volume $=\frac{4}{3}\pi r^3$.

(Area of curved surface $=4\pi r^2$.)

Pyramid, volume $=$ area of base $\times\frac{1}{3}$ perpendicular height.

Cone, volume $=$ area of base $\times\frac{1}{3}$ perpendicular height.

(Area of curved surface $=$ circumference of base $\times\frac{1}{2}$ slant height.)

Right-angled triangle ABC (Fig. 16); $AC^2=AB^2+BC^2$.

Measurement of Angles.—Angles may be measured in *degrees*, or in *radians*.

A **degree** is the angle at the centre of a circle subtended by an arc of $\frac{1}{360}$th of the circumference.

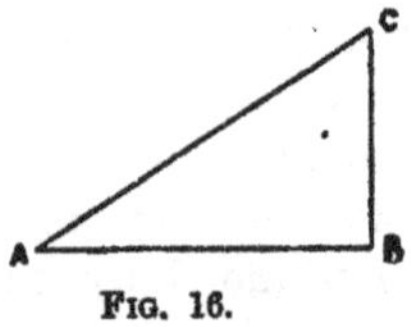

FIG. 16.

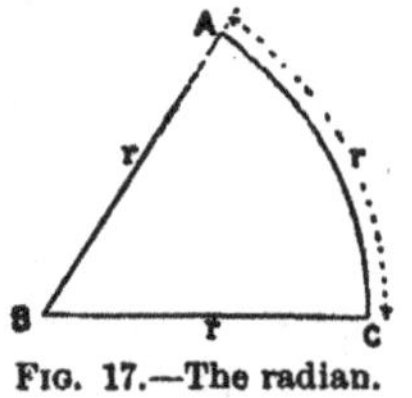

FIG. 17.—The radian.

A **radian** is the angle at the centre of a circle subtended by an arc equal to the radius of the circle; in Fig. 17, ABC is one radian.

In a complete circle there are 360 degrees and 2π radians, therefore

$$2\pi \text{ radians}=360 \text{ degrees},$$

or

$$\pi \text{ radians}=180 \text{ degrees}.$$

An angle expressed in radians can be transformed to degrees by multiplying by $\frac{180}{\pi}$; or if expressed in degrees, can be transformed to radians by multiplying by $\frac{\pi}{180}$.

An angle may be expressed in radians by dividing the length of the circular arc subtending it by the radius of the arc, both being in the same units.

TRIGONOMETRICAL RATIOS.

The following definitions should be understood. Given an angle *POM* (Fig. 18), take any distance *OP* and draw *PM* perpendicular to *OM*.

Then the following ratios of the sides will be independent of the length *OP* and will depend only on the magnitude of the given angle.

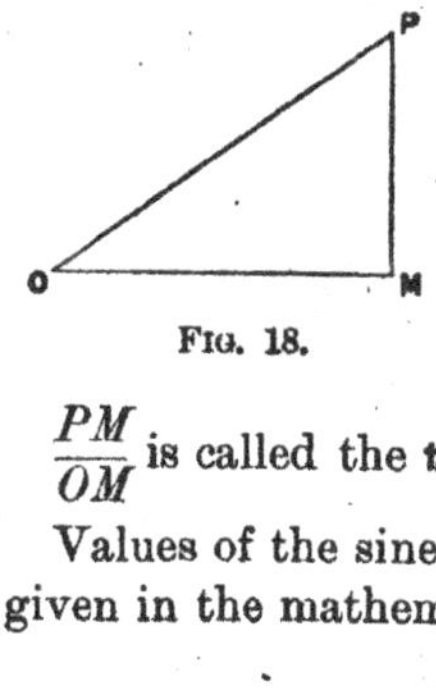

FIG. 18.

$\frac{PM}{OP}$ is called the **sine** of the angle *POM*, written sin *POM*.

$\frac{OM}{OP}$ is called the **cosine** of the angle *POM*, written cos *POM*.

$\frac{PM}{OM}$ is called the **tangent** of the angle *POM*, written tan *POM*.

Values of the sine, cosine, and tangent of angles up to 90° are given in the mathematical tables at the end of the book.

USEFUL CONSTANTS.

1 inch	=2·54 centimetres.
1 metre	=39·37 inches.
5280 feet	=1 mile.
6 feet	=1 fathom.
1 chain	=66 feet.
80 chains	=1 mile.
1 knot	=6080 feet per hour.

1 square inch =6·45 square centimetres.
1 square metre=1550 square inches.
1 cubic inch =16·39 cubic centimetres.
1 cubic metre =61,025 cubic inches=1·308 cubic yard.
1 litre =1000 cubic centimetres=1·7617 pint.
1 gallon =0·1605 cubic foot=4·541 litres.
1 bushel =1·2837 cubic feet.
One pound avoirdupois=7000 grains=453·6 grams.
One kilogram=2·205 pounds.
One gallon of pure water at 62° F. weighs 10 lbs.

EXERCISES ON CHAP. I.

1. Convert 9 ft. 6½ in. to metres.

2. Convert 2·94 metres to feet and inches.

3. Convert 3 miles 15 chains to kilometres.

4. Convert 53·7 millimetres to inches.

5. A rectangle has sides 4⅜″ and 2⅞″. Calculate its area.

6. A triangle, base 8 cms.; perpendicular height, 13·25 cms. Calculate its area.

7. Draw carefully to scale a triangle having sides respectively 4½″, 3¼″, and 5⅜″. Measure its perpendicular height from your drawing and calculate the area of the triangle from this and the length of the base.

8. What is (*a*) the circumference and (*b*) the area of a circle whose diameter is 14 cms.? Take $\pi=\frac{22}{7}$.

9. Calculate the volume of a ball 9″ diam.

10. Draw full size a 5-sided figure *ABCDE* from the following particulars. Take measurements from your drawing and calculate its area.

Sides, $AB=4''$, $BC=3''$, $CD=2''$, $DE=1\frac{1}{2}''$, $EA=1''$.
Diagonals, $BD=3''$, $AD=1\frac{1}{2}''$.

11. A figure stands on a base *AB* 5″ long. Its heights at intervals of 1″, starting from *A*, are 2″, 4″, 2½″, 3½″, 1″, 5″. Straight lines joining the tops of these ordinates bound the top of the figure. Calculate, using the trapezoidal rule, the area of the figure.

12. Draw at random any figure bounded by a curve at the top,

and find its area by applying (*a*) Simpson's rule, (*b*) the ordinary engineering rule.

13. Describe any micrometer screw gauge with which you are acquainted, suitable for measuring to the $\frac{1}{1000}$th of an inch. Sketch and describe carefully the method of graduation and the position of the gauge when set to measure ·374 inch. (1899.)

CHAPTER II.

MATTER, FORCE, WEIGHT.

Definition of terms.—Applied mechanics treats of those laws of force and the effects of force upon matter which apply to works of human art. As science stands at present, it is impossible to state exactly what **force** and **matter** really are, and we are compelled to explain them by reference to some of their properties. Matter is anything which our senses tell us exists. Matter exists in many different forms, and can often be changed from one form to another, but man cannot create it, nor can he annihilate it. Matter always occupies space, and a given piece of matter, occupying a definite space, is called a **body**.

Force may exert push or pull on a body, or may set it in motion, or bring it to rest. The most familiar conception we have of force is obtained from the manner in which our muscles must be exerted when we support a body.

All bodies are measured, as regards the quantity of matter, or **mass**, they contain, by comparison with a standard body. The standard for this country is the quantity of matter contained in a lump of platinum preserved in the Exchequer Office. This quantity of matter is called one **pound**. In countries using the metric system the standard mass is the **gram**, and is the quantity of matter contained in a cubic centimetre of water at the temperature of 4° C.

When two given bodies, of different materials but having the same volume, are found to contain differing quantities of matter, that which contains the greater quantity is said to be more *dense* than the other. The **density** of a material is stated by the quantity of matter, or mass, of a cubic unit of it. Thus, the density of water is about 62·3, there being 62·3 lbs. mass in one cubic foot of water; wrought iron has a density of about 480.

Specific density is the density of a substance when compared with that of a standard, water being taken for this purpose. Thus, the specific density of water being 1, the specific density of wrought iron is 7·7.

The most familiar force we have is **weight**, which all bodies possess. Weight is due to gravitation, which is manifest in the attraction which all bodies have for one another. Gravitational effort is proportional to the product of the masses of the two bodies and inversely proportional to the square of the distance between them. It is very small for bodies of ordinary size, but is perfectly evident when one, or both bodies, possesses a large quantity of matter. Thus, for bodies near the surface of the earth, the gravitational effort is seen by what we call the weight of the body. Weight means the tendency towards the earth's centre possessed by all bodies.

The weight of a body may vary.—Weight, as we have seen above, is proportional to the product of the masses of the earth and of the body. Assuming these to be constant, a given body will always have the same weight at the same place on the earth. Weight, however, is inversely proportional to the square of the distance from the earth's centre to the centre of the body, and therefore any change in this distance will produce a change in the body's weight. Thus, the weight of a given body is slightly less at the top of a mountain than at sea level.

The earth is not perfectly spherical, but is flattened towards the poles. Consequently a body at sea level near the poles will be nearer to the earth's centre than one at sea level near the equator. Therefore a body near the poles has a greater weight than it would possess near the equator.

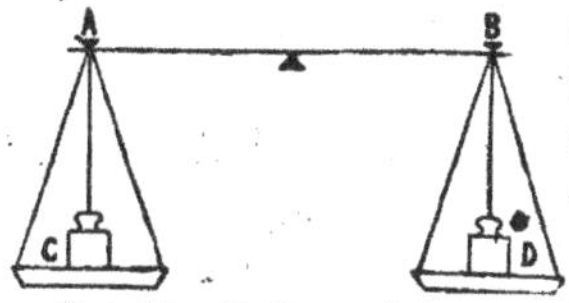

Fig. 19.—Ordinary balance.

The effect caused by the whirling of the earth on its axis also makes the apparent weight of a body near the poles slightly greater than it would possess near the equator.

Measurement of mass.—Quantities of matter can be measured by comparing their weights, using an ordinary **balance** for this purpose. The balance beam will become horizontal when equal vertical forces acting downwards are applied to its ends A and B (Fig. 19). These forces are produced by the weights of the bodies placed in the pans, and when the weights, as shown by the beam, are equal, we have equal masses in the pans. Using a standard lb. mass in the pan C, we can obtain another lb. mass by this means in pan D, and this can be done at any place without variation in the mass measured, as equal masses have equal weights when both are at the same place.

Spring balances (Fig. 20), which measure force applied to them by the extensions of a spring, show the actual weight of bodies placed in their pans. These appliances, therefore, will indicate different readings with the same body placed in the pan at different places on the earth's surface. Thus, it can be shown that a body, the weight of which is 32,088 lbs. at the equator, will have a weight of 32,252 lbs. at the poles. Engineers use as their unit of force, in most cases, the weight of the standard lb. mass.

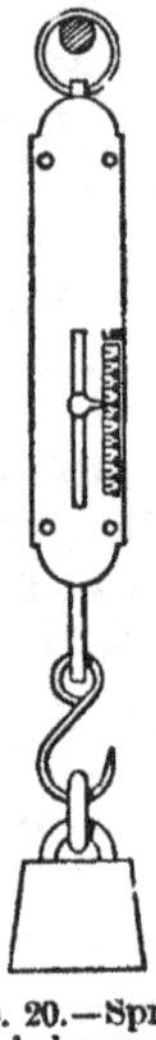

FIG. 20.—Spring balance.

This, as we have seen, is indefinite unless we also state the place where the mass has to be weighed. Thus, if we say, the weight of the lb. mass at sea level at Greenwich, we have a perfectly definite force and one which is used by many people.

It will be observed from the above figures that the alteration in the weight of a body by transference from the equator to the poles is too small to affect engineering work, being about 0·5 per cent. It is, therefore, generally neglected in engineering calculations, although this is no reason why the student should be ignorant of the fact that such alteration exists.

Specific gravity.—Specific gravity is the weight of a given volume of a substance when compared with the weight of an equal volume of water. It is usual in engineering work to measure specific gravities at a temperature of 60° F. The specific gravity of water being 1, the specific gravity of wrought iron would be 7·7, and of lead 11·4. It will be noticed that the number giving the specific gravity of a body will be the same as that giving its specific density. Specific gravity, however, refers to weight, and specific density to quantity of matter.

Some important relations.—The quantity of matter in, or mass of, a given body may be calculated thus:

Let V=its volume in cubic feet,

δ=its specific density.

Then $V \times 62{\cdot}3$ would be its mass if it were water,

and $V \times 62{\cdot}3 \times \delta$ will be its actual mass, in lbs.

The weight of a body may be calculated thus:

Let V=its volume in cubic feet,

ρ=its specific gravity.

Then $V \times 62{\cdot}3$=its weight if it were water,

and $V \times 62{\cdot}3 \times \rho$=its actual weight, in lbs.

The specific gravity of a body can be found roughly by first weighing it and then measuring it and calculating its volume from the dimensions.

Suppose V=its volume in cubic feet.

W=its weight in lbs.

Then $W = V \times 62{\cdot}3 \times \rho$,

or $$\rho = \frac{W}{62{\cdot}3\,.\,V}.$$

EXAMPLE.—A piece of flat bar iron, 12″ long, section $2'' \times \frac{1}{2}''$ is found to weigh 3·38 lbs. Find its specific gravity.

Here $$V = \frac{12 \times 2 \times \frac{1}{2}}{1728} \text{ cubic feet.}$$

$$\therefore \rho = \frac{3{\cdot}38 \times 1728}{12 \times 2 \times \frac{1}{2} \times 62{\cdot}3}$$

$$= \frac{3{\cdot}38 \times 144}{62{\cdot}3} = \underline{7{\cdot}8.}$$

Practical Applications.—An important part of the routine work of the engineer is the calculation from its drawings of the weights of various parts of a structure or machine. This he does by first calculating the volume of the part either in cubic inches or cubic feet and then multiplying this volume by the weight of the material per cubic inch or per cubic foot. Or, he may proceed, after having obtained the volume in cubic feet, to multiply this by 62·5,* which gives the weight of the part if made of water, 62·5 lbs. being the weight of 1 cub. ft. of water. If this result be now multiplied by the specific gravity of the material, the

* 62·3 more nearly, but 62·5 is near enough for almost all engineering purposes.

result will be the weight of the part. This procedure has to be followed out, especially in finding the weights of castings.

In finding the **weights of plates**, it is advantageous, if much work has to be done, to tabulate for reference the weights of plates one inch thick, of the substances used per square foot superficial area. If the actual area of the plate in square feet be now calculated, this, multiplied by the tabular number and the thickness of the plate in inches, gives the total weight.

For **bar iron** and **rolled sections** of different materials, the weight of each shape and size of section per foot running length is tabulated; the actual length, in feet, of stuff used multiplied by the tabular number will give its weight.

In estimating the **weights of castings**, it is customary to divide up the drawing of the casting into numerous parts, so chosen as to simplify the necessary mensuration work of finding the volume. Fillets, small bosses, etc., are omitted in this cutting up and allowed for afterwards. In structural work where the parts are riveted together or secured by bolts or pins, the heads of rivets, bolts and pins and nuts are allowed for separately.

The following table gives the weights and specific gravities of some common substances:

WEIGHTS AND SPECIFIC GRAVITIES.

Material.	Weight of		Weight of a sheet 1″ thick, 1 sq. foot area.	Specific Gravity.
	One cub. foot.	One cub. inch.		
	lbs.	lb.	lbs.	
Wrought iron,	480	0·28	40	7·7
Steel, - -	490	0·28	41	7·8
Cast iron, -	450	0·26	$37\frac{1}{2}$	7·2
Copper, - -	550	0·32	46	8·8
Brass, - -	525	0·30	44	8·6
Gun metal, -	540	0·31	45	8·7
Aluminium, -	165	0·095	14	2·6
Zinc, - -	450	0·26	$37\frac{1}{2}$	7·2
Tin, - -	465	0·27	39	7·4
Lead, - -	710	0·41	59	11·4
Fresh water, -	62·5	0·036	—	1·0
Sea water, -	64	0·037	—	1·024

A few examples are appended to show some of the methods adopted.

EXAMPLE 1. A cast-iron pipe 3" bore, 6 ft. long, metal of body ½" thick, circular flange at each end 7½" diam. × ¾" thick (Fig. 21). Calculate its weight.

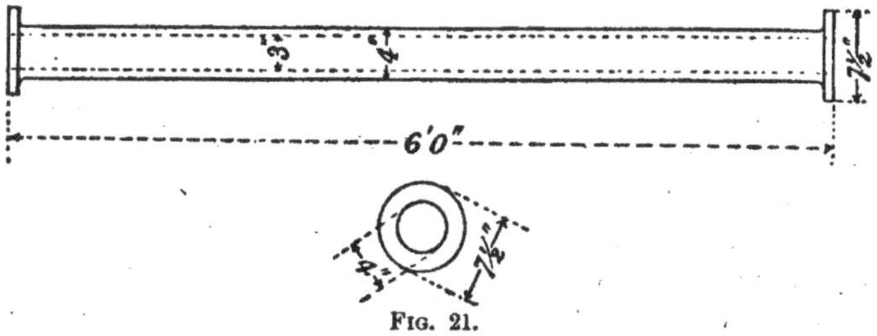

FIG. 21.

Remove the flanges and calculate their volume separately from the volume of the body of the pipe.

$$\begin{aligned}\text{Volume of body of pipe} &= \text{cross sectional area} \times \text{length}\\ &= \pi(2^2 - 1\tfrac{1}{2}^2) \times 72\\ &= 396 \text{ cubic inches.}\\ \text{Volume of each flange} &= \pi(3\tfrac{3}{4}^2 - 2^2) \times \tfrac{3}{4}\\ &= 23{\cdot}7 \text{ cubic inches.}\\ \text{Total volume of metal} &= 396 + 23{\cdot}7 + 23{\cdot}7\\ &= 443{\cdot}4 \text{ cubic inches.}\end{aligned}$$

Now, cast iron weighs 0·26 lb. per cubic inch;

$$\begin{aligned}\therefore \text{ weight of pipe} &= 443{\cdot}4 \times 0{\cdot}26\\ &= \underline{115{\cdot}3} \text{ lbs.}\end{aligned}$$

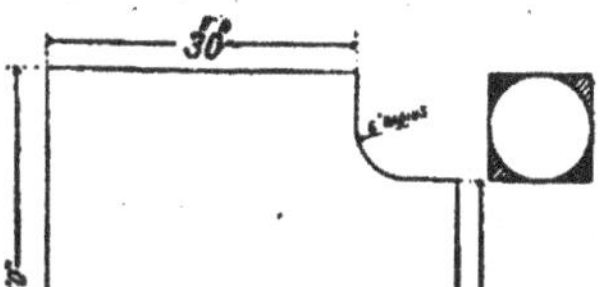

FIG. 22.

EXAMPLE 2. A wrought-iron plate, ½" thick, originally square, has a piece cut out of the corner as shown (Fig. 22). Calculate its weight.

(i) Area of original square
$= 4 \times 4 = 16$ square feet.

(ii) Area of circle 1 ft. diam.
$= 0{\cdot}7854$ square foot.

(iii) Area of square 1 ft. side
$= 1{\cdot}0$ square foot.

Difference between (iii) and (ii) $= 0{\cdot}2146$ square foot.

This difference is made up of the four shaded pieces, and three of these are cut out together with the circle 1 ft. diam.

$$\text{Area of piece cut out} = 0{\cdot}7854 + (\tfrac{3}{4} \times 0{\cdot}2146)$$
$$= 0{\cdot}9463 \text{ square foot};$$
$$\therefore \text{ Area of actual plate} = 16 - 0{\cdot}9463$$
$$= 15{\cdot}054 \text{ square feet.}$$

Now, wrought iron weighs 40 lbs. per superficial foot if the plate is 1″ thick;

$$\therefore \text{ weight of plate} = 15{\cdot}054 \times 40 \times \tfrac{1}{2}$$
$$= \underline{301} \text{ lbs.}$$

EXAMPLE 3. A copper float ball 14″ diam. is made of metal $\frac{1}{16}$″ thick. Calculate its weight.

In this case, as the metal is thin compared with the diameter of the ball, we may find the volume of metal near enough for practical purposes by multiplying the spherical area by the thickness of metal. To be quite exact we should have to calculate the volume of metal by taking the volume of a sphere 13⅞″ diam. from the volume of a sphere 14″ diam.

$$\text{Spherical area} = 4\pi r^2$$
$$= (4 \times \tfrac{22}{7} \times 7 \times 7) \text{ square inches.}$$

Approximate volume of metal $= (4 \times 22 \times 7 \times \frac{1}{16})$ cubic inches, and as copper weighs 0·32 lb. per cubic inch,

$$\text{weight of ball} = 4 \times 22 \times 7 \times \tfrac{1}{16} \times 0{\cdot}32$$
$$= \underline{12{\cdot}3} \text{ lbs.}$$

EXERCISES ON CHAP. II.

1. Find the weight of a piece of flat bar iron 24″ long, section 2″ × ½″.

2. Find the weight of a wrought-iron bar, 13·16″ long, section 1·5″ × 0·4″.

3. A piece of angle iron, section as in Fig. 23, is 30 ft. long. Calculate its weight, neglecting rounded corners.

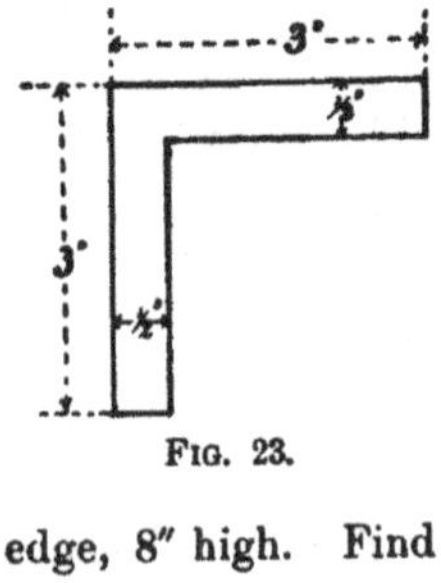

FIG. 23.

4. A circular brass plate, 2 ft. diam., is ⅛″ thick. Calculate its weight.

5. A hollow cylinder of wrought iron is 4″ inside diameter, 4⅜″ outside diameter and 10 feet long. Calculate its weight.

6. A solid pyramid of lead, square base 4″ edge, 8″ high. Find its weight.

7. A copper cone, 10″ diam. of base, 12″ high, metal 0·05″ thick, no bottom. Find its weight.

8. A hollow conical vessel, 6″ inside diam. at top, 9″ deep inside, is full of water. Calculate the weight of water.

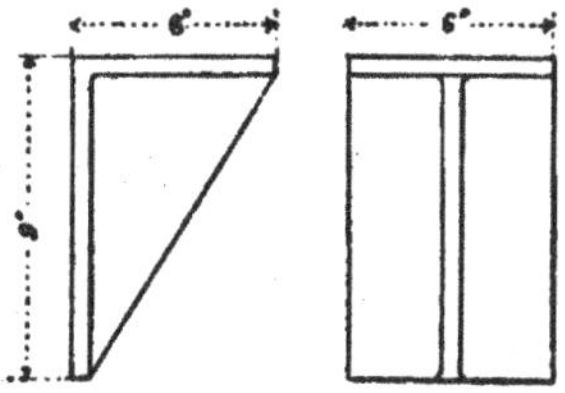

FIG. 24.

9. A cast-iron bracket, metal all over $\frac{1}{2}''$ thick, has dimensions as shown in Fig. 24. Find its weight.

10. A Lancashire boiler is 7 ft. diam. and 30 ft. long. The two internal tubes are each 3 ft. diam. Plates of cylindrical outside portion $\frac{5}{8}''$ thick. End plates $\frac{3}{4}''$ thick. Internal tubes $\frac{7}{16}''$ thick. Neglect lapping at joints, rivets, etc., and find weight of material if of wrought iron throughout.

11. Calculate what weight of sheet lead, 0·1″ thick, will be required for lining a timber tank, the internal dimensions of which are 6 ft. long, 4 ft. broad, 3 ft. deep.

12. A solid ball of cast iron is to have a weight of 90 lbs. Calculate its diameter.

CHAPTER III.

TWO AND THREE FORCES ACTING AT A POINT.

Representation of a force.—To describe completely a force acting on a body we require to state the following particulars, (*a*) its magnitude, (*b*) its point of application, (*c*) its direction, (*d*) its sense, *i.e.* to state whether the force is pushing or pulling at the point of application.

A straight line may be employed to represent a given force, for it may be drawn of any length and so represent to a given scale the magnitude of the force, the end of the line shows the point of application, the direction of the line gives the direction, and an arrow point on the line will indicate the sense of the force.

Thus, a pull of 5 lbs. acting at a point *O* in a body (Fig. 25) at 45° to the horizontal, would be completely represented by the line *OA* and arrow point as shown.

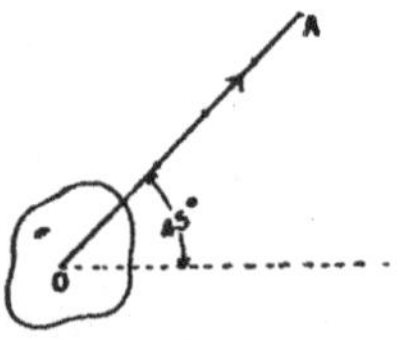

FIG. 25.—Graphical representation of a force.

We often speak, as above, of a force "acting at a point." Of course this must not be understood literally, for no material is so very hard that it would not be penetrated by even a very small force applied to it at a mathematical point. What is meant by the statement, is that the force may be imagined to be concentrated at the point in question without thereby affecting the condition of the body as a whole.

Forces acting in the same straight line.—A body is said to be in equilibrium if the forces acting on it balance one another. Thus, if two equal opposite pulls P, P (Fig. 26) be applied at a point O in a body both in the same straight line, they will evidently balance one another and the body will be in equilibrium.

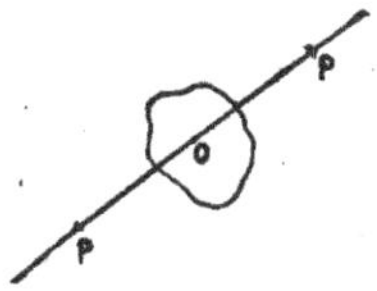

FIG. 26.

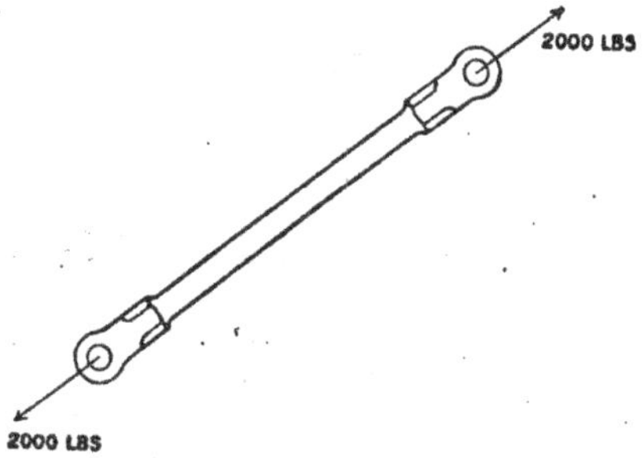

FIG. 27.—Tie bar under pull.

A tie bar subjected to two equal opposite pulls of 2000 lbs. each (Fig. 27) acting in the direction of its length will be in equilibrium. If one only of these pulls were reduced or increased even by very little the bar would move. This bar could not possibly be imagined pulled with a force of 2000 lbs. at one end only and yet to remain at rest, any more than a pull of 5 lbs. could be applied by the hand to one end of a piece of string unless the other end were pulled with a force of 5 lbs. in the opposite direction.

FIG. 28.—Column under push.

In the same way, if a column or strut (Fig. 28) be pushed at one end and remain in equilibrium, there must be an equal opposite push acting in the same straight line at the other end.

It is impossible for a single force to act alone. To every force there must be an equal opposite force, or what is exactly equivalent to an equal opposite force. This equal opposite force is often called a **reaction**.

If several forces in the same straight line act at a point, the point will be in equilibrium if the sum of the forces of one sense is equal to the sum of those of opposite sense. If these sums are not equal, then a force is required to balance the point, and its magnitude will be equal to the difference of these sums and the force must have the same sense as the smaller sum. In the given case (Fig. 29) these sums are

$$2+3+5=10 \text{ lbs., sense from } A \text{ to } B;$$
$$8+1=\ 9 \text{ lbs., sense from } B \text{ to } A.$$

And a force of $(10-9)=1$ lb. of a sense B to A will produce equilibrium.

Two intersecting forces.—If two forces are given acting at a point, their lines of direction intersecting, a single force may

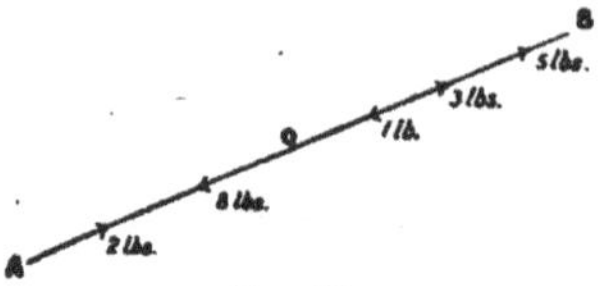

FIG. 29.

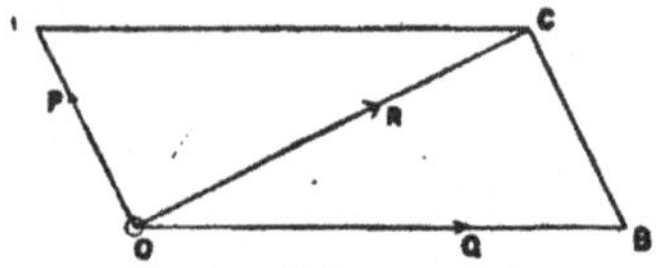

FIG. 30—Parallelogram of forces.

be found which would have the same effect, if applied alone, as the two forces together have. This single force is called the **Resultant** of the given forces, and may be found by the following construction.

Let P and Q be two pulls applied to a nail at O (Fig. 30); their joint tendency will be to carry the nail upwards to the right. Set off OA, to some suitable scale, equal to P, and OB, to the same scale, equal to Q. Complete the parallelogram $OACB$ and draw its diagonal OC. Measure off OC to the same scale of force and this will give the magnitude of R. If a pull R be now applied to the nail along the line OC, it will have the same effect as P and Q together have. This method is called the **Parallelogram of Forces**; P and Q are called **Components** of R.

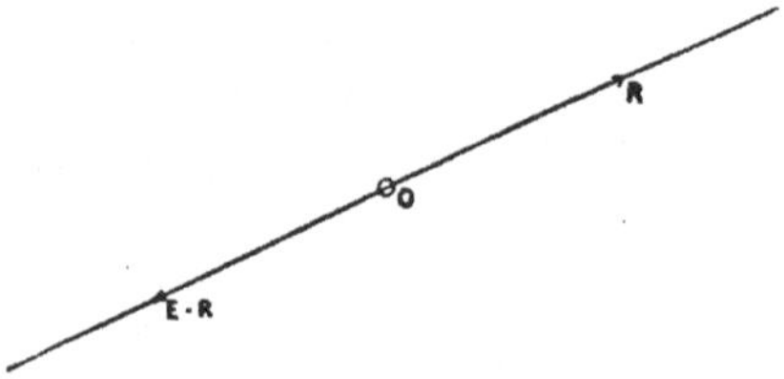

FIG. 31.—E and R balance.

Let us now remove the forces P and Q, and instead apply R alone to the nail. We may balance R by applying another pull E, equal and opposite to R and in the same straight line (Fig. 31), and if we do so, there will now be no tendency to disturb the nail. But since R is exactly equivalent to P and Q together, we may replace R again by P and Q (Fig. 32), thereby giving three pulls on the nail which will balance one another without any tendency to disturb the position of the nail in the board. E is generally called the **Equilibrant**, meaning the force required to keep the other forces in equilibrium.

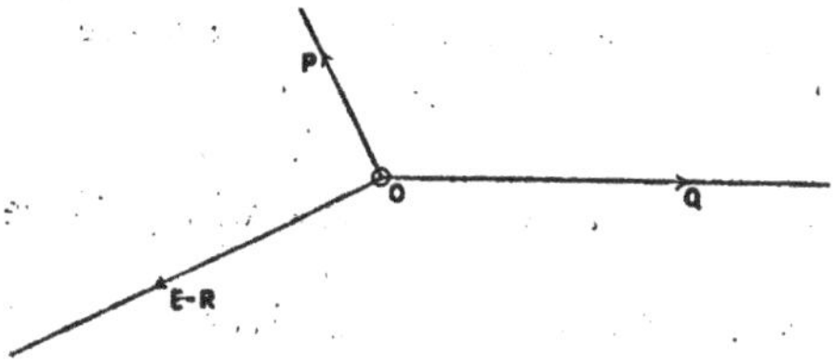

FIG. 32.—*P*, *Q*, and *E* balance.

Experimental verification.—The most satisfactory way for the beginner to verify the truth of the above principle is for him to make an experiment illustrating it.

EXPT.—Procure three wooden pulleys about 2″ or 3″ diameter, having their edges grooved to receive string, and with holes through their centres so that they will run freely on bradawls. Pin a sheet of paper to a vertical board and mount two pulleys at A and B by means of bradawls (Fig. 33). Tie two strings to a small split key ring, pass a bradawl through the ring into the board at O, and lead the strings over the pulleys at A and B. Fasten any bodies of known weights W_1, W_2, to the ends of the strings.

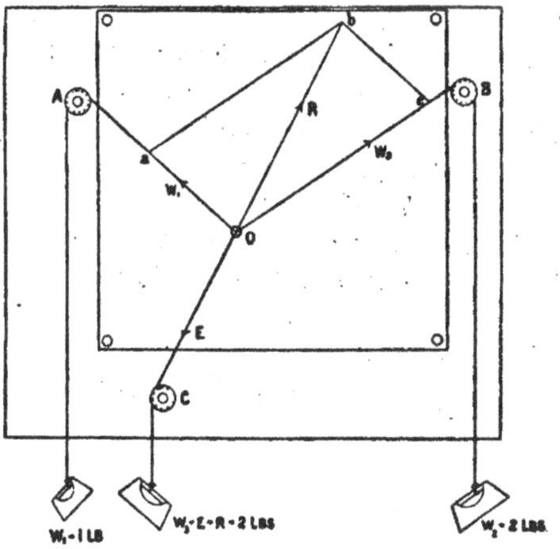

FIG. 33.—Verification of the parallelogram of forces by experiment.

We have now two forces W_1 and W_2 acting on the bradawl at O along the strings OA and OB. Mark the directions of these strings carefully on the paper and remove them to construct the parallelogram of forces *Oabc*. *R*, the resultant of W_1 and W_2 acting at *O*, will thus be found. Produce *bO* and replace the strings. By means of another pulley and bradawl at *C*, arrange a third string tied to the ring to lie along *bO* produced. Tie a body of weight W_3 to the end of this string, W_3 being equal to *R* in magnitude. This will give a third force $E = W_3 = R$ acting at *O*. If what has been done is correct, the three forces W_1, W_2, and *E* should now balance one another, and if they do, we should be able to remove the bradawl at *O* without the ring changing its position. Try if this is so.

You will probably notice that after the bradawl is removed from *O* the ring can be made to take up positions some little distance from *O*. This is due to the stiffness of the strings and the friction of the pulleys on the bradawls, and these causes prevent the perfect success of the experiment when regarded as a means of testing the truth of the parallelogram of forces.

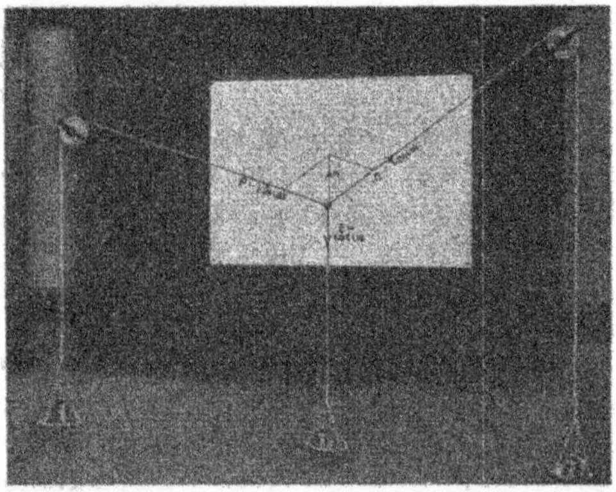

FIG. 34.—Students' apparatus for experiments on forces.

Fig. 34 is reproduced from a photograph of an apparatus arranged for students' use. The pulleys used are of aluminium, with pivot bearings, and may be clamped to any part of the edge of the board. The apparatus may be used also for testing several of the following principles concerning forces. The student should not forget in using it, that scale pans possess weight, and that such weights should be added to those put into them to get the total pull in the cords. Several experiments should be made by each student, using different weights and pulley positions each time.

Notice that before attempting to apply the parallelogram of forces to find the resultant of two forces, both given forces must be made to act either towards or from the point of application. Thus, given P' pushing and Q pulling at O (Fig. 35),

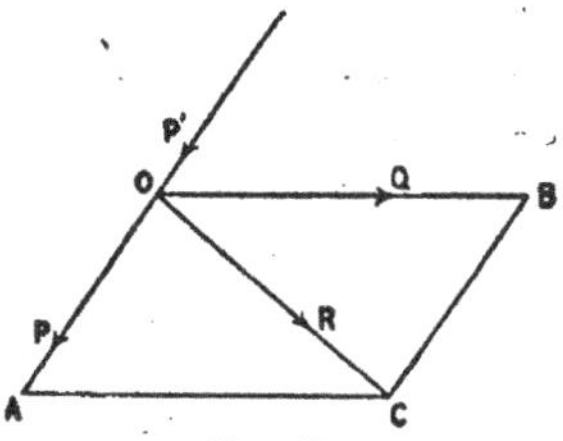

FIG. 35.

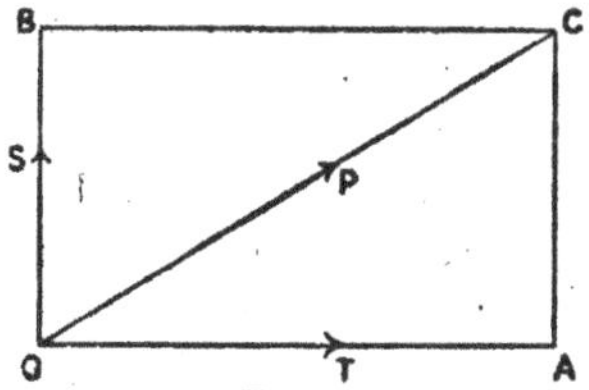

FIG. 36.

the tendency will be to carry O downwards to the right. Substitute $P=P'$, pulling at O, for P' and take OA to scale to represent P, also OB to represent Q. The parallelogram $OACB$ may now be drawn, giving R, the resultant of P' and Q.

Substitution of components for resultant.—Since the resultant produces exactly the same effect as its components, we may use either resultant or components in working out a question. It is often convenient to substitute for a given force its components along two given lines, which are usually taken perpendicular to one another. Thus, if we are given P acting at O, and it would be more convenient instead of P to have forces in OA and OB (Fig. 36), then, by setting off $OC=P$ and completing the parallelogram $OBCA$, two forces $S=OB$ and $T=OA$ are found, which if substituted for P, would have the same effect on O.

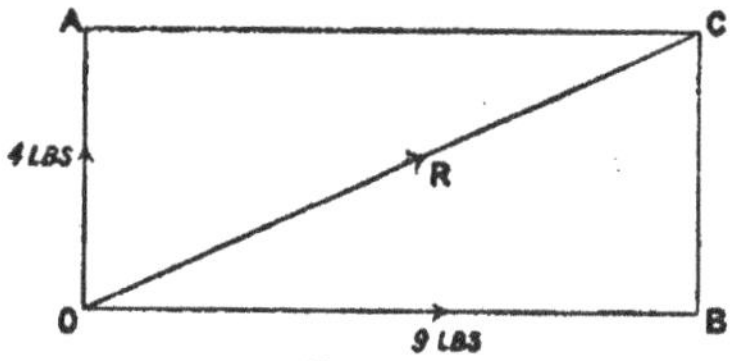

FIG. 37.

EXAMPLE 1. Two forces of 4 lbs. and 9 lbs. pull a nail in directions at right angles to one another. Find their resultant.

Draw OA and OB to scale (Fig. 37) to represent the given forces. Complete the parallelogram; its diagonal OC will give R.

R may also be found by calculation. Thus

$$OC^2 = OB^2 + BC^2$$
$$= OB^2 + OA^2 = 9^2 + 4^2;$$
$$\therefore\ OC = \sqrt{81+16} = \sqrt{97},$$

or

$$R = \underline{9 \cdot 8} \text{ lbs.}$$

In solving questions on forces by construction, take care to use a large scale. By doing so you will secure much more accurate results. Thus, in the above case a scale of ½″ to a lb. would be suitable.

EXAMPLE 2. A horse exerts a pull of 200 lbs. on a tram car at 30° to the direction of the rails as seen in plan. Find the force urging the tram forward, and that tending to pull it off the rails.

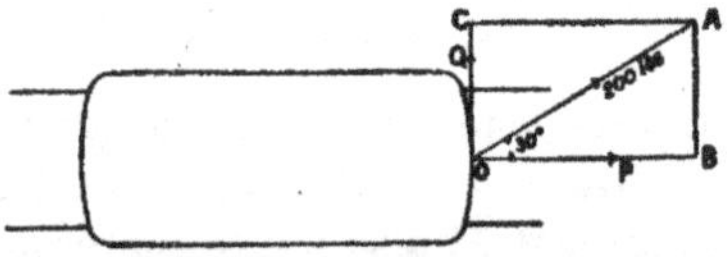

FIG. 38.

Set off OA (Fig. 38), to scale to represent 200 lbs. acting at 30° to OB, the direction of the rails. Draw OC perpendicular to OB. Complete the parallelogram $OBAC$; then OB, to scale, gives the force P, tending to urge the tram along the rails, and OC gives Q, the force tending to pull it off.

By calculation, since in the triangle OAB, the angle AOB is 30° and ABO is 90°, the sides have the following proportion:

$$BA : AO : OB = 1 : 2 : \sqrt{3},$$

or

$$OB : AO : OB = 1 : 2 : \sqrt{3};$$
$$\therefore\ Q : 200 : P = 1 : 2 : \sqrt{3};$$
$$\therefore\ Q = \tfrac{1}{2} \times 200 = \underline{100} \text{ lbs.}$$

and

$$P = 100 \times \sqrt{3} = \underline{173 \cdot 2} \text{ lbs.}$$

EXAMPLE 3. A load of 14 lbs. is hung by a cord 10 ft. long from an overhead beam, the arrangement being that of a pendulum. Find what horizontal force applied to the load will keep it 2 ft. from the vertical through the point of support.

Draw the figure to scale, as shown at OAB (Fig. 39). Let P be the required force; then the three forces acting on the load, P, W, and the pull of the cord, T, keep it balanced, so that T must be equal and opposite to the resultant of P and W. Set off AC to a suitable scale of force to represent 14 lbs., and complete the parallelogram *ACDE*. *T* will be represented by *AD*, and *P* by *AE*, which being measured gives $P=2 \cdot 86$ lbs.

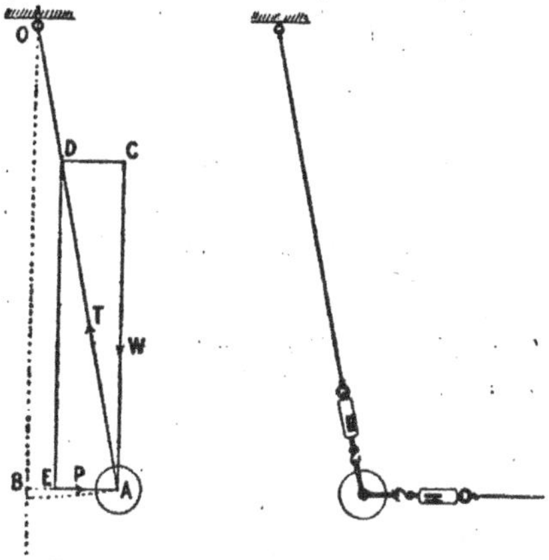

Fig. 39.—Forces acting on a pendulum.

Expt.—Test your construction by actually arranging a cord and weight. Put a spring balance in the cord and apply *P* by means of another spring balance (Fig. 39). *P* and *T* can now be measured directly. If readings of *P* are taken when *W* is at different distances from the vertical, a curve may be plotted on squared paper which will show the values of *P* for all positions of *W*. The results for intervals of ½ ft. are given in the table, the curve shown (Fig. 40) being plotted by using the values of *AB* for abscissae or horizontal distances, and the corresponding values of *P* for ordinates or vertical distances.

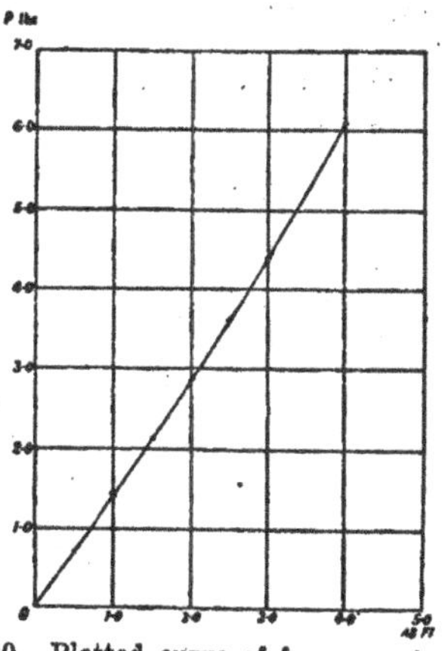

Fig. 40.—Plotted curve of force required to keep the pendulum out of the vertical.

AB	*P*
ft.	lbs.
0·5	0·7
1·0	1·41
1·5	2·12
2·0	2·86
2·5	3·6
3·0	4·4
3·5	5·2
4·0	6·1

EXAMPLE 4. The forces in the parts of a simple roof truss may be easily found by applying the parallelogram of forces.

Thus, let AB and AC be the rafters and BC the tie bar (Fig. 41). If a load W is applied at the top, the pushes in BA and CA may be found at once by considering these as the components of W. Thus, aA and cA are the pushes in BA and CA respectively. Now, if the bar BA is exerting a push aA at A, it must be exerting an equal contrary push at B, and it may be assumed that the wall is pushing vertically upwards to support the truss at B, with a force, say, P; this is called the **reaction** of the support. There are, therefore, three forces in equilibrium at B, viz., the push from AB, P, and a force in the tie bar. Take this last as being equal and opposite to the resultant of the other two, and find the resultant by setting off $dB=aA$ and completing the parallelogram $dBef$. Bf will give the force in the tie bar BC and eB will give the reaction of the support. In the same way AC is pushing at C with a force equal and opposite to cA; hence by the same method find the reaction Q. It will be found that $Ch=Bf$, showing that the tie bar is pulling equally at both ends, as we know must be the case, and therefore if Ch and Bf be not equal we know that something is wrong with our work.

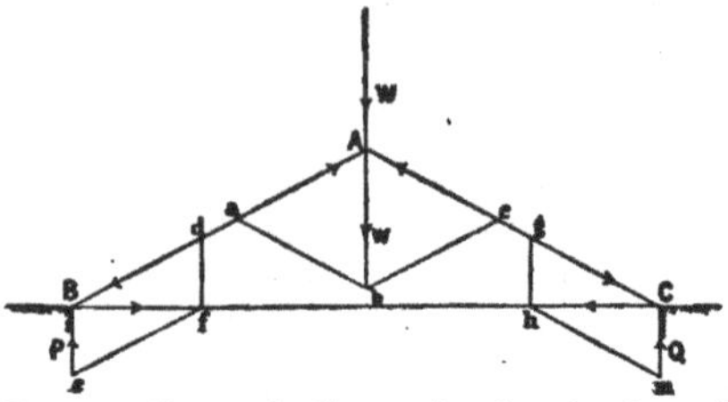

FIG. 41.—Forces in the parts of a simple roof truss.

FIG. 42.—Experimental model of a simple roof truss.

Fig. 42 shows a suitable apparatus for the student to use in order to test the above constructions. It consists of two wooden bars, connected loosely at the top by means of a bolt. The lower end of one bar is pivoted to a bracket secured to the base board, and the lower end of the other is mounted on a roller to minimise friction. A cord, with a spring balance, serves for a tie, and a compression spring balance, consisting of two brass tubes sliding one in the other, with a spring inserted, forms part of each inclined bar. The forces in each rafter, and in the tie, can be read from these three balances.

EXERCISES ON CHAP. III.

1. Represent graphically a pull of 10 lbs. acting at a point, its direction being N.E. Scale ¼″ to a lb.

2. Represent graphically two pulls acting at a point, one of 5 lbs., direction S.W.; one of 8 lbs., direction E. Find their resultant. Scale ½″ to a lb.

3. Represent graphically a push of 7 lbs. acting at a point, direction N.; also a pull of 4 lbs. acting at the same point, direction S.E. Find their resultant. Scale ½″ to a lb.

4. A pull of 25 lbs. and a push of 54 lbs. act at a point along the same straight line in opposition to one another. Represent them graphically, and find their resultant.

5. Draw a horizontal line, and mark a point O in it near its centre. Pulls of 2 lbs., 5 lbs., and 9 lbs. act at O in the right-hand portion of the line, and pulls of 4 lbs. and 6 lbs. together with pushes of 3 lbs. and 12 lbs. in the left-hand portion. Find the equilibrant.

6. Two pulls of 6 lbs. and 10 lbs. act on a point (*a*) at 90°, (*b*) at 120°, (*c*) at 60°. Find their resultant in each case both by construction and calculation.

7. A push of 20 lbs. and a pull of 30 lbs. act at the same point, their lines making 40° with each other. Find their resultant.

8. The resultant of two forces whose lines are perpendicular to one another is 15 lbs. One is a force of 4 lbs. Find the other.

9. A force of 100 lbs., acting in a horizontal line has to be balanced by two forces, one of 50 lbs. and the other of 120 lbs. Show their lines of action.

10. Three cords are attached to a ring; one cord carries a weight of 10 lbs. and hangs vertical. The other cords are attached to an overhead beam and are inclined one at 45° and one at 60° to the vertical. Find the pull in each.

11. An overhead pulley has a chain passing over it from a winch, and a load of 5 cwts. is being hoisted. The chain carrying the load hangs vertical and the chain leading to the winch makes 30° with the vertical. Suppose the force in each part of the chain to be 5 cwts., and find the resultant force on the pulley.

12. Prove by diagrams and by experiment that the force required to balance two given equal and opposing forces becomes smaller as the given forces approach, being finally nearly in the same straight line.

13. A barge is pulled along the centre of a canal 60 ft. wide by a horse on the tow-path whose centre is 4 ft. from the bank. The horse pulls the rope, which is 80 ft. long, with a force of 120 lbs. Find, by construction, the force urging the barge along the canal and the force urging it towards the bank.

14. A horse draws a load up an incline of 1 in 20. The traces are inclined at 30° to the horizontal and the pull of the horse on them is 180 lbs. Find by construction the backward pull on the horse taken parallel to the incline and the downward pull on the horse taken at 90° to the incline.

15. A man pulls a nail by means of a string in a direction at 30° to the board. If he exerts a force of 20 lbs., calculate the force tending to draw the nail and that tending to bend it.

CHAPTER IV.

TRIANGLE AND POLYGON OF FORCES. SIMPLE STRUCTURES.

Triangle of forces.—Let us now consider what conditions must be fulfilled in order that three forces acting at the same point all in the same plane may balance one another.

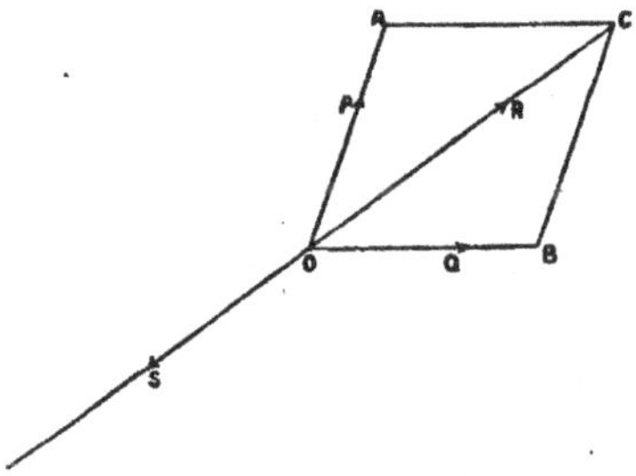

FIG. 43.—*P*, *Q* and *S* balance.

It has been seen already that if three forces, such as *P*, *Q*, and *S*, act at a point *O* (Fig. 43), one of them must be equal and opposite to the resultant of the other two. Find, by the parallelogram of forces, *R*, the resultant of *P* and *Q*, then *S* must be equal and opposite to *R*.

This proportion will evidently be true :

$$Q : P : R = OB : OA : CO.$$

Now $\qquad R = S$ and $OA = BC$;

$$\therefore \quad Q : P : S = OB : BC : CO,$$

that is, the three given forces are proportional to the sides of the triangle *OBC*.

The equilibrium of *P*, *Q*, and *S*, may therefore be tested by seeing whether a triangle can be drawn with sides proportional to these forces. Thus, in Fig. 44, *Ob* is parallel and proportional to *Q*, *bc* to *P*, and *cO* to *S*. If the lines so drawn give a closed triangle, then the given forces will be in equilibrium. This triangle *Obc* is called the **triangle of forces** for the given forces *P*, *Q*, and *S*.

Notice in drawing the triangle of forces that the sides *must be drawn in the proper order* to represent the sense of the forces. Thus, *Ob* is drawn to the right to indicate that *Q* acts to the right, *bc* upwards as *P* acts upwards, and *cO* down to the left to indicate the sense of *S*. So long as attention is paid to this the triangle of forces may be begun with a line parallel and proportional to any one of the given forces. The student should test this fact for himself by actual construction.

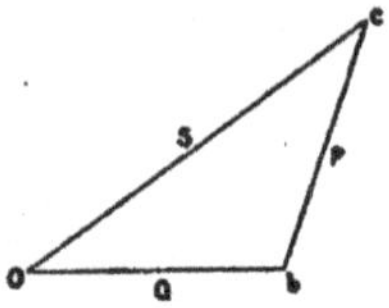

FIG. 44.—Triangle of forces for *P*, *Q* and *S*.

Resultant of several forces.—By means of the parallelogram of forces, the resultant may be found of any number of forces acting at a point, all being in one plane. Thus, given *P*, *Q*, *S*, *T*, acting at *O* (Fig. 45), in the plane of the paper; to find their resultant *R*, first find the resultant of any pair, such as *P* and *S*, by the parallelogram. Call this R_1. Then, take the other pair, *Q* and *T*, and find R_2, their resultant. Finally, find the resultant *R*, of R_1 and R_2. This will clearly be the resultant of the given forces.

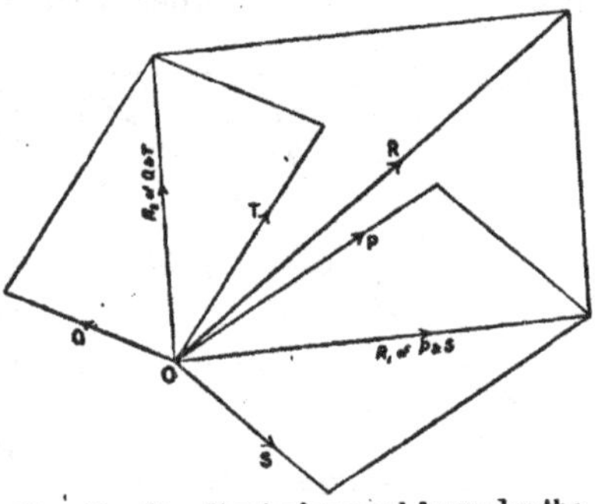

FIG. 45.—Resultant of several forces by the parallelogram of forces.

If we apply *E*, equal and opposite to *R*, *R* would be balanced at *O*, and therefore, of course, the given forces would also be balanced by *E*. So that by this method we may also find the equilibrant of a given number of forces.

EXPT.—Try one or two examples of this on your experimental board. Use five or six cords attached to a ring and led over pulleys, their ends being provided with scale pans. Have different weights in the pans and fix the ring by means of a bradawl at a given position on the board. Find the resultant of these forces, and by means of another cord and pulley apply a force equal and opposite to R to the ring. Now remove the bradawl and see if the ring remains in equilibrium in its original position.

Polygon of forces.—By means of an extension of the triangle of forces, we may ascertain whether a given number of forces acting at a point, all in the same plane, are in equilibrium. Thus, given P, Q, S, T, acting at O (Fig. 46), all being in the plane of the paper. Draw AB (Fig. 47) parallel and propor-

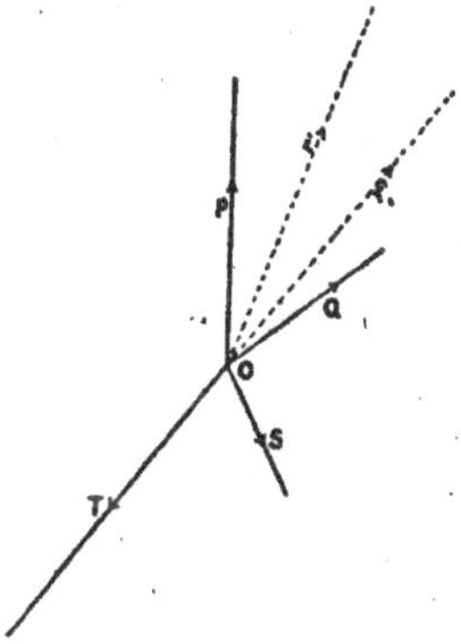

FIG. 46.

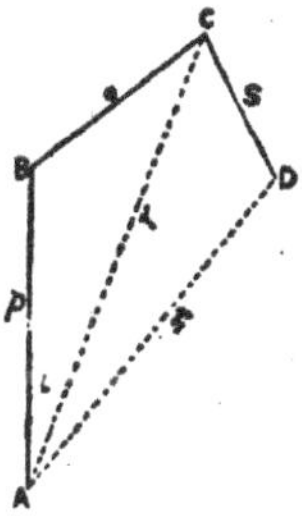

FIG. 47.—Polygon of forces.

tional to P, and BC in the same way to represent Q. Then, by the triangle of forces, CA will give a force r, which, if applied as a push downwards to O, will balance P and Q. Reverse it as shown at r' (Fig. 46), and r' will be the resultant of P and Q. Remove P and Q and let r' be applied instead.

Now, taking r' and S; AC (Fig. 47) represents r', draw CD to represent S, and, by the triangle of forces, DA will give a force r_1 which if applied at O (Fig. 46) will balance r' and S. It therefore follows that P, Q, and S, will be balanced by the application of r_1 to O, and if r_1 is reversed, as shown by r'_1, then r'_1 will be the resultant of P, Q, and S. Now if T, the last of the given forces, is found to be equal and opposite to r'_1, that is, if it is represented by the line DA in Fig. 47, T and r'_1 will balance one another, and if this is the case, the given forces P, Q, S, and T must be in equilibrium.

The test of equilibrium therefore is: Can a **closed polygon** be drawn with its sides parallel and proportional to the given forces respectively, taken as before in proper order? If so, then the forces are in equilibrium.

This polygon is called the **polygon of forces** for the given forces. It may be used to find the resultant of a number of

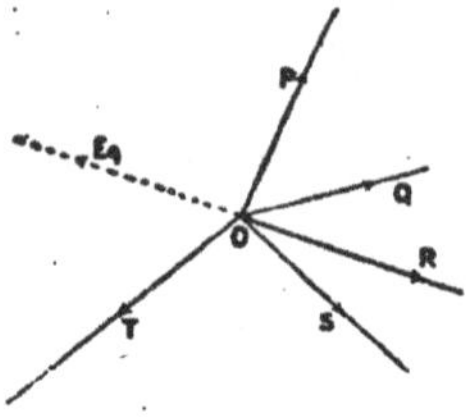

Fig. 48.

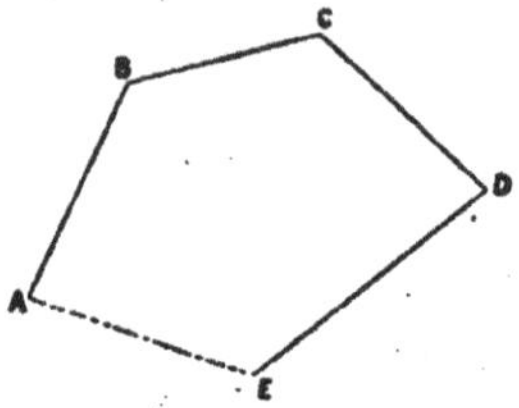

Fig. 49.

forces acting at a point, in one plane, which are not in equilibrium.

Thus, if on drawing the polygon of forces for the given forces *P*, *Q*, *S*, *T* (Fig. 48), as at *ABCDE* (Fig. 49), it is found that the

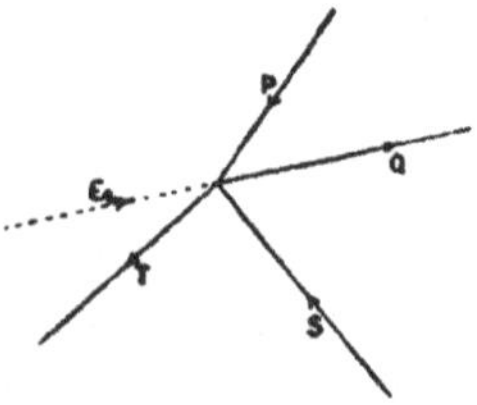

Fig. 50.

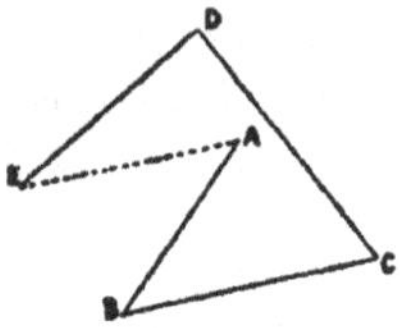

Fig. 51.

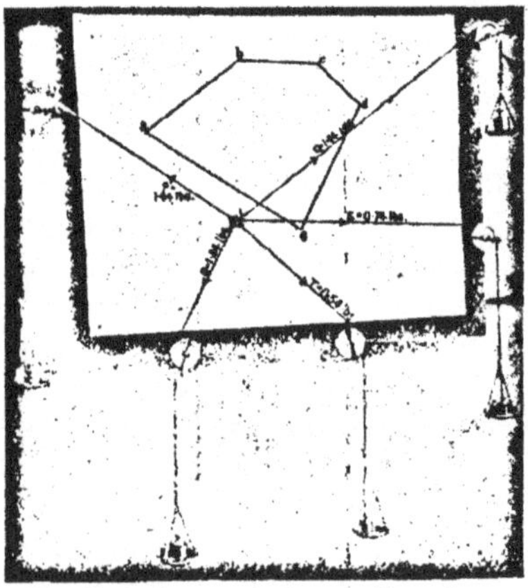

Fig. 51A.—An experiment on the polygon of forces.

polygon does not close, then the closing line *EA* will give a force *Eq*, which, if applied as a pull at *O* in the sense from

E to *A*, will balance the given forces, and will therefore if reversed, as at *R*, be their resultant.

Notice that it need not concern us if the forces are not all pushes or all pulls, provided we take them in their proper order. Figs. 50 and 51 show a case of this, and in Fig. 51A an example is shown experimentally worked out.

The student should work out some similar cases for himself, and verify the fact that it does not affect the result which force is taken first in drawing the polygon, or the numerical order in which they are taken, provided their senses are properly considered.

Experimental models of simple structures.—A model **derrick crane** is shown in Fig. 52 having arrangements provided for experimentally finding the forces in its parts. It consists of an upright post firmly secured to a base board, an inclined jib, with a compression spring balance fitted to it, and a cord to serve as a tie and having an ordinary spring balance to indicate the pull. The jib can be fitted with a pulley at the end for the cord supporting the weight to run over, or it may be used without the pulley, the supporting cord being then simply secured to its upper end. The method of using the model is as follows.

FIG. 52.—Experimental model of a derrick crane.

EXPT.—Place a known weight in the scale pan and then measure the height of the post from the junction of the jib to the junction of the tie, the length of the jib and the length of the tie. From these dimensions make an outline diagram of the crane and show the vertical line of the weight. This is shown at *ABC* (Fig. 53). If *W* is simply hung from *A*, then by the parallelogram of forces *ADEF*, $AD = W$ being first set off, the pull *T* of the tie and the thrust *Q* of the jib will be found.

Or, the triangle of forces, as at *abc* (Fig. 54), may be used. If the pulley is fitted to the jib end, and the supporting cord

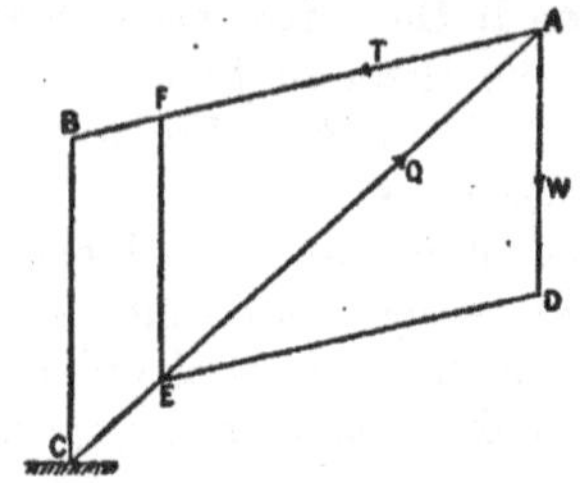

FIG. 53.—Solution of the derrick crane by the parallelogram of forces.

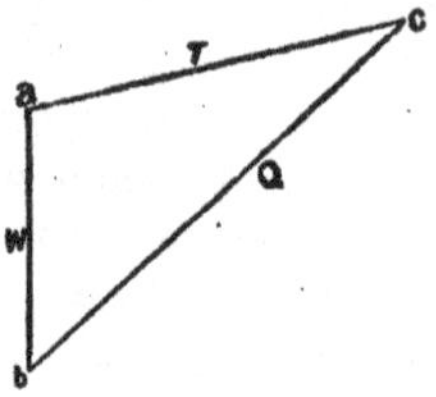

FIG. 54.—Solution of the derrick crane by the triangle of forces.

passed over it and led to a point D on the post (Fig. 55), this will give an additional force W which will be nearly equal to W, the stiffness of the cord and the friction of the pulley making it slightly different. Taking $W' = W$, the polygon of forces will give T and Q. This is shown at *abcd* (Fig. 56). In either case,

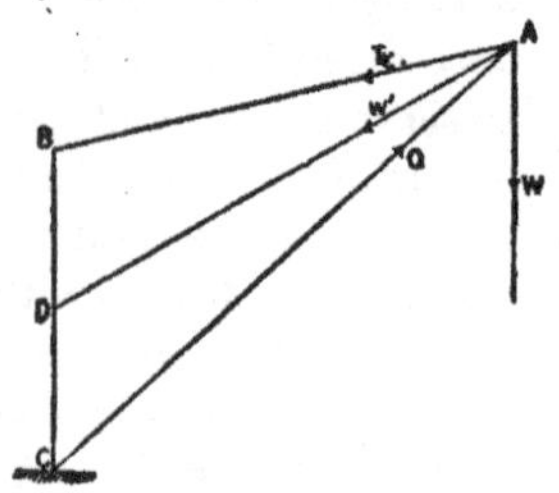

FIG. 55.—Derrick crane, the chain passing to D on the post.

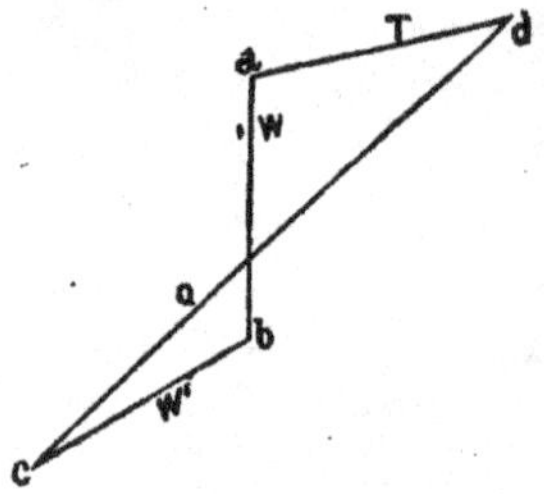

FIG. 56.—Solution by the polygon of forces.

after the force diagram is drawn and Q and T determined thereby, the spring balances should be read and compared with the scaled results. Generally they will read more than those found by construction, because the weight of the parts has been neglected and these will cause forces in the tie and jib when W is removed from the pan. Take W away and read the balances again, these readings, subtracted from the previous ones, should give results agreeing very closely with the scaled ones.

A **wall crane** (Fig. 57) can easily be arranged, using the jib from the derrick crane. AB is the jib, arranged horizontally,

BC an inclined tie, fitted with a spring balance. The cord suspending *W* may either be secured to *B* or passed over a pulley there and led to any point *D* between *A* and *C*. The method of using this model and the construction for the results are precisely similar to those for the derrick crane.

Sheer legs are used for moving very heavy weights, such as for placing engine parts and the boilers into position on board a vessel. A model sheer legs is shown in Fig. 58. It

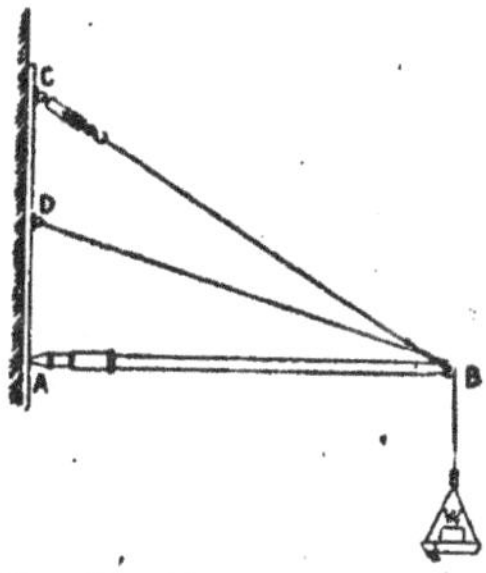

FIG. 57.—Experimental model of a wall crane.

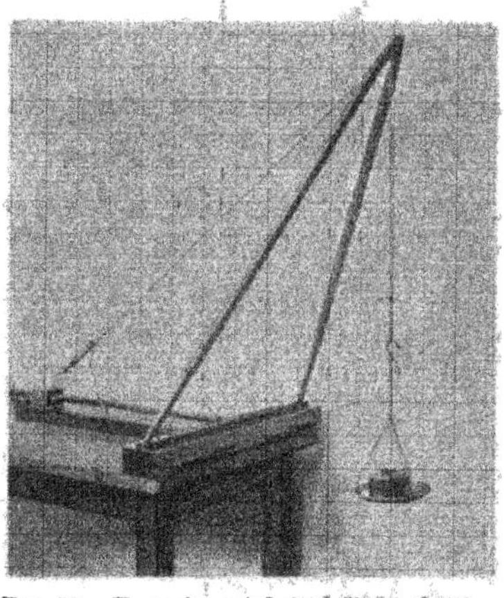

FIG. 58.—Experimental model of a sheer legs.

consists of two jibs hinged together at the top, and each fitted with a compression spring balance. A tie, with spring balance, passes from the top of the legs to the rear of the base board. The bottom ends of the legs bear in notches cut in the base board, there being a number of these, so that the legs may be spread out more or be brought closer together. The cord supporting the weight may be attached to the top of the legs or passed over a pulley there and brought down to some point on the base board.

Examining the forces acting at the top of the legs, we see that in this case these are not all in the same plane. If, however, the resultant push of the two legs combined is known, this resultant would fall in the same plane as the other forces, and the solution would then be exactly the same as for the derrick crane. The method of carrying out an experiment is as follows.

Expt.—Put a known weight in the scale pan, measure all the dimensions required in order to make an outline diagram of the model, and show the line of W on the diagram (Fig. 59). Draw in addition to the outline plan and elevation of the model, a view showing the true shape of the legs. Taking the directions from the elevation, and setting off $ab = W$ and $bc = W' = W$, first, the polygon of forces may be drawn (Fig. 59), giving $cd = R$

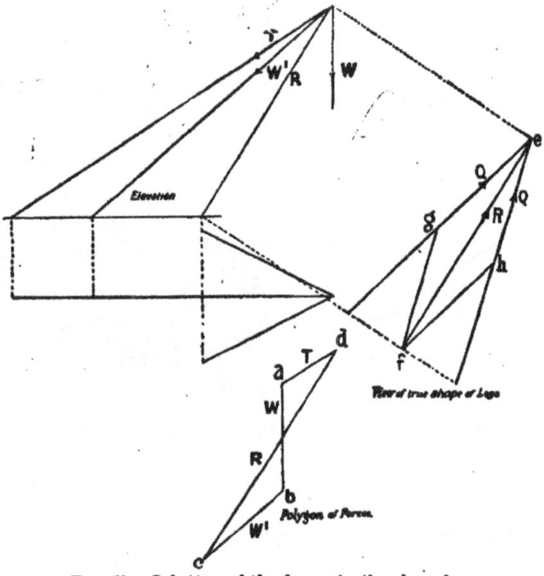

Fig. 59.—Solution of the forces in the sheer legs.

and $da = T$. Setting off $ef = R$ midway between the legs as shown in the view giving their true shape, draw the parallelogram of forces $ehfg$, and the thrusts QQ of each leg will be found.

This construction gives all the required forces, which, when measured to scale, should be found to agree closely with those shown by the spring balances, after allowance has been made for the effects of the weights of the parts, in the same manner as for the derrick crane.

Inclined Plane.—A model of a small carriage resting

FIG. 60.—Model of an inclined plane.

on an inclined plane is shown in Fig. 60. The forces required to support it if the plane were removed are shown experimentally by the pulls of the two cords, one arranged parallel to the plane and the other at 90° to it. These, with the weight of the carriage, give three forces acting on it and keeping it in equilibrium. Measure the height and length of the plane and make an outline diagram as shown at *ABC* (Fig. 61). Find these forces by construction, using the triangle of forces *abc*. Experiments should be made and force diagrams drawn when *P* is acting parallel to the base and also when *P* is applied at any angle to the plane.

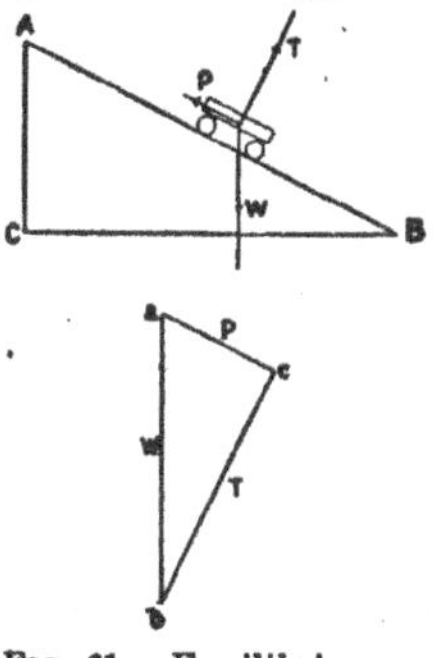

FIG. 61.—Equilibrium of a carriage on an inclined plane.

EXERCISES ON CHAP. IV.

1. *AB* and *AC* are two scaffold poles in the same vertical plane, lashed together at their tops. *AB* is 20 feet and *AC* 15 feet long. The distance *BC* between their feet is 15 feet. Find by construction the push in each pole when a load of 1 ton is hung from the top.

2. The jib of a model derrick crane is 47″ long, the tie 38″, and the post 31″. Find by construction the push in the jib and the pull in the tie when a load of 4·7 lbs. is simply hung from the end of the jib.

3. A crane jib measures 19 ft., the tie 17½ ft., and the post 9 ft. A load of 50 cwts. is attached to a chain which passes over a single pulley at the top of the jib, then along the tie. Find the push in the jib and the pull in the tie by construction.

4. Answer Question 3 supposing the chain, after leaving the pulley at the top of the jib, to pass along the jib.

5. *A* is a hinge fixed to a vertical wall 6 ft. vertically over another, *B*. A triangular frame *ABC*, $AC=8$ ft., $BC=10$ ft, is attached to *A* and *B*, the arrangement forming a wall crane. A load of ½ ton is attached to a chain which passes over a pulley at *C*, then along *CA* to a winding arrangement on the other side of the wall. Find by construction the forces in *AC* and *BC*, indicating whether they are push or pull.

6. In Question 5, turn the frame upside down and answer the same.

7. Draw to scale a frame *ABCD* from these particulars: $AB=4$ ft., $AD=4$ ft., $BC=5$ ft., $DC=6$ ft. Diagonal bar $BD=5$ ft. The frame is attached to a vertical wall at *A* and *D*, *A* being uppermost. Find by construction the forces in all the bars, marking push or pull, when a load of 2 tons is hung from *C*.

8. A boiler weighing 25 tons is placed on board a ship by means of a pair of sheer legs. The bottom pivots of the legs are 30 feet apart and each leg is 50 feet long. The distance, measured horizontally from the pivots to the centre of the ship hatch, is 25 feet, and the back leg of the sheers is 100 feet long. Assume that the load is simply hung from the top of the legs and find the push in each leg and the pull in the back leg when the boiler is going through the hatch.

9. A load *W* of 2000 lbs. is hung from a pin *P*, at which pieces *AP* and *BP* meet like the tie and jib of a crane. The angles *WPB* and *WPA* are 30° and 60° respectively. Show by a sketch how to find the forces in *AP* and *BP*. Distinguish as to each piece being a strut or a tie. (1897.)

10. Two pieces in a hinged structure meet at a pin, and a load is applied at the pin. Show how we find the pushing or pulling forces in the pieces. Describe an apparatus which enables your method to be illustrated. (1898.)

11. A carriage mounted on frictionless wheels rests on a plane inclined at 25° to the horizontal. If the carriage weighs 10 lbs., find by construction the force required to keep it in equilibrium, (*a*) when the cord is horizontal, (*b*) when parallel to the plane, (*c*) when at an angle of 30° to the plane.

CHAPTER V.

MOMENTS. PARALLEL FORCES. COUPLES.

Moments.—The **moment of a force** means the tendency of the force to turn the body on which it acts about a given axis. The moment of a force is measured by the product of the magnitude of the force and the length of a line drawn from the axis perpendicular to the line of action of the force.

Suppose a rod AB (Fig. 62) to be suspended by means of a bradawl pushed through a hole at A into the experimental board, so that it hangs vertically. Attach a cord to the rod at C, and lead it over a pulley at D, so that the portion CD is horizontal. If a weight W_1 be attached to the end of the cord, this will give a horizontal pull, $P = W_1$, to the rod at C. The effect will be to turn the rod in the same direction as the hands of a clock. By doubling or trebling the load W_1, the tendency to rotate the rod is doubled or trebled. By increasing the distance AC, the turning tendency is increased in the same proportion, so that the turning tendency, or moment of the force P, is proportional jointly to the magnitude of P, and the perpendicular distance AC from the axis at A to the line of action of P.

FIG. 62.—Moment of a force.

Notice it is not merely the distance from the axis to the point of application of P that is taken. For if this were so it is easily seen that the calculated moment of P about A would remain the same no matter in what direction P is applied, provided it always acts at C, whereas, actually, inclining the line of P's action diminishes the turning tendency, until finally if W_1 be hung direct from C so that P is vertical, there will be no tendency whatever to turn the rod. Hence in calculating moments the perpendicular to the line of action of the force must always be taken.

Now suppose a weight $W_1 = 5$ lbs. hung to the cord, and that AC is 10″ (Fig. 63). The moment of P will be measured by taking the product of 5 and 10; thus the moment of P about $A = 5 \times 10$ lb.-inch units. In giving the numerical value of a moment, the units of force and distance employed must always

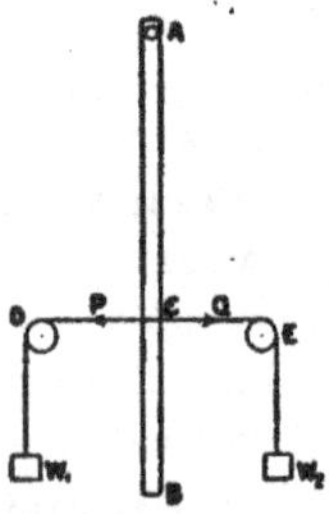

FIG. 63.—Two equal forces, giving equal opposing moments.

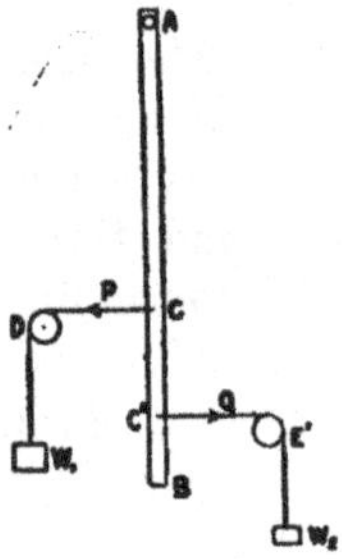

FIG. 64.—Two unequal forces, giving equal opposing moments.

be stated. Thus, ton foot, cwt. inch, gram centimetre, are units of moment. The moment of P will be clockwise. The rod AB may be balanced against rotation if another weight $W_2 = W_1$ be applied as shown (Fig. 63), so as to produce a force Q equal to P pulling horizontally at C in the opposite sense to P. Thus, moment of $Q = 5 \times 10 = 50$ lb.-inch units, and will tend to turn AB in the opposite direction to the hands of a clock, that is, anti-clockwise.

It will be found also, by trial, that the rod will be balanced if the weight W_2 be altered, say diminished, provided at the same time the distance from A at which Q is applied be altered, in the present case increased, viz. AC to AC' (Fig. 64). It will be found in all these cases that $Q \times AC'$ must always amount to 50 lb.-inch units. That is, for AB to be in equilibrium:

Clockwise moment of P = anticlockwise moment of Q.

If the direction of P be altered by raising the left-hand pulley to D' (Fig. 65), the moment of P is now $P \times AM$, clockwise. The rod will be balanced by W_2, provided that matters are so arranged that the product $Q \times AN$, which measures the anticlockwise moment of Q, is equal to $P \times AM$.

The result may be stated thus : **Two forces which act on a body free to rotate, having equal moments of opposite sign about the axis of rotation, will balance the body.**

The student should test the truth of this statement by working carefully several experiments similar to that described above, using different forces and distances in each case.

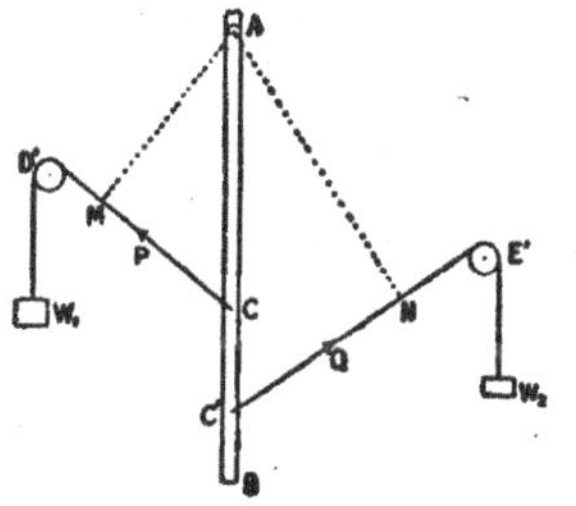

FIG. 65.—Two inclined forces, having equal opposing moments.

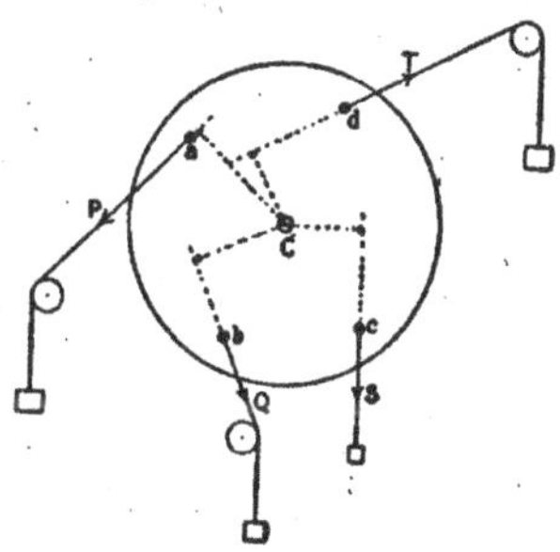

FIG. 66.—Disc in equilibrium under the action of several forces.

Principle of Moments.- EXPT.—By means of a screw at C, mount a circular wooden disc on the vertical experimental board (Fig. 66). Apply any number of forces, such as P, Q, S, T by means of cords attached to the disc at a, b, c, d, led over pulleys and having weights at their ends. Let the disc find its position of equilibrium. Calculate the moment of each force about C by multiplying the magnitude of the force by the length of the perpendicular from C to its line of action, producing this last if necessary. Arrange these moments in two columns, one for clockwise, one for anticlockwise moments. Take the sum of each column, and we should expect to find that these are equal, for, **if the disc is in equilibrium, the total clockwise turning tendency must be equal to the total anticlockwise turning tendency, in order that rotation may not take place.** This principle may be used to solve a great many problems.

EXAMPLE 1. A beam 12 feet long, supported at its ends, carries a load of 2 tons, 4 ft. from one end. Find the reactions of its supports. Neglect meanwhile the weight of the beam itself.

FIG. 67.

Let AB be the beam (Fig. 67) and P and Q the reactions of its supports. Imagine the beam to be free to rotate about B, and take moments about B of the forces acting on it. Find P thus:

Clockwise moment of P about $B = P \times 12$ ton-foot units.

Anticlockwise moment of W about $B = W \times 8 = 2 \times 8 = 16$ ton-foot units.

Q has no moment about B, as its line of action passes through B, and therefore the perpendicular from B to the line of action has no length.

Now the clockwise moment must equal the anticlockwise moment.

$$\therefore \quad P \times 12 = W \times 8,$$
$$12P = 16, \quad P = 1\tfrac{1}{3} = \underline{1{\cdot}33} \text{ tons.}$$

In the same way, take moments about A in order to find Q; then

Anticlockwise moment of $Q = Q \times 12$ ton-foot units.

Clockwise moment of $W = W \times 4 = 2 \times 4 = 8$ ton-foot units.

Anticlockwise moment = clockwise moment.

$$\therefore \quad Q \times 12 = W \times 4,$$
$$12Q = 8, \quad Q = \tfrac{2}{3} = \underline{0{\cdot}66} \text{ ton.}$$

EXAMPLE 2. A beam 20 feet long, supported at its ends, carries loads at intervals as shown (Fig. 68). Find the reactions of the supports, neglecting meanwhile the weight of the beam.

FIG. 68.

Taking moments about B.

Clockwise moment: $P \times 20$ ton-foot units.

Anticlockwise moments:

$$\begin{aligned} W_1 \times 17 &= \tfrac{1}{2} \times 17 = 8{\cdot}5 \text{ ton-foot units.} \\ W_2 \times 13 &= \tfrac{1}{4} \times 13 = 3{\cdot}25 \quad ,, \quad ,, \\ W_3 \times 7 &= 1 \times 7 = 7{\cdot}0 \quad ,, \quad ,, \\ W_4 \times 4 &= 2 \times 4 = 8{\cdot}0 \quad ,, \quad ,, \end{aligned}$$

Total anticlockwise moment $= \underline{26{\cdot}75}$ ton-foot units.

Clockwise moments = anticlockwise moments.

$$\therefore \quad P \times 20 = 26{\cdot}75, \quad P = \underline{1{\cdot}3375} \text{ tons.}$$

Taking moments about A,

Anticlockwise moment $= Q \times 20$ ton-foot units.

Clockwise moments:

$$
\begin{aligned}
W_1 \times 3 &= \tfrac{1}{2} \times 3 = 1{\cdot}5 \text{ ton-foot units.}\\
W_2 \times 7 &= \tfrac{1}{4} \times 7 = 1{\cdot}75 \quad ,, \quad ,,\\
W_3 \times 13 &= 1 \times 13 = 13{\cdot}0 \quad ,, \quad ,,\\
W_4 \times 16 &= 2 \times 16 = 32{\cdot}0 \quad ,, \quad ,,
\end{aligned}
$$

Total clockwise moment $= 48{\cdot}25$ ton-foot units.

Anticlockwise moments = clockwise moments.

$$\therefore \quad Q \times 20 = 48{\cdot}25,$$

$$Q = 2{\cdot}4125 \text{ tons.}$$

Notice in these questions, that a system of forces exists in each case the lines of which are parallel to one another, and that by the above solutions, the sum of the downward forces

FIG. 69.—Apparatus for determining the reactions of the supports of a beam.

equals the sum of the upward forces. Thus, in Example 1, the downward force is 2 tons and the two upward reactions have a sum of $1\frac{1}{3} + \frac{2}{3} = 2$ tons.

In Example 2, the downward forces give

$$\tfrac{1}{2} + \tfrac{1}{4} + 1 + 2 = 3\tfrac{3}{4} = 3{\cdot}75 \text{ tons.}$$

The upward reactions give

$$1{\cdot}3375 + 2{\cdot}4125 = 3{\cdot}75 \text{ tons.}$$

Or, in a given system of parallel forces the sum of the forces of one sense must in every case be equal to the sum of the forces of the

other sense, otherwise the body will be displaced as a whole in the same sense as the greater sum. Thus, if in the beam examples the sum of the reactions falls below the sum of the loads, the beam will move downwards and *vice versa*. Fig. 69 shows an apparatus for experimentally finding the reactions of a loaded beam.

EXPT.—Hang the same rod AB as before vertically in front of the experimental board using this time a long piece of cord for the suspension (Fig. 70). Attach cords at C and D and apply horizontal forces P and Q by means of pulleys and weights. To balance the rod AB, a force acting to the right, of magnitude, by the foregoing, equal to the sum of P and Q, must now be applied. Attach a weight $W_3=P+Q$ to another cord and by means of a pulley, apply a horizontal force $E=P+Q$ to the rod at F. This will prevent bodily movement of the rod to the left, but, very likely, the rod will not now hang vertically. In this case shift the cord at F, up or down, until such a position is found that the rod hangs vertical under the combined actions of the horizontal forces P, Q and E. E is now the equilibrant of P and Q.

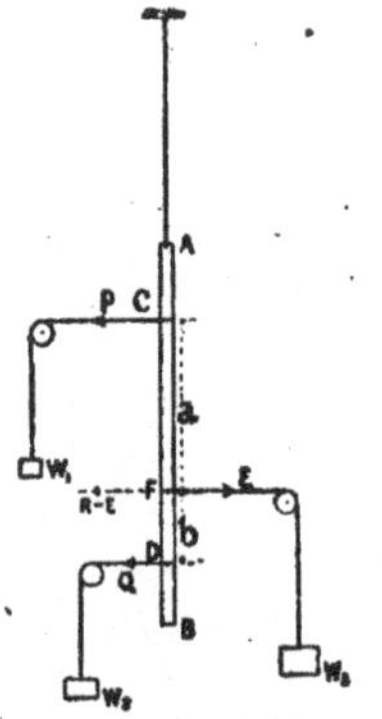

FIG. 70.—Equilibrant of two parallel forces of the same sense.

The position of F may be found thus: Imagine the rod to be free to turn about C and take moments of the horizontal forces (which alone tend to rotate the rod) about C.

Clockwise moment, $Q\times CD=Q\times(a+b)$.

Anticlockwise moment, $E\times CF=E\times a$.

P passes through C and has therefore no moment.

Clockwise moment = anticlockwise moment;

$$\therefore\ Q(a+b)=E\times a, \text{ and } E=P+Q,$$

$$\therefore\ Q(a+b)=(P+Q)a;$$

$$\therefore\ Qa+Qb=Pa+Qa,$$

or
$$Qb=Pa,$$

or
$$\frac{P}{Q}=\frac{b}{a},$$

or
$$P:Q=b:a.$$

The point F therefore divides the distance CD between the forces P and Q in inverse proportion to the forces.

Since E is the equilibrant of P and Q, if its sense be reversed, it will be the resultant of these forces.

Resultant of parallel forces.—We have, therefore, the following means of finding the resultant of two parallel forces of the same sense: $R=P+Q$; the line of R divides the distance between P and Q in inverse proportion to P and Q.

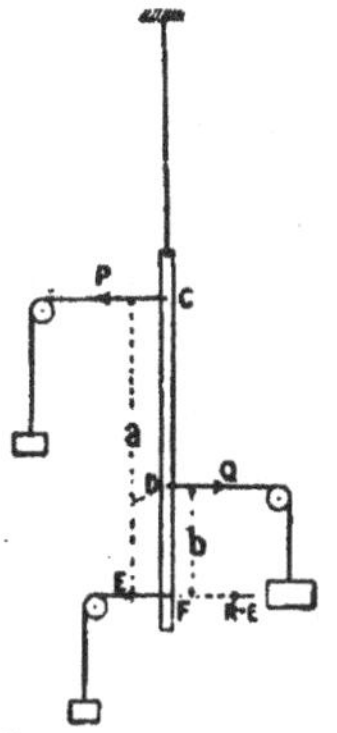

FIG. 71.—Equilibrant of two parallel forces of opposite sense.

If the forces are of opposite sense, the student should verify experimentally the following statements:

(1) R is equal in magnitude to the difference between P and Q; thus, if Q is the greater force, $R=Q-P$.

(2) R acts in the same sense as the greater force.

(3) $P:Q=b:a$. (Fig. 71.)

EXPT.—Do this by actually applying forces P and Q to the suspended rod. Calculate the magnitude of E from (1) and its point of application F from (3). Apply E and see if the rod balances vertically. Notice in this figure, that P, E and Q are arranged in an exactly similar manner to P, Q and E in Fig. 70. So that since P, E and Q balance in both cases, the same results apply to both. Thus:

Given forces of same sense.	Given forces of opposite sense.
(1) $R=E=P+Q$.	(1) $Q=P+E=P+R$; $\therefore R=Q-P$.
(2) R acts in same sense as given forces.	(2) R acts in same sense as larger given force.
(3) $P:Q=b:a$.	(3) $P:Q=b:a$.

In (3) reference must be made respectively to Figs. 70 and 71. It is seen that in each case, a is the distance from P to R, and b is the distance from Q to R.

It may now be inferred from the above statements, that **the resultant of any number of parallel forces has a magnitude equal to the algebraic sum of the forces, and its position may be calculated by taking moments about any fixed point in the body.**

Couples.—It has now been seen that the resultant of two parallel forces of opposite sense may be found from

$$R=P-Q \quad \text{.............................} \quad (1)$$

$$P:Q=b:a \quad \text{.................................} \quad (2)$$

and that if R be reversed, giving $E=R$, then P, Q and E will balance.

Notice that if P and Q are nearly equal to one another, that R will become very small [from (1)], and that b and a must be nearly equal to one another and therefore both must be very large. So that as P and Q become more nearly equal to one another, E will become smaller and smaller and will move further away from P and Q. In the particular case of P becoming equal to Q, E will become zero and its distance from P and Q will be infinitely great. In this case P and Q are called a **couple**, and it follows from the above that no single force can balance a couple.

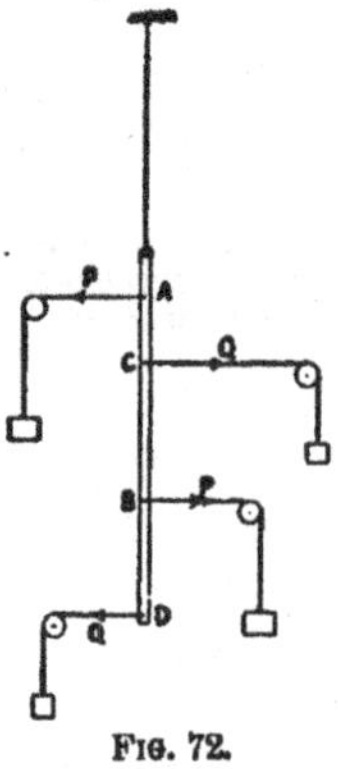

FIG. 72.

The **moment of a couple** is measured by the product of one of the forces and the perpendicular distance between them, called the *arm* of the couple.

EXPT.—Using the suspended rod, Fig. 72, apply two equal horizontal forces P, P, at A and B, thus causing an anticlockwise couple of moment $P \times AB$ to act on the rod. It will now be found impossible to keep the rod vertical by the application of any single force.

To balance, apply Q, Q, at C and D, thereby giving a clockwise couple to the rod. Select the values of Q and the arm CD so that the moments of the clockwise couple and the anticlockwise couple are equal, that is :

$$P \times AB = Q \times CD.$$

The rod now remains balanced vertically.

EXAMPLE. Suppose in the above experiment that P, P, are forces of 6 lbs. each, $AB=10''$ and $CD=15''$. Find the forces Q, Q.

$$P \times AB = Q \times CD,$$
$$6 \times 10 = Q \times 15,$$
$$Q = \underline{4} \text{ lbs.}$$

Fig. 72 shows all the forces horizontal, but this is not essential. **Any two couples of equal moment and opposite turning tendencies will balance the rod.**

EXPT.—Test this statement by inclining the lines of the P couple and also the lines of the Q couple, but at different angles for the two couples, and make their moments equal again by adjusting the weights hung on.

Couples have many interesting and useful properties, but most of them must be reserved until the student is more advanced. One thing in particular should be noticed—a couple applied to a body will not displace it as a whole from its given position, but will only cause it to rotate. Conversely, a body which is beginning to rotate must have a couple acting on it. This principle is made use of in the example following.

EXAMPLE. Given the total force transmitted from the piston to the crosshead of an engine, and the lengths of the crank and connecting rod, to find the pressure on the guides and the turning moment on the crank, neglecting frictional effects.

Make an outline diagram of crank, connecting rod, and piston rod tó scale (Fig. 73). Let P=force along piston rod, T=force along connecting rod, and Q = reaction of guide, which would be perpendicular to the piston rod if there were no friction. P, T, and Q, acting at the crosshead A, balance one another. Set off $Aa=P$ to any convenient scale of force. By drawing the parallelogram of forces $Aabc$, we obtain T, represented by Ac, and Q represented by Ab. The pressure on the guide will be equal and opposite to Q. The connecting rod exerts a force equal and opposite to T at its crank pin end B. This force acting on the crank pin has a moment about C, and, in consequence, the crank rotates.

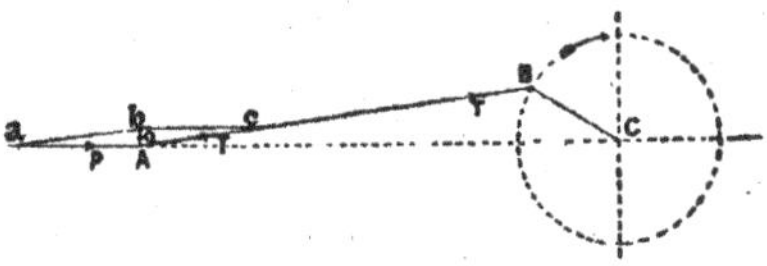

FIG. 73.—Forces on guide, and along connecting rod.

Considering the crank separately, the force T (Fig. 74), applied at B, requires an equal opposite force, which in the actual crank can only be applied at C, viz., the reaction from the crank shaft bearings. These two parallel equal forces T, T, of opposite sense, form a couple, of moment $T \times Bd$, gives the turning moment on the crank. This couple is balanced by the couple due to the resistance of the machinery driven by the engine, not shown on the diagram.

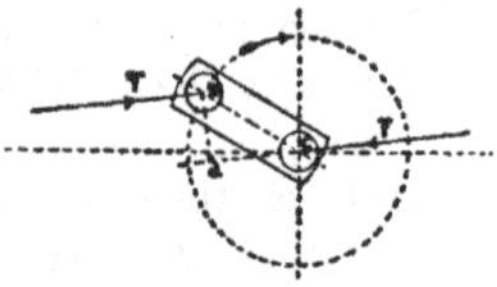

FIG. 74.—Forces on crank.

The force P transmitted by the piston rod does not remain constant during the stroke, but it is a very useful exercise at this stage to work out completely an example in which it is assumed to be uniform. The actual practical problem can be dealt with later.

The method which may be adopted is to draw outline diagrams of the crank and connecting rod when the crank is at intervals of 30° (Fig. 75), and then to find, by the parallelogram of forces, the forces Q and T for each position, assuming a constant value for P, say 1000 lbs. These values of

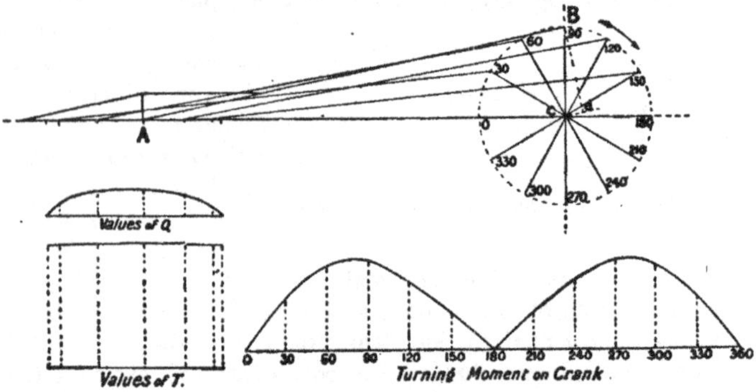

FIG. 75.—Turning moment diagrams for an engine.

Q and T set off as ordinates on a base line taken to represent the stroke, and a curve drawn through their tops (Fig. 75) will show the values of Q and T for all positions of the crosshead. Work out also the value of $T \times Bd$ for each position of the crank. The values of this set off as ordinates on a base taken to represent the revolution, and divided into intervals of 30°, will give the turning moment on the crank at any crank position (Fig. 75).

Consider again the rank in this position (Fig. 76). We notice that the whole tendency of *T* is partly to exert a push on the crank along *BC* and partly to rotate it. This can be easily seen if we resolve *T* into two forces, one along the crank and one perpendicular to it; thus, *V* gives the push along *BC* and *S* produces the turning moment. When the crank is in any position between 90° and 180°, and also between 270° and 360°, *V* is a pull, and in other positions it is a push.

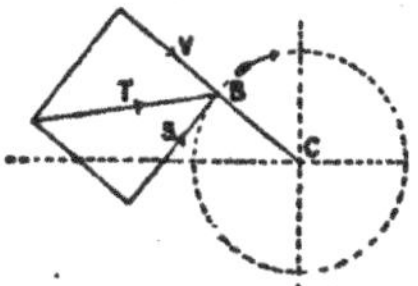

FIG. 76.—Force acting along the crank.

Experiment on forces in an engine.—It is easy to show these forces experimentally. Fig. 77 shows a flat piece of wood

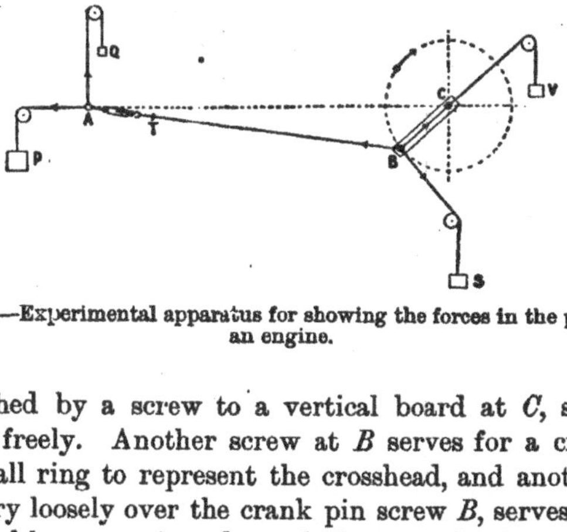

FIG. 77.—Experimental apparatus for showing the forces in the parts of an engine.

BC attached by a screw to a vertical board at *C*, so that it can turn freely. Another screw at *B* serves for a crank pin. *A* is a small ring to represent the crosshead, and another ring, fitting very loosely over the crank pin screw *B*, serves for connecting rod brasses. A cord attached to *A* and led over a pulley gives *P*. A cord hanging vertically from *A* gives *Q*. A spring balance introduced into a cord connecting *A* and *B* gives *T*, and cords led from the ring at *B* over pulleys give *V* and *S* respectively. The weights used should be fairly heavy, say 10 lbs. at *P* and the others to correspond with the crank position taken, in order to minimise as far as possible the disturbing effect produced by the weight of the spring balance.

EXERCISES ON CHAP. V.

1. AB is a uniform bar pivoted at C, its centre of length. W is a load of 5 lbs. placed at D, CD being 15″. If we have to restore balance by means of a 3 lb. weight, where must it be placed?

2. A bent lever ACB is pivoted at C; arm AC is horizontal and 9″ long; arm BC is vertical and 39″ long. A load of 300 lbs. is hung from A. Find what horizontal force at B will produce equilibrium. Neglect the weight of the lever.

3. The arms of a bent lever ACB are perpendicular to one another, and the lever is pivoted at C. Arm AC is 6″ long, and BC is 27″ long and inclined 30° to the vertical. Find what horizontal force P at B will balance a force $Q=250$ lbs. applied at A at 90° to AC. Neglect the weight of the lever.

4. A rod 5 ft. long has a weight of 2 lbs. at one end and 3 lbs. at the other, also a weight of 5 lbs. at its centre. Find the point about which it will balance. Neglect the weight of the rod.

5. Give a dimensioned sketch of a practicable arrangement of levers whereby a weight of ½ ton may be balanced by one of 10 lbs.

6. A beam 12 ft. long, supported at its ends, carries a load of 1¼ tons at a point 4 ft. from one end. Find the reactions of the supports, neglecting the weight of the beam.

7. A beam 20 ft. long, supported at its ends, has a load of 2 tons at the centre of its span, another of 1 ton at 3 ft. from one end, and another of 3 tons at 4 ft. from the other end. Neglect the weight of the beam and find the reactions of the supports.

8. AB is a beam 16 ft. long. It is supported at the end A and at C 4 ft. from the end B. A load of 4 tons is placed 6 ft. from A and another of 2 tons at the end B. Neglect the weight of the beam and find the reactions of the supports.

9. A handle used for turning a machine is 15″ radius. A man exerts a constant force of 30 lbs. (a) continually in a horizontal direction, (b) continually in a direction tangential to the circle of rotation of the handle. Draw diagrams to show in each case the turning moment for all positions of the handle.

10. A man whose weight is 160 lbs. can lift, unaided, a load of 3 cwts. Suppose he uses a lever 4 ft. long, the fulcrum being 3″ from one end, find what weight he can raise (a) if his end of the lever is moving down, (b) if his end of the lever is moving up, the fulcrum and weight changing places with each other.

11. The diameter of the safety valve of a steam boiler is 3 inches. The weight on the end of the lever is 55 lbs., and the distance from the centre of the valve to the fulcrum is 4·5 inches. What must be the length of the lever from the centre of the valve to the point of suspension of the weight, in order that the valve will just lift when the pressure of steam in the boiler is 80 lbs. per square inch? Neglect the weight of the lever and the valve. (1896.)

CHAPTER VI.

CENTRE OF GRAVITY. FORCES NOT ALL APPLIED AT THE SAME POINT. HANGING CHAINS. ARCH.

Centre of parallel forces.—Let two forces P and Q act on the rod AB (Fig. 77a), their directions being perpendicular to AB. The resultant R, found as before, will pass through C, and will divide AB in the proportion

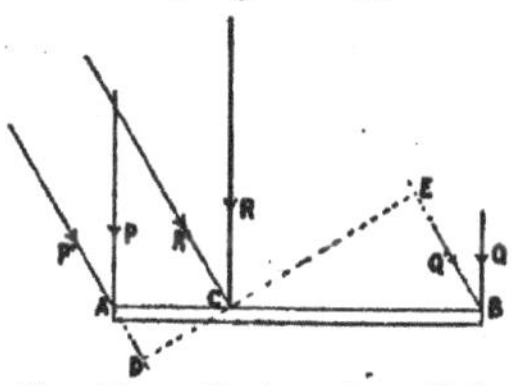

Fig. 77a.—Centre of parallel forces.

$$P : Q = BC : AC.$$

Suppose we incline the directions of P and Q, as at P' and Q', without altering their magnitudes. R will now act parallel to P' and Q', its magnitude will be unaltered, and it may be seen, if we draw a line through C, perpendicular to P' and Q', that this line DE is also divided inversely proportional to P and Q, that is,

$$P' : Q' = EC : CD.$$

It therefore follows that R', the resultant of P' and Q', will also pass through C, and it may be shown in the same way, that no matter how P' and Q' are inclined, provided their magnitudes are unaltered and that they are kept parallel to one another, that their resultant always acts through the same point C. This point is called the **centre** of the parallel forces P and Q.

If there are a number of parallel forces it will be easily seen that their resultant also always passes through the same point whatever may be the inclination of the forces.

Centre of gravity.—Suppose now we have a sheet of thin metal. Every particle of the metal is being pulled towards the earth's centre, so that we have a large number of forces acting on the body in lines which are practically parallel to one another (Fig. 78). The resultant of these forces is what we call the weight of the plate, *W* say. Now no matter how the plate may be turned (which is equivalent to inclining the forces on the particles to their first direction) there will be a point in the plate through which *W* always acts, this point being the centre of the parallel forces acting on the particles. Let *G* be this point, then *G* is called the **centre of weight, or centre of gravity** of the plate.

FIG. 78.—Centre of gravity.

In a great many bodies, we may see by inspection where the centre of gravity is. Thus, a *uniform rod* will have its centre of gravity at the middle of its length. A *square*, at the intersection of its diagonals, so also a *rectangle* and a *parallelogram*. A *circle* will have its centre of gravity at its geometrical centre. In the case of a *triangle*, the centre of gravity will be found one-third way up a line from the centre of the base to the opposite corner. *Pyramids* and *cones* have their centres of gravity $\frac{1}{4}$ way up a line from the centre of the base to the apex. In *prisms*, with ends perpendicular to their axes, the centre of gravity will lie at the geometrical centre of the middle cross section.

When we are considering the equilibrium of a given body we may regard its whole weight as concentrated at its centre of gravity.

Centre of gravity by experiment.—Suspend a thin plate of any irregular outline and of any material by a cord attached at *A* (Fig. 79). The pull in the cord will be *P* equal to *W*, the weight of the plate. If the plate be at rest, these forces, being the only two acting on it, must be in the same straight line. It follows therefore that *G*, the centre of gravity of the plate, must fall vertically under *A*.

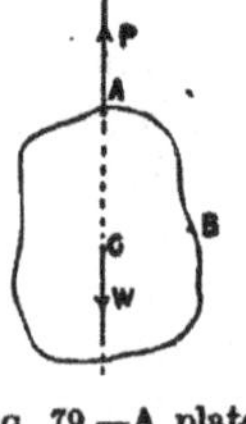

FIG. 79.—A plate hung from *A*.

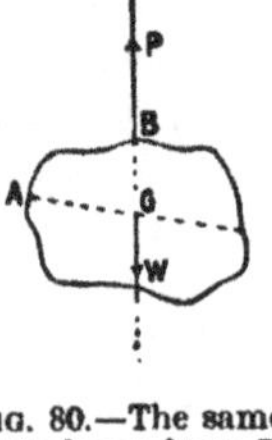

FIG. 80.—The same plate hung from *B*.

Produce the line of P downwards on the plate as shown, then G is in this line. Now hang the plate from another point in it such as B (Fig. 80). G again must be vertically under B, so if the line of P be again drawn on the plate, the intersection of this line with that first drawn will give G.

EXPT.—Find by experiment the centres of gravity of pieces of card board cut into the following shapes: triangle; square with a piece cut off one corner; shape of a letter **H**; shape of a letter **L**; shape of a letter **T**.

Centre of gravity by calculation.—For plates of fairly regular outline, the position of the centre of gravity may be easily calculated by using the principle of moments. Thus, to find the centre of gravity of the plate shown in Fig. 81. Divide it up into rectangles as shown. Then, the centre of gravity of *abcg* is at *m*, and of *gdef* at *n*.

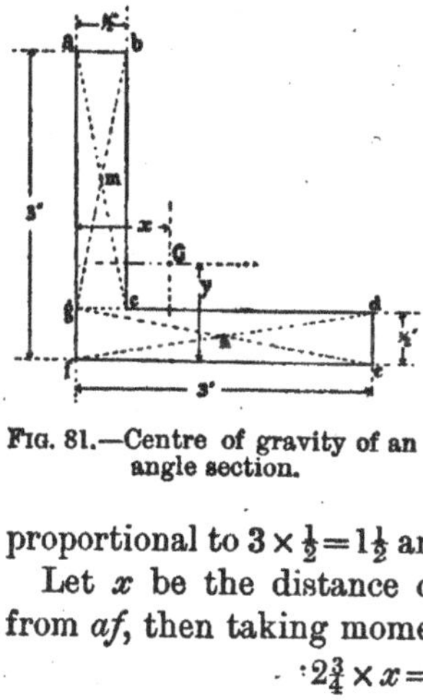

FIG. 81.—Centre of gravity of an angle section.

Also the weights of the rectangles will be proportional to their areas. So that weight of *abcg* is proportional to $2\frac{1}{2}\times\frac{1}{2}=1\frac{1}{4}$, and weight of *gdef* proportional to $3\times\frac{1}{2}=1\frac{1}{2}$ and the weight of the whole plate to $2\frac{3}{4}$.

Let x be the distance of the centre of gravity of the plate from *af*, then taking moments about *af*,

$$\begin{aligned} 2\tfrac{3}{4}\times x &= (1\tfrac{1}{4}\times\tfrac{1}{4})+(1\tfrac{1}{2}\times 1\tfrac{1}{2}) \\ &= \tfrac{5}{16}+\tfrac{9}{4} \\ &= \tfrac{41}{16}; \\ \therefore\ x &= \tfrac{41}{16}\times\tfrac{4}{11}=\tfrac{41}{44}=0{\cdot}932''. \end{aligned}$$

Now take moments about *fe*, and let y be the distance of the centre of gravity from *fe*.

$$\begin{aligned} 2\tfrac{3}{4}\times y &= (1\tfrac{1}{2}\times\tfrac{1}{4})+(1\tfrac{1}{4}\times 1\tfrac{3}{4}) \\ &= \tfrac{3}{8}+\tfrac{35}{16} \\ &= \tfrac{41}{16}; \\ \therefore\ y &= \tfrac{41}{16}\times\tfrac{4}{11}=\tfrac{41}{44}=0{\cdot}932''. \end{aligned}$$

The centre of gravity is therefore a point 0·932" from each of the sides *af* and *fe*.

Centre of gravity by a graphical method.—We may proceed in another way. Thus, to find the centre of gravity of the plate shown in Fig 82, divide it into triangles by the line *bd*. Find the centre of gravity of each by construction, *i.e.* bisect *bd* at *e*, join *ea* and *ec* and measure $\frac{1}{3}$rd up each of them from *e*. This construction will give c_1 and c_2, the centres of gravity of *abd* and *bcd*. The centre of gravity of the whole plate must be in the line joining c_1 and c_2. Now divide the plate again by the line *ac*. Find as before c_3 and c_4, the centres of gravity of *abc* and *acd*. Join c_3, c_4. The centre of gravity of the whole plate must be in the line c_3c_4. *G* will therefore be the point where c_1c_2 and c_3c_4 intersect.

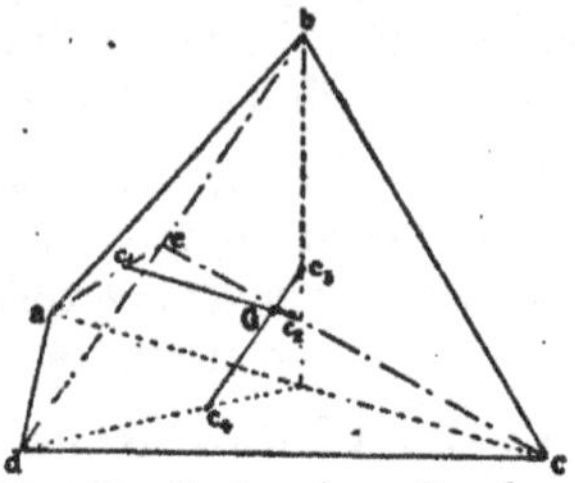

FIG. 82.—Centre of gravity of an irregular plate.

State of equilibrium of a body.—A body is said to be in **stable equilibrium** if, on being slightly disturbed, it tends to return to its original position; **unstable** if it tends to go over further, and **neutral** if it will remain at rest indifferently in any position. We may easily test for a body's equilibrium.

Thus, suppose a **cone** to stand on its base on a horizontal surface (Fig. 83). Its weight being *W*, then *R*, the reaction of the

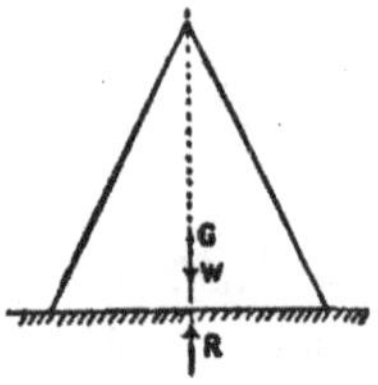

FIG. 83.

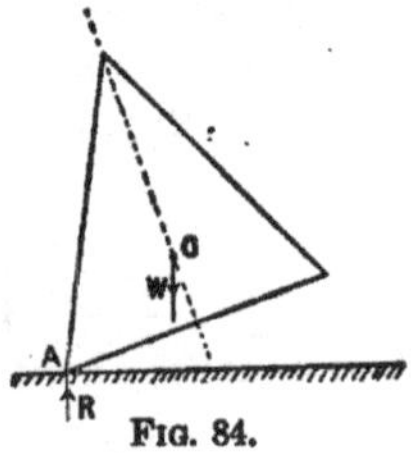

FIG. 84.

Stable equilibrium of a cone.

surface, will be equal to *W* and in same straight line. Disturb the cone slightly. *R* shifts along to *A* (Fig. 84), and *R* and *W* now form a couple tending to bring the cone back to its original position. This is therefore a case of stable equilibrium.

Now stand the cone on its apex (Fig. 85). If we disturb it slightly, *R* and *W* form a couple tending to upset it (Fig. 86). The position is therefore unstable.

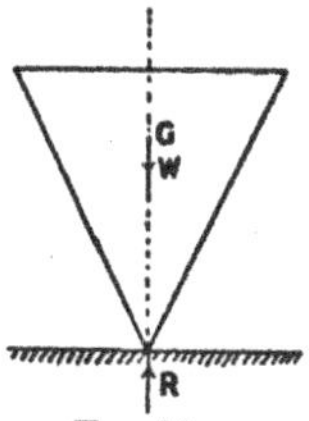

FIG. 85.

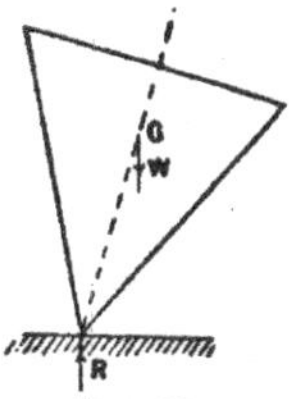

FIG. 86.

Unstable equilibrium of a cone.

A **ball** resting on a horizontal table will be in neutral equilibrium, because *R* and *W* (Fig. 87) will always be in the same straight line no matter how the ball is disturbed. It will therefore remain at rest in any position.

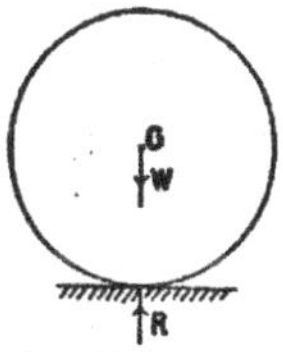

FIG. 87.—Neutral equilibrium of a ball.

A **cylinder** lying on its side (Fig. 88) will also be in neutral equilibrium, but if we cut a slice off the top, the equilibrium will be found to be stable. For the centre of gravity *G′* (Fig. 89) will be brought below *G* by cutting off the slice, and consequently, if the cylinder be slightly disturbed, as in Fig. 90, *R* and *W* will give a couple tending to bring it back again to its original position.

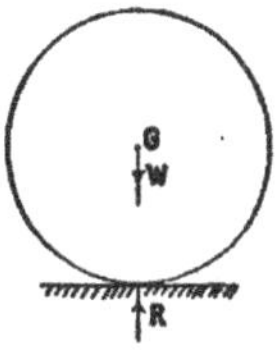

FIG. 88.—Neutral equilibrium of a cylinder.

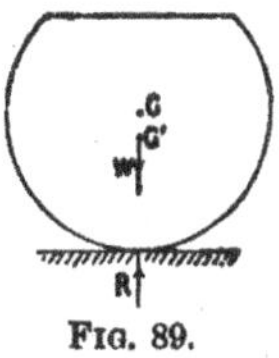

FIG. 89.

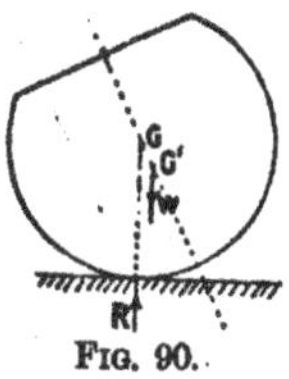

FIG. 90.

Portion of a cylinder in stable equilibrium.

It may be seen, from some of the above examples, that another property of the centre of gravity of a body is that, if the body is free to be moved by its weight from its given position, it will always do so in such a way that the centre of gravity is lowered thereby.

Any forces acting in a plane.—We may now consider what conditions must be fulfilled if a number of forces, all in the same plane, but not necessarily passing through the same point, act on a body and produce equilibrium.

There are three conditions to be satisfied:

(1) *There must be no tendency to produce vertical movement of the body, either up or down.*

(2) *There must be no tendency to produce horizontal movement, either to right or left.*

(3) *There must be no tendency to rotate the body.*

The easiest way of testing whether these conditions are satisfied is to take components of each force in horizontal and vertical directions. Then, that there may be no vertical movement, the sum of the upward components must be equal to the sum of those acting downward; for no horizontal movement, the sum of the components acting towards the right must equal the sum of those acting towards the left; and for no rotation, the sum of all the clockwise moments, about any point, produced by the components must equal the sum of the anticlockwise moments about the same point.

Graphical solution by the link polygon.—The equilibrium of a number of given forces all in one plane may be tested by this method. Given any number of forces such as *P*, *Q*, *S*, *T* (Fig. 91), all in one plane. Take any one of them, *e.g.* *Q*, and balance it by applying any two forces, p_1, p_2, at any point such as *B*, on *Q*'s line. Applying the triangle of forces to p_1, *B*, and p_2, the magnitudes of p_1 and p_2 can be found, as at *Oab*. Now imagine that *BC* is a rod, pushing at *B* with force p_2, and its other end pushing at *C*, on the line of *S*, with equal force p_2. Balance *S* and p_2 at *C* by a third force, p_3, acting along a rod *DC*. The direction and magnitude of this force may be found from the triangle of forces *Obc*. Let

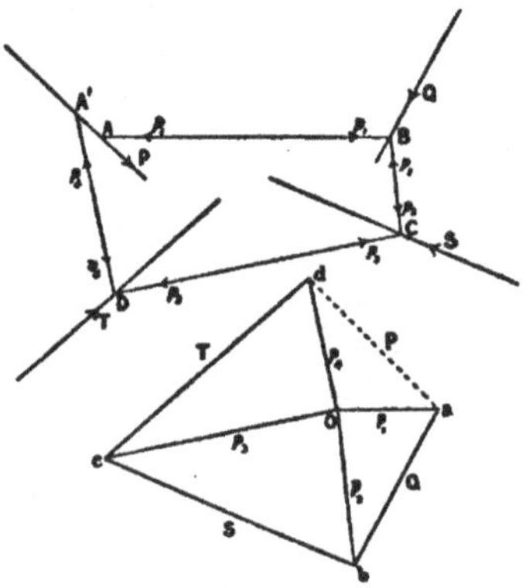

FIG. 91.—The link polygon.

the end D of the rod push at D on T's line with force p_3. Balance T and p_3 at D by a force p_4, the line and magnitude of which will be found from the triangle of forces Ocd. We have now balanced all the given forces except P, and this would be balanced also provided the push p_4 acting at A' on P's line, the push p_1 acting at A on the same line, and P, balance one another.

For this to be possible the following conditions must be satisfied.

(1) A and A' must coincide, for the three forces p_1, p_4, and P must pass through the same point.

(2) The triangle of forces for them must close, as shown at aOd.

If these conditions are fulfilled, we have a closed system of rods or links $ABCD$, and also a closed force polygon $abcd$ (Fig. 91). Notice that the closed force polygon $abcd$ will have the same shape no matter which force we select to begin with, for its sides are parallel and proportional to the given forces. On the other hand the **link polygon**, as it is called, will have a shape depending on how we select the directions of the first two links AB and BC. This will alter the position of O in the force polygon, but it will not alter the essential conditions.

In practice we proceed thus. First draw the force polygon $abcd$ for the given forces. *For equilibrium this must close.* Then select any point O and join it to each corner of the force polygon. Draw the links between the given forces parallel to these lines Oa, Ob, etc., on the force polygon. Notice that Oa, on the force polygon, comes between da representing P, and ab representing Q; this means that the link connecting P and Q must be drawn parallel to Oa; attention paid to this point will save mistakes. *If the link polygon closes also the given forces are in equilibrium.*

The method is suitable for use in many cases where graphical solution is desirable.

Tensions in a hanging cord.—Expt.—Arrange an experimental link polygon by attaching a cord $ACDEFB$ to the ends A and B of a rod (Fig. 92); hang the rod up by two cords attached to fixed supports H and K on a vertical board, and to A and B. If weights are now hung from C, D, E, and F, the cord will take a definite shape. To find its tensions at any part of its length, we may apply the triangle of forces. Mark on a piece of drawing paper, stretched on the board, the directions of the cord and of the applied weights, and remove the paper.

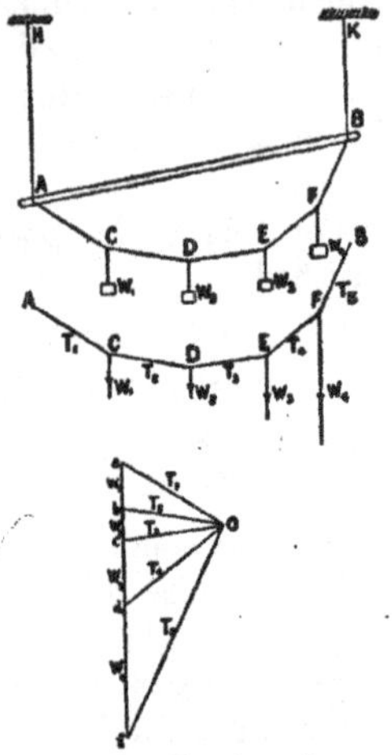

As W_1, W_2, W_3, and W_4, are known, the triangle of forces can be now drawn for each of the points C, D, E, and F. Calling the tensions in the different parts of the cord T_1, T_2, T_3, T_4, and T_5, abO will be the triangle of forces for W_1, T_2, and T_1 acting at C; bcO that for W_2, T_3, and T_2 acting at D; cdO that for W_3, T_4, and T_3 acting at E; dfO that for W_4, T_5, and T_4 acting at F. The tensions T_1, T_2, T_3, T_4, and T_5, will therefore be given by the lines Oa, Ob, Oc, Od, Of respectively, measured to the scale of force.

FIG. 92.—Tensions in a hanging cord.

Tensions at the ends of a stretched chain.—This construction enables us to solve some important practical problems. Thus, noticing that af in the force diagram represents the total load applied to the cord, and that Oa and Of represent the tensions in the two end portions of the cord and are drawn in the same direction as these portions, we can easily find the tensions at the ends of a hanging chain if we only know the inclination of the chain there, and the total applied weight.

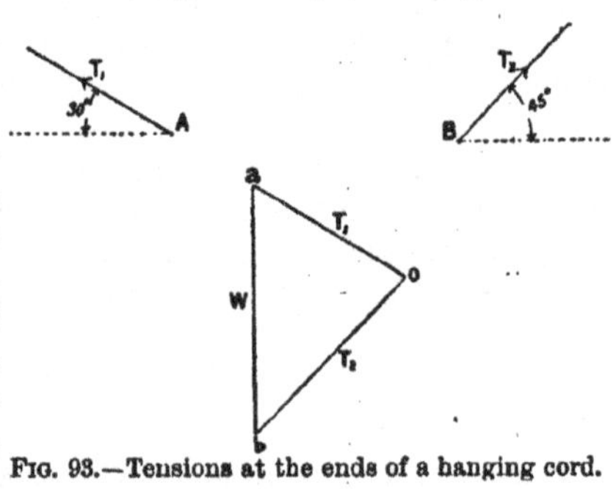

FIG. 93.—Tensions at the ends of a hanging cord.

Thus, suppose a chain (Fig. 93) hanging from two supports

A and *B* has its ends making angles of 30° and 45° respectively with the horizontal, and that the total applied weight is 50 lbs. Set off *ab* to scale to represent 50 lbs., and draw *aO* and *bO* at angles of 30° and 45° respectively to the horizontal. Then *aO*, to the scale of force, gives T_1 equal to 36½ lbs., the tension at the 30° end of chain and *Ob* to the same scale gives T_2 equal to 44½ lbs., the tension at the 45° end. The form taken by the hanging chain is not required for the solution, and is therefore not shown in the figure.

Calculation of same problem from dip and span.—*AB* (Fig. 94) represents a fine wire, or a chain stretched between two supports on the same level; to find its tensions at the middle of the span and at the supports, we may proceed as follows. Imagine the wire cut at *C*, the centre of its span, and instead of the pull of the left-hand portion, apply a horizontal force *H* of the same amount; this will keep the right-hand portion of the chain or wire in equilibrium. We apply a horizontal force at *C*, because the chain or wire will be horizontal there as the supports are on the same level, and also because the wire or chain is only capable of resisting pull, consequently its tension at any point will be always in the same direction as the chain at that point.

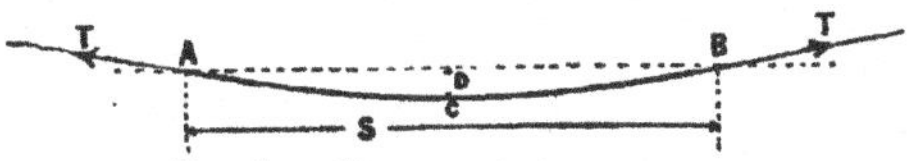

FIG. 94.—Hanging chain or wire.

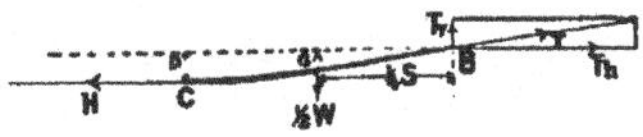

FIG. 95.—Relations between the dip, span, etc.

Consider now the equilibrium of the right-hand portion of the chain or wire. *T*, its tension at the end *B* (Fig. 95), may be split into horizontal and vertical components T_v and T_h. If *W* is the whole weight of the chain or wire, ½ *W* will be the weight of the portion under consideration and may be placed at its centre of gravity *G*, which we may assume (without serious error if the dip of the chain is small compared with the span) to be midway between the verticals through *C* and *B*. Let *D* be the dip of the chain and *S* its span, both in the same units of length. Notice now that two horizontal forces and two vertical forces only act on the balanced half chain, consequently the vertical forces must be equal and opposite and so also the horizontal forces; two opposite equal couples are therefore formed by these forces, giving

$$T_h = H$$

$$T_v = \frac{W}{2}$$

Moment of clockwise couple $= H \times D$

Moment of anticlockwise couple $= \frac{1}{2} W \times \frac{S}{4}$

$$\text{and } H \times D = \frac{1}{2} W \times \frac{S}{4}$$

$$\text{or } H = \frac{1}{8} W . \frac{S}{D}$$

Having found H, and knowing that T_h is equal to H, we may now find T from its known components T_h and T_v, thus

$$T = \sqrt{T_h^2 + T_v^2},$$

and acts at an angle to the horizontal the tangent of which is $\frac{T_v}{T_h}$.

An actual example.—An experiment was tried on a chain 20′ 2″ long, weight 3 lbs., when hung from two supports, 19′ 9″ apart, both on the same level. Instead of fixing the ends of the chain, each end was supported by cords lead over pulleys shown at A and C, (Fig 96) the pulls in these giving T_v and T_h. By the experiment it was found that

$$T_v = 1{\cdot}5 \text{ lbs.}$$

$$\text{and } T_h = H = 4{\cdot}24 \text{ lbs.}$$

The dip at the centre of the span was measured and found to be 1·73 feet.

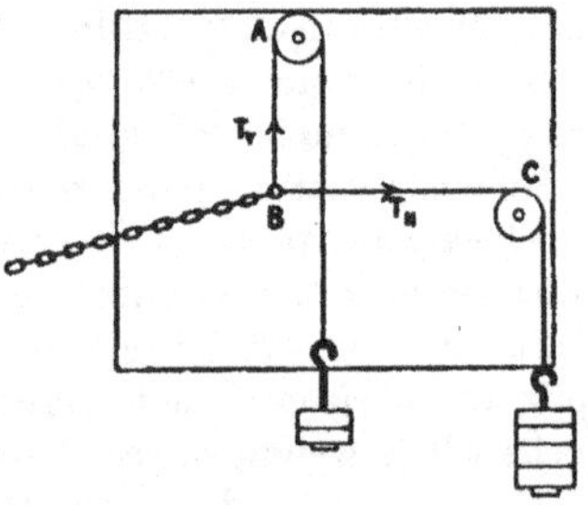

Fig. 96.—The forces at the end of a chain, determined experimentally.

By calculation, from $H = \frac{1}{8} W \frac{S}{D}$

$$H = \frac{3 \times 19{\cdot}75}{8 \times 1{\cdot}73} = 4{\cdot}28 \text{ lbs.},$$

showing an error of less than one per cent when compared with the observed value of H obtained in the experiment.

The arch.—Supposing we have several vertical loads supported by means of a cord pulled at its ends (Fig. 97), and that we know the shape taken up by the cord. Imagine the whole apparatus to become rigid for a moment and turn it upside down; we should then have a structure in equilibrium (Fig. 98) under

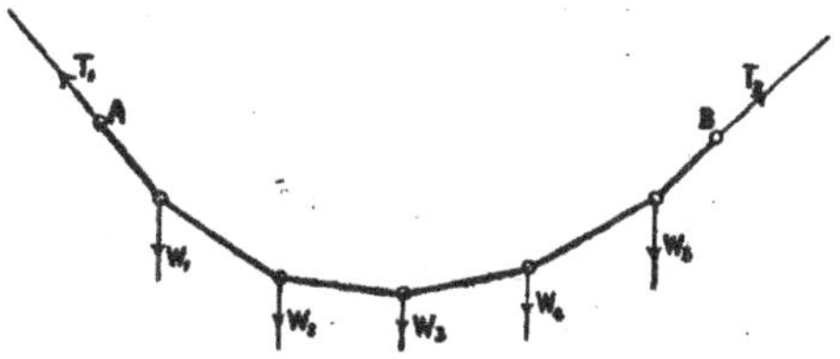

Fig. 97.—Hanging cord, supporting loads.

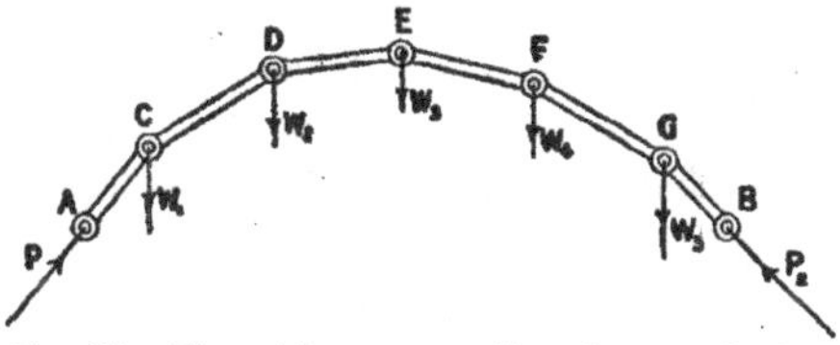

Fig. 98.—Hinged bars, supporting the same loads.

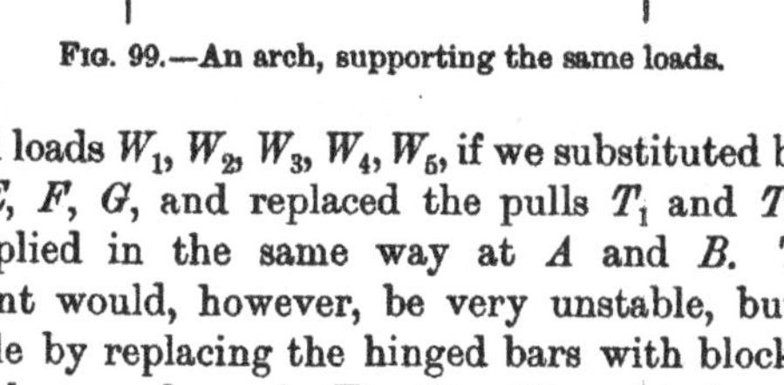

Fig. 99.—An arch, supporting the same loads.

the applied loads W_1, W_2, W_3, W_4, W_5, if we substituted bars hinged at *C*, *D*, *E*, *F*, *G*, and replaced the pulls T_1 and T_2 by equal pushes applied in the same way at *A* and *B*. The whole arrangement would, however, be very unstable, but could be made stable by replacing the hinged bars with blocks properly fitted together, as shown in Fig. 99. These blocks would push one another at their joints in just the same way as the bars did, only now we have the structure called an **arch**. The curve formed by the hinged bars is called the *line of resistance of the arch.* It is interesting and instructive to watch the changes produced in the line of resistance of a model arch such as that shown in Fig. 100. Here the joints, instead of being flat, are

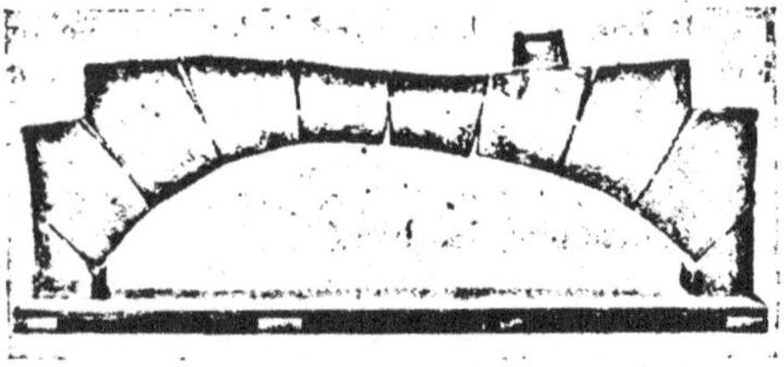

FIG. 100.—Experimental model of an arch.

made rounded so that the blocks may roll on one another, thereby giving by their places of contact, points on the curve of resistance. The places of contact of the blocks have been marked in white chalk on the model and are easily seen in the figure. In an actual arch, the shape of the line of resistance alters, when loads are applied, in the same way as it does on the model, only, of course, as the joints are flat, the shape of the arch itself does not change unless rupture occurs. In practice, an arch is considered safe when the line of resistance for any possible load passes through any part of the middle third of every joint.

EXERCISES ON CHAP. VI.

1. A uniform plank, 20 ft. long, weight 90 lbs., rests on supports at its ends. A load of 500 lbs. rests 8 ft. from one end. Find the reactions of the supports.

2. A uniform beam 12 ft. long, supported at its ends, carries a distributed load, including its own weight, of $\frac{1}{2}$ ton per foot run. A concentrated load of 1 ton rests 5 ft. from one end, and another of 3 tons, 4 ft. from the other end. Calculate the reactions of the supports.

3. A uniform beam 16 ft. long weighs 300 lbs. It is supported at one end and at a point 4 ft. from the other end. Calculate the reactions of the supports.

4. A cone, 3 ft. diameter of base, 4 ft. high, stands on its base on a horizontal surface. Specific gravity of material = 3. What horizontal force at the top will turn the cone over?

5. A triangular plate, ABC, of wrought iron 1" thick, lies on a horizontal surface. $AB=3$ ft., $BC=3\frac{1}{2}$ ft., $CA=4$ ft. Find what vertical lifting force applied at A will raise that corner of the plate.

6. A wall 8 ft. high, 14" thick, is built of material weighing 130 lbs. per cubic foot. The normal wind pressure on the face of the wall is 50 lbs. per square foot of vertical surface. Consider a piece of the wall one foot long, and calculate the overthrowing moment of the wind on it and also the resisting moment of the weight of the wall. Will the wall stand or fall?

7. Show in a diagram the couples acting on a hinged door, 7 ft. high, 3 ft. wide, weight 90 lbs. There are two hinges, placed one foot from top and bottom of the door.

8. The jib of a crane is 40 feet long and weighs $\frac{1}{2}$ ton. The tie is 30 ft. long and the post is 25 ft. high. Make an outline diagram to scale and calculate the pull on the tie produced by the weight of the jib. Take the centre of gravity of the jib at 16 ft. from the lower end.

9. A symmetrical roof weighs 16 tons. On one side of it there are 7 tons of snow equally distributed. Find the pressures on the supports.

10. A horizontal beam 10 feet long weighs $\frac{1}{2}$ ton and is pivoted 4 ft. from one end. Its centre of gravity lies in the longer part, 1 foot from the pivot. Find where a load of 500 lbs. must be placed to keep the beam balanced.

11. A ladder 24 ft. long weighs 50 lbs. and has its centre of gravity 8 ft. from one end. A bag of tools, weight 100 lbs., is slung at the centre of length of the ladder. A lad and a man carry the whole between them, the lad being at the lighter end of the ladder. Find where the man must be if his share of the load is 90 lbs.

12. A chain of weight 20 lbs. is stretched between two points on the same level, 40 feet apart. If the dip is 4 ft., calculate the pulls at the middle and at each end of the chain.

13. The weight of a chain hanging from two points of support is 220 lbs.; its inclinations to the horizontal at the points of support are 25° and 42° respectively; what are the tensions in the chain at the points of support? (1900.)

14. A symmetrical pair of steps, hinged together at the top and connected together by a string at the bottom, stands on a smooth horizontal plane. If the length of each side be 3 feet 3 inches, and the string be 3 feet in length, find the tension of the string when a person of 140 lbs. in weight stands on the steps at a height of 2 feet from the ground. How is the tension in the string affected as the person ascends the steps? (1896.)

CHAPTER VII.

STRESS. STRAIN. ELASTICITY. ULTIMATE TENSILE STRENGTH. STRENGTH OF SHELLS. SHEAR.

Stress.—What happens to a piece of material when forces are applied to it in the direction of its length? This is a question which now requires to be studied. Suppose AB (Fig. 101) is a bar the ends of which are subjected to equal and opposite pulls P, P. Imagine the bar to be cut at CD at 90° to its axis, and consider what must be done to preserve the equilibrium of the left-hand portion. In order to balance the force P it will be necessary to apply an equal and opposite force at CD (Fig. 102); let Q be this force. As the cut is only imaginary, the portion to the right of CD in the actual bar must have supplied this force Q, in order to keep the left-hand portion balanced. In the same way, the left-hand portion of the bar exerts a pull Q equal to P, acting towards the left (Fig. 103), on the right-hand portion of the bar.

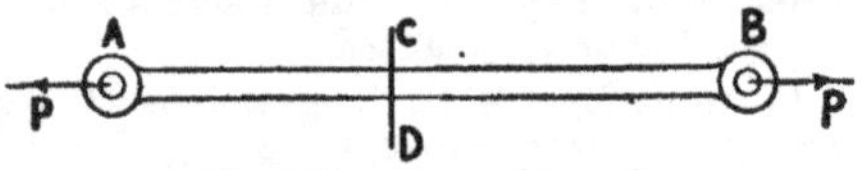

FIG. 101.—Bar under pull.

FIG. 102.—Equilibrium of the left-hand portion.

FIG. 103.—Equilibrium of the right-hand portion.

The force Q in the actual bar will be distributed in some manner over the section of the material, the sum of the pulls on every part of the section being equal to Q (Fig. 104). For sections taken near the ends of the bar, where the forces P, P

are applied, the distribution cannot be definitely stated, but at some little distance from the ends, and right along the bar, the distribution is probably uniform, each square unit of the sections bearing equal forces. This force per unit area of the section is called **stress.** The stress may be easily calculated by dividing the total force on the section by the area of the section. Thus, if the bar is pulled at its ends with forces of 10 tons each, and its sectional area is 2½ square inches, the stress will be 10 divided by 2½, or 4 tons per square inch. If the bar is of uniform section, a stress equal to this will be found on any cross section, except those very near the ends where the distribution is unknown.

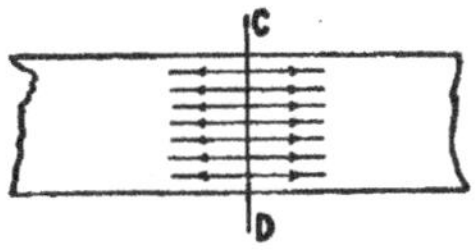

FIG. 104.—Distribution of the stress.

Ties and Struts.—Those portions of a structure which are intended to be under *pull* are called **ties**, and if intended to be under *push*, are called **struts.** Ties are said to be under **tensile stress** when pulled, and struts under **compressive stress** when pushed.

If we consider the case of a tie slightly bent at first and then pulled, we can easily see that the tendency is to straighten it (Fig. 105); in the same way a pulled string becomes straight. Also, if the tie is straight to begin with, there will be no tendency to bend it when pulls are applied. It follows therefore that since no lateral stiffness is required in ties to resist bending, that the shape of the cross section is immaterial. It is very different, however, in struts.

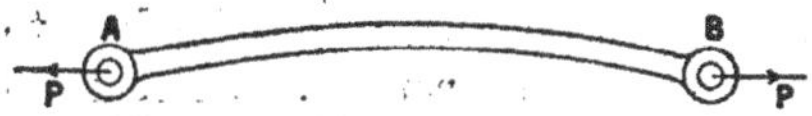

FIG. 105.—Tie bar, originally bent.

If the strut is originally bent, the tendency will be to bend it still more (Fig. 106), consequently the section must be chosen so as to give the necessary lateral stiffness to resist this bending action. If the strut is straight to begin with, and is kept straight when the pushes are applied, then the stress on any section not too near its ends will be uniformly distributed, and will be found as before by dividing the total force on the section by the area of the section.

FIG. 106.—Strut, originally bent.

Both ties and struts should be made straight to begin with, as we have seen that ties tend to become straight if originally bent, and struts will bend more if not at first straight. Horizontal or inclined ties of great length compared with their cross sectional dimensions will have a bending tendency due to their own weight. Such ties should have a suitable cross section to give stiffness, or should be supported at intervals by suspending rods. Struts are of many different forms depending on their lengths and the magnitude of the loads. A few common forms are shown in Fig. 107.

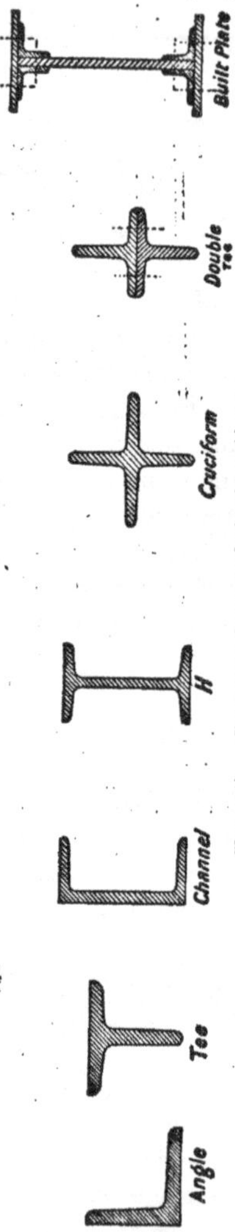

FIG. 107.—Common forms of strut sections.

Columns are vertical pieces designed for the purpose of carrying weights and come under the heading of struts as they are

FIG. 108.—Flanged column, load applied centrally.

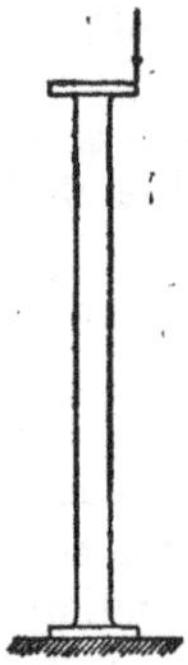
FIG. 109.—Flanged column, load applied at edge of flange.

subjected to push forces. Lateral stiffness must be arranged for in columns as in struts. Short blocks used as columns fail by *crushing*. Very long columns fail by *bending* and thus *breaking*. Columns are often made stiffer by putting flanges on the ends, the effect being that the column bends as shown by the dotted lines in Fig. 108, instead of as a whole. But if the load is not centrally applied at the ends, the effect may be worse than if there were no flanges. As an extreme case, think of W applied at the edge of the flange (Fig. 109). It is easy to see that the bending tendency is much greater than would be the case if the flanges were absent.

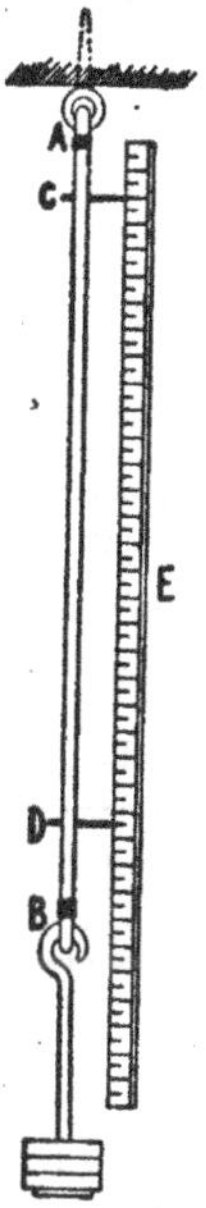

FIG. 110.—Apparatus for measuring the extensions of rubber.

Change of length of a loaded bar.—All materials stretch when pull forces are applied and become shorter with push forces. With sufficiently delicate apparatus, these changes of length can be measured in metals. With material such as rubber, an ordinary scale is sufficient to measure the differences. Fig. 110 shows a round rod of rubber about ½ inch diameter and 3 or 4 feet long tied to a support at A and having a hook for carrying weights at B. C and D are two needles pushed through the rubber; these needles will move on a scale E when loads are applied, and from their readings the changes in length of the portion CD can be obtained.

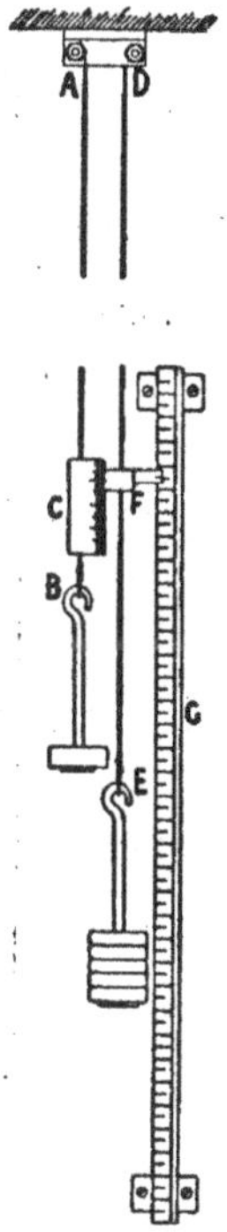

FIG. 111.—Apparatus for measuring the extensions of a pulled wire.

A simple apparatus for measuring the **extensions of wires** under various loads consists of two wires hung side by side from the same support. One wire AB (Fig. 111) carries a constant load sufficient to keep it taut, and has a scale of inches divided into tenths fixed at C. The other wire DE is the wire under test. The load on it can be varied, and when this is done a vernier fixed at F will move over the scale C, and will give the changes in length of the portion DF. The arrangement for carrying the scales prevents any drooping of the support at AD from being measured as an extension of the wire under test. The support for the wires should, in order to have a long portion of the wire available for testing, be fixed at the ceiling, or as high up the wall as possible.

Extensions in ordinary test pieces.—Test pieces of ordinary bars or plates used for engineering work must be subjected to great loads before a measurable extension is produced. This is done in large testing machines and the extensions are measured by means of instruments called **Extensometers.** In Prof. Ewing's Extensometer, the extension of the piece is measured by the movement of a fine wire over the scale of a microscope. It is possible with the instrument to measure a change of length of $\frac{1}{50,000}$th inch on a test piece 8″ long. The following results were obtained by its use, and are given in illustration of an important fact.

Test bar of flat iron 1·501″ wide, 0·492″ thick, 8″ long between the test points. The load was applied in steps of 1000 lbs. until a maximum of 10,000 lbs. was reached. The bar carried this load for 2 or 3 minutes, and then the load was taken off 1000 lbs. at a time.

An Experiment with Ewing's Extensometer.

Load, lbs.	Load increasing.		Load diminishing.	
	Scale reading.	Differences per 1000 lbs. load.	Scale reading.	Differences per 1000 lbs. load.
0	3·00		3·00	
		0·18		0·16
1000	3·18		3·16	
		0·18		0·18
2000	3·36		3·34	
		0·18		0·19
3000	3·54		3·53	
		0·19		0·19
4000	3·73		3·72	
		0·21		0·20
5000	3·94		3·92	
		0·19		0·21
6000	4·13		4·13	
		0·19		0·19
7000	4·32		4·32	
		0·20		0·20
8000	4·52		4·52	
		0·20		0·20
9000	4·72		4·72	
		0·19		0·19
10,000	4·91		4·91	

The scale reading was such that one part on the scale corresponds to an extension of $\frac{1}{500}$ inch. Consequently the total stretch for 10,000 lbs. load was $\frac{1 \cdot 91}{500}$ inch.

Plotting on squared paper column 1 for ordinates and column 2 for abscissae, we see that all the points lie practically on a

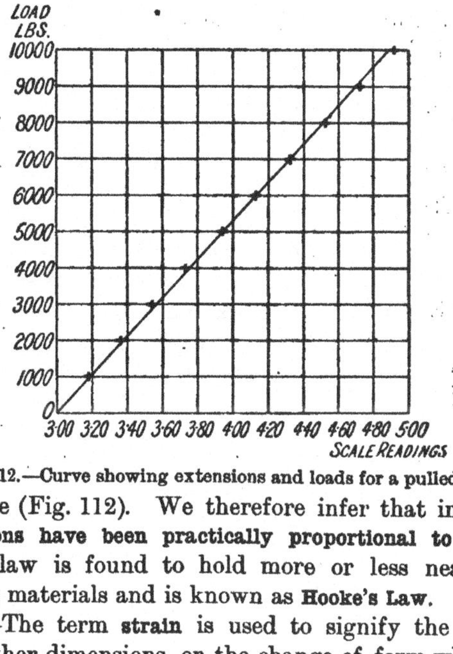

FIG. 112.—Curve showing extensions and loads for a pulled bar.

straight line (Fig. 112). We therefore infer that in this piece **the extensions have been practically proportional to the loads.** The same law is found to hold more or less nearly in all engineering materials and is known as **Hooke's Law.**

Strain.—The term **strain** is used to signify the change of length or other dimensions, or the change of form which occurs in a material when loads are applied. Strain is measured in a pulled or pushed bar by stating the change of length per unit of original length of the bar. To obtain it in any particular case, divide the total extension by the original length of the bar. Thus, in the above experiment,

$$\text{tensile strain} = \frac{1 \cdot 91}{500 \times 8} = 0 \cdot 0004778.$$

Strain is not measured in any units, as it is simply the ratio of two lengths.

Some important definitions.—If we go on loading a test piece, we presently reach a point where Hooke's Law breaks down, and the extensions cease to be proportional to the load, but increase at a more rapid rate. Up to this point, if the load be removed, the piece will come back to its original length, but if the loading is carried beyond this point the bar will be found to be permanently extended when the load is removed.

Elasticity is that property of matter by virtue of which it tends to return, or spring back, to its original shape and dimensions when the applied forces are removed.

Elastic Limit is the name given to the stress at which Hooke's Law breaks down, and at which the bar takes up a permanent extension. The permanent extension is called **Permanent Set,** and material loaded beyond the Elastic Limit is said to be **overstrained.**

Modulus of elasticity.—Notice that both strain and stress are proportional to the load producing them up to the elastic limit, and that therefore they are proportional to one another for a given material. It follows that if the stress be divided by the corresponding strain, a constant number for the same material will be produced. The fraction $\frac{\text{stress}}{\text{strain}}$, is called **modulus of elasticity.** *Young's Modulus of Elasticity,* is that which refers to a bar pushed or pulled ; it is usually written E. For calculation of E we have the following relations :—

Let W = load applied,
A = sectional area of bar in square inches,
L = length of bar,
e = change of length produced by W,

both lengths being in the same units. Then

$$\text{stress} = \frac{W}{A},$$

$$\text{strain} = \frac{e}{L},$$

and

$$E = \frac{\text{stress}}{\text{strain}} = \frac{W}{A} \div \frac{e}{L}$$

$$= \frac{W \cdot L}{A \cdot e}.$$

The modulus of elasticity will be described as tons per square inch or lbs. per square inch depending on how the stress is measured. The average values of Young's Modulus for some common materials are given in the following table :

YOUNG'S MODULUS OF ELASTICITY.

Material.	E. (tons per square inch).
Wrought iron, - - -	13,000
Steel, - - - - -	13,500
Cast iron, - - - -	6,000
Rolled copper, - - -	6,200
Brass, - - - - -	5,700
Gun metal, - - - -	5,000
Phosphor bronze, - - -	6,000
Aluminium bronze, - -	6,500

Phenomena beyond the elastic limit.—In **ductile** materials, such as iron and steel, if further loading be applied after the elastic limit is passed, a point is reached where the material draws out considerably with no, or very little, increase to the load. This point is called the **yield point.** Further loading produces considerable extension throughout the material which can easily be seen even in a short piece, and it will be also observed, if the diameter of the piece is measured from time to time, that contraction is going on all over the specimen. Presently the load is reached at which the wire is about to break. At the place where fracture is about to occur considerable contraction will be observed, until finally rupture occurs. In materials like cast iron and brass, there is no yield point, and the total extension of specimens of these materials before breaking is much smaller than for iron or steel. In general, the *absence of a yield point and small extension show hard material lacking ductility* ; this is further shown by small local contraction at the fracture. Ductile material, carrying a constant load beyond the elastic limit, goes on extending, or **creeping,** during a long period of time.

It is usual to estimate the **ductility** of a material from the extension on a measured length of the test piece, and also from the local contraction at fracture. Plastic material showing considerable extension before rupture is more suitable for withstanding shocks when worked into a structure than hard stuff which breaks off short with little extension. The ductility of a material is often tested by bending a bar of it to a given radius, and its ability to resist shocks is tested by repeated blows applied by a weight falling on the middle of the bar, the bar being supported at its ends and turned over after each blow. *In practice the elastic limit of the material should not be approached when the loads are applied*, but the information gained during tensile testing after the elastic limit is reached is valuable for determining the qualities of the material. During testing, the loads should be applied at a fairly uniform rate and without shocks; the material should not be given a rest at any time after loading has once started, as this is liable to alter its qualities.

Ultimate strength.—The **breaking stress, or ultimate strength** of a material is measured by dividing the breaking load by the original sectional area of the piece. This is always done for engineering purposes, although it is fictitious, as owing to the contraction at the fracture, the actual area over which the breaking load is distributed is smaller than the original area, and therefore the stress on the section at the fracture is higher than would be shown by the above calculation. It is convenient, however, in practice, to measure the ultimate strength as stated; for we wish to know what load would break a given piece in order, by making the actual load a certain fraction of the breaking load, to prevent that occurring. Thus, if we know that the ultimate strength of mild steel is 30 tons per square inch, then a load of 37·5 tons would be required to break a bar $2\frac{1}{2}''$ by $\frac{1}{2}''$ section. To prevent this, we arrange that the actual load shall be, say, one fifth of this, or 7·5 tons.

Factor of safety.—The **factor of safety** is the ratio of the breaking load to the working load. Thus,

$$\text{Factor of safety} = \frac{\text{breaking load}}{\text{working load}}.$$

The magnitude of the factor of safety to be used in any given case depends on the nature of the loading. A low factor of safety may be employed where the load is steady, or is applied and removed very gradually. High factors of safety are employed where the load is suddenly applied, or where the loads are pushes and pulls alternating. The factor of safety allows a margin for shocks and for our ignorance of the possible loads a structure may have to bear.

Fatigue of materials.—It is well known that a load which would not break a piece if gradually applied, will ultimately fracture it if applied and taken off many times. A still smaller load will produce the same result if continually applied as alternately push and pull. Material treated in this way is said to get into a state of **fatigue**, although it is difficult to say exactly what happens to it. Bauschinger found that a piece of Bessemer mild steel plate which had an ultimate strength of 28·6 tons per square inch with the load gradually applied, broke when a stress of 15·7 tons per square inch was continually applied and removed, and that a stress of 8·6 tons per square inch applied alternately as push and pull effected the same result. Another specimen of the same plate having a constant stress of 14·3 tons per square inch broke when an additional stress of 9·5 tons per square inch was continually applied and taken off. This last case approximates to the conditions in an actual structure. Other materials showed similar results.

Factors of safety used in practice.—In practice, live loads, such as the weight of a locomotive and train on a bridge, are usually doubled and added to the dead load which consists of the weight of the bridge itself. The working stress allowed in bridge work of steel having an ultimate tensile strength of 30 tons per square inch ranges from 4 to 7½ tons per square inch, the lower value being used for parts which are alternately pushed and pulled. These stresses correspond respectively to factors of safety of 7½ and 4. Pieces of wrought iron or steel subjected to shocks should have a factor of safety of from 10 to 12. For cast iron the factor of safety ranges from 5 to 15; this material is not suitable for withstanding shocks and the latter factor is used for such cases. Timber is liable to sag under loads, consequently factors of safety of from 8 to 20 are used.

Effect of heating and cooling.—The injurious effects of overstraining, or of repeated applications of loads may be got rid of by **annealing**. This consists of heating to redness and then cooling slowly. Bars and plates are usually partly annealed on coming from the makers. This is due to them leaving the rolls hot and then cooling down fairly slowly. Cold rolling and wire drawing produces hardness by setting up over-strain in the material. This can be got rid of by subsequent annealing. Crane chains are occasionally *passed through the fire* so as to anneal them and restore their original qualities. Copper is hardened by mechanical treatment such as wire-drawing or bending; it may be softened and its ductility restored by being heated to redness and plunged into cold water. Steel containing more than 0·2 per cent. of carbon is made very hard by the same treatment.

Tensile tests on wires.—Experiments on pulling wires of various materials until they break are very instructive and are easily performed by students. It has already been shown how the extensions of a wire may be measured for varying loads within the elastic limit. The same apparatus (Fig. 111) may be used, after the elastic limit is passed, if a long scale, divided in inches and tenths, is clamped to a firm support and arranged so that the testing weights may move past it as they descend while the wire is extending. Use the vernier until the elastic limit is reached and the scale after. The slight error produced by any drooping of the support after the material begins to draw out can be safely neglected as it will be only a very small percentage of the total extension.

EXPT.—Arrange apparatus as described above and carry out tensile tests on wires of copper, brass, iron and steel. In each case, note throughout the test the extensions produced by gradually increasing loads. Plot these on squared paper and work out the results of the tests, using the same method as has been employed in the following record of a test on a copper wire.

TENSILE TEST ON COPPER WIRE.

Length, 9′ 8½″.

Diameter, 0·036″.

The load was applied in 2-pound increments. Up to the elastic limit the extensions were measured by a vernier on the test wire moving over a scale on another wire hung from the same support. The vernier read to 0·01″. After the elastic limit the extensions were measured by a long boxwood scale, clamped to a fixed support.

Load, lbs.	Scale reading, inches.	Extensions, inches.	Remarks.
1	3·00	0·00	This load was the weight of the hook.
3	3·02	0·02	
5	3·03	0·03	
7	3·05	0·05	
9	3·06	0·06	
11	3·08	0·08	
13	3·09	0·09	
15	3·2	0·20	Elastic limit reached.
19	3·4	0·4	Scale changed to boxwood rule.
21	3·9	0·9	
23	4·8	1·8	
25	6·0	3·0	
27	7·6	4·6	
29	9·6	6·6	
31	12·0	9·0	
33	14·9	11·9	
35	19·0	16·0	
37	26·0	23·0	
39	32·0	29·0	Wire broke.

Columns 1 and 3 plotted give the complete curve as shown in Fig. 113. The curve up to 15 lbs. load has also been plotted in Fig. 114 to an increased scale of extensions.

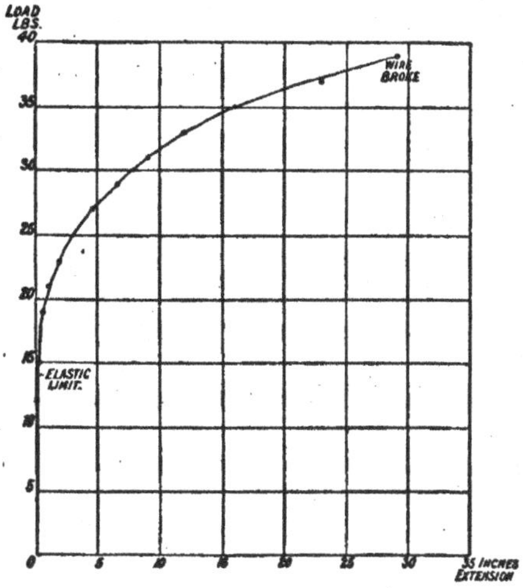

FIG. 113.—Plotted load-extension diagram for a copper wire.

Calculation of results.—Area of wire $= \pi \dfrac{d^2}{4} = 0{\cdot}001018$ sq. inch.

Load at elastic limit = 13 lbs.

$$\text{Elastic limit} = \frac{13}{0{\cdot}001018} = 12{,}770 \text{ lbs. per sq. inch}$$

$$= \underline{5{\cdot}7} \text{ tons per sq. inch.}$$

Breaking load = 39 lbs.

$$= \frac{39}{0{\cdot}001018} = 38{,}310 \text{ lbs. per sq. inch}$$

$$= \underline{17{\cdot}1} \text{ tons per sq. inch.}$$

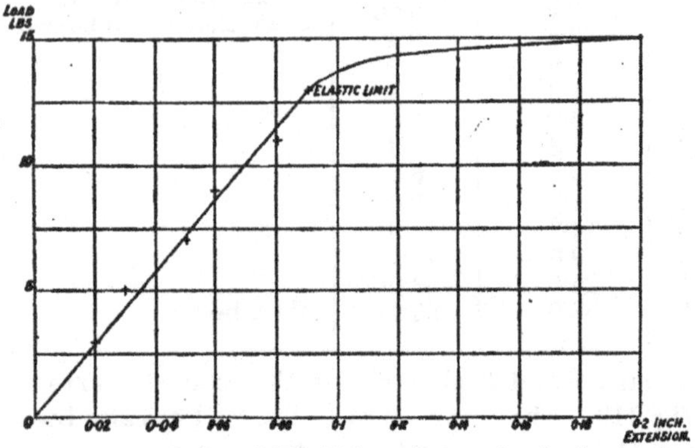

FIG. 114.—Elastic portion of Fig. 113 to an increased scale of extensions.

For **Young's modulus**, 12 lbs. load gives extension 0·09″ on a length of 9′ 8½″.

$$\text{Stress} = \frac{12}{0{\cdot}001018} = 10{,}800 \text{ lbs. per sq. inch.}$$

$$\text{Strain} = \frac{0{\cdot}09}{116{\cdot}5} = 0{\cdot}000772.$$

$$E = \frac{\text{stress}}{\text{strain}} = \frac{10800}{0{\cdot}000772} = 13{,}990{,}000 \text{ lbs. per sq. inch}$$

$$= \underline{6250} \text{ tons per sq. inch.}$$

The total extension on 116·5″ was 29″, or

$$\text{Extension} = \frac{29}{116{\cdot}5} \times 100 = \underline{24{\cdot}9} \text{ per cent.}$$

Autographic records.—Wire testing machines can be conveniently arranged so as to automatically draw a diagram similar to that in Fig. 113. The diagram reproduced in

Fig. 115 was drawn by such a machine while a copper wire was under tensile test. These diagrams are called **autographic records**. In Fig. 116 copies of autographic records for some well known materials have been traced on the same sheet. The loads have been plotted as stresses in tons per square inch of sectional area; the extensions are those on 10″ of the specimens. The comparative qualities of the materials for which the curves are given in this diagram can be understood by inspection.

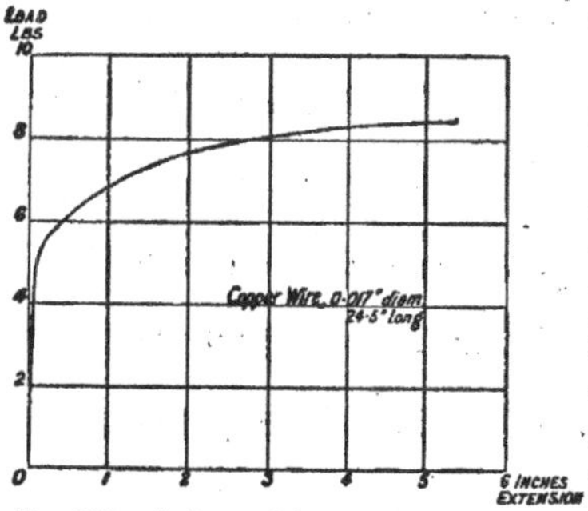

FIG. 115.—Autographic record for a copper wire.

Stresses in a cylindrical shell.—It is interesting to consider the case of a thin cylindrical closed vessel, or **shell**, as it is called, subjected to fluid pressure internally. As we shall see in Chap. XVII., the fluid pressure everywhere on the internal surface is perpendicular to the surface of the vessel. Referring to Fig. 117,

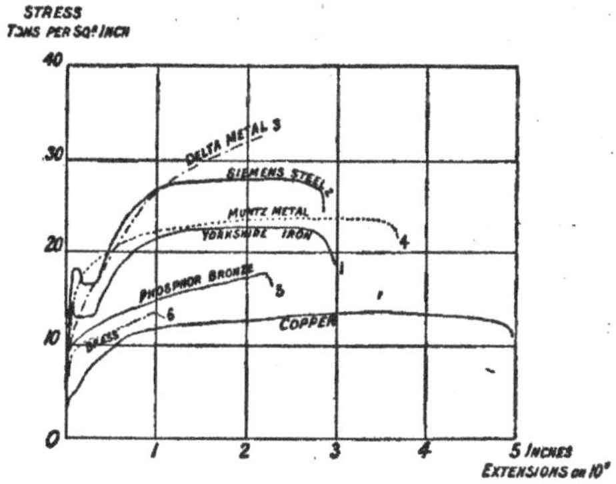

FIG. 116.—Autographic records for some well-known metals.

Let d = diameter of shell, inches;
p = fluid pressure, pounds per square inch;
t = thickness of the plate, in inches;
P = total pressure on end of vessel, then

$$P = p \times \frac{\pi d^2}{4}.$$

The material of the cylindrical part of the vessel is therefore subjected to pull forces equal to P, and will consequently be

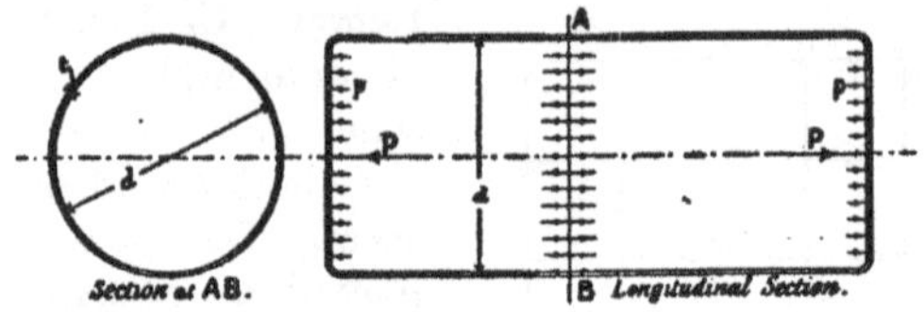

FIG. 117.—Cylindrical shell under internal fluid pressure.

under tensile stress. To find the amount of this stress, the total pull P is divided over the area of the section AB. This area is given approximately by

$$\text{Sectional area} = \text{circumference of shell} \times \text{thickness of plate}$$
$$= \pi d \times t.$$

$$\therefore \text{ stress on section } AB = \frac{P}{\pi dt}$$
$$= \frac{p \times \frac{\pi d^2}{4}}{\pi dt}$$
$$= \frac{pd}{4t} \text{ lbs. per square inch.}$$

The stress on a longitudinal section of the shell must now be determined. The whole length of the shell need not be considered, but only the portion between two cross sections taken one inch apart, for if this portion is taken sufficiently far from the ends of the shell, the *staying action* of the ends becomes negligible, and the stresses on all such rings will be alike. The fluid pressure on the ring is represented by the arrows shown (Fig. 118), everywhere directed perpendicular to the curved surface of the ring. Take components of these, as shown, parallel and perpendicular to the diameter AB, and consider the portion of the ring above AB. The horizontal components acting towards the right and left will balance one another, and need not be further considered. The vertical

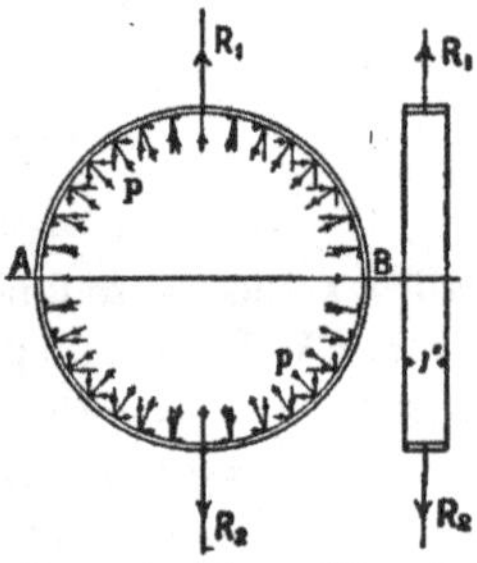

FIG. 118.—Ring, 1″ broad, cut from the shell.

components will have a resultant R_1. Similarly, on the portion below AB, a resultant force R_2 will act, equal and opposite to R_1. These two forces put the material of the ring at A and B under tensile stress, which must now be calculated.

First to obtain R. There will be no difference experienced in the equilibrium of the ring if we imagine it to be filled up to the level of AB with cement. The pressure on the surface of

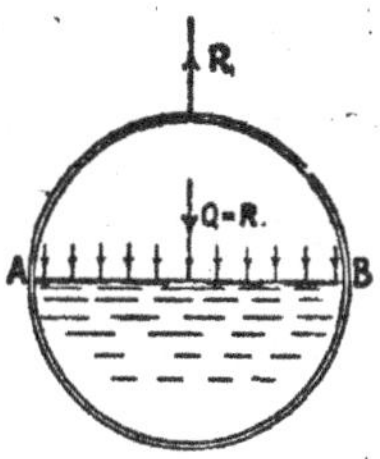

FIG. 119.

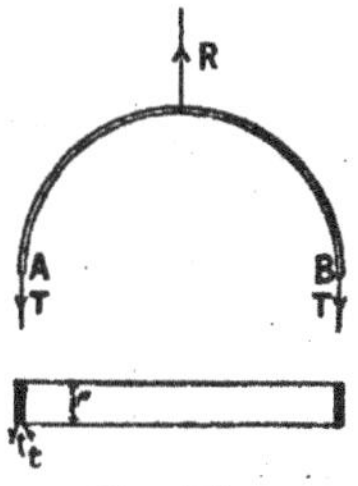

FIG. 120.

the cement will be perpendicular to AB (Fig. 119) and the resultant force due to this will be

$$Q = p \times \text{area of surface of } AB$$
$$= p \times d \times 1.$$

R and Q now preserve the equilibrium of the ring, therefore

$$R = Q$$
$$= p \times d.$$

Imagine the material at A and B to be cut, and that forces T, T, are supplied at the sections in order to balance R (Fig. 120), then if the stress on the sections at A and B be called q,

$$T = q \times \text{area cut at } A \text{ or } B,$$
$$= q \times t \times 1.$$

Also $$R = 2T.$$

$$\therefore \; pd = 2 \times q \times t;$$

$$\therefore \quad q = \frac{pd}{2t} \text{ lbs. per square inch.}$$

It has already been found that a circumferential seam has a tensile stress equal to $\frac{pd}{4t}$, consequently **the stress on a longitudinal seam is just double that on a circumferential seam.** For this reason, boilers are usually made with the longitudinal joints much stronger than the circumferential joints.

EXAMPLE. A boiler shell is 6 feet in diameter, and the metal is $\frac{1}{2}''$ thick. If the steam pressure is 100 pounds per square inch, calculate the stress on circumferential and longitudinal sections.

$$\text{Total pressure on end of shell} = 100 \times \frac{\pi d^2}{4}$$

$$\text{Area of circumferential section} = \pi dt\,;$$

$$\therefore\ \text{Stress on circumferential section} = \frac{100 \times \pi d^2}{4\pi dt} = \frac{100 \times d}{4 \times t}$$

$$= \frac{100 \times 72}{4 \times \frac{1}{2}}$$

$$= \underline{3600} \text{ pounds per square inch.}$$

$$\text{Stress on longitudinal section} = \underline{7200} \text{ pounds per square inch.}$$

Thin flat plates, such as are used for boilers, are not able to withstand pressure without *bulging*. This is easily understood when we consider that the plates can practically only withstand tension, not bending, and it must be remembered that a comparatively small force P (Fig. 121) suffices to put two large forces T, T, out of line with one another when P is applied transversely to their lines of action. Flat surfaces subjected to fluid pressure consequently require to be *stayed*. The student is referred to books on boilers for descriptive sketches showing how this is done in different types. A spherical shell would be *self-staying*, but practical considerations prevent this shape being used for boilers, although it is occasionally used for other vessels where no heat is to be applied.

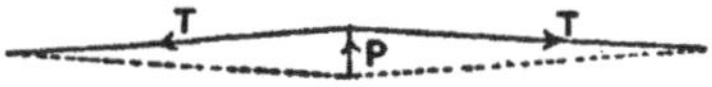

FIG. 121.

The results obtained above are only applicable to vessels the walls of which are thin compared to their diameters. Thick walled vessels, such as hydraulic pipes and cylinders, have to be considered from a different point of view, as the stresses in such cases are not uniformly distributed over the sections of the material as has been assumed for thin shells. The theory involved is too complicated for beginners, and is therefore not given here.

Shearing action.—Another kind of stress to which material is often subjected, is called **shear stress.** Shear stress acts tangentially to the surface or section of the material. A body subjected to shear stress tends to break at any section by the two portions of the body sliding past one another. In Fig. 122, a piece of material is shown between the blades of a shearing machine. When the pressure is applied, the material will fracture by its two portions sliding past one another at *AB*. The section *AB* will therefore be under shear stress.

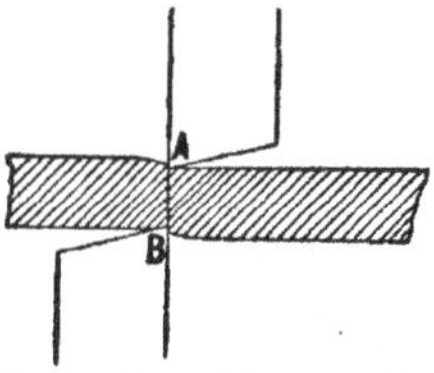

FIG. 122.—Plate under shear stress at *AB*.

In punching machines (Fig. 123), the piece punched out separates from the surrounding metal by shearing. Again, in the common knuckle joint (Fig. 124), when pulls or pushes are applied to the rod, the joint may break by the pin shearing at two sections, the left-hand rod then carries away the centre portion of the pin with it.

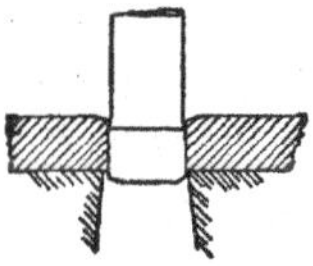
FIG. 123.—Shear stress produced in a punching machine.

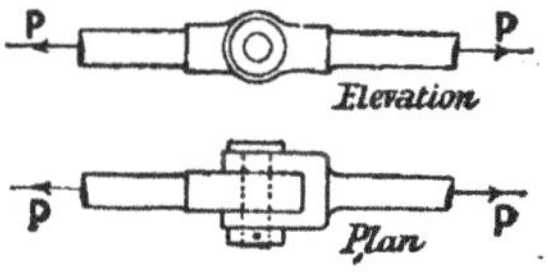

FIG. 124.—Knuckle joint.

In all these cases, the shear stress is calculated by dividing the total force producing the shearing action by the area of the section of the material over which the shear is distributed. Thus, in the punching machine,

Let force pushing the punch down $= P$ tons;

diameter of hole punched $= d$ inches;

thickness of plate $= t$ inches; then

area under shear $=$ circumference of hole $\times\, t$

$= \pi d \times t$ square inches.

$$\text{Shear stress} = \frac{P}{\pi d t} \text{ tons per square inch.}$$

In the knuckle joint, if d is the diameter of the pin in inches,

$$\text{Area under shear} = 2 \times \frac{\pi d^2}{4} = \frac{\pi d^2}{2}.$$

$$\text{Shear stress} = \frac{P}{\frac{\pi d^2}{2}} = \frac{2P}{\pi d^2}.$$

EXAMPLE. Suppose two bars connected by a knuckle joint to be pulled with forces of 2 tons. Find the diameter of the pin if the safe shear stress is 4 tons per square inch.

$$\text{Shear stress} = \frac{2P}{\pi d^2};$$

$$\therefore\ 4 = \frac{2 \times 2}{\frac{22}{7} \times d^2},$$

$$d^2 = \frac{2 \times 2 \times 7}{4 \times 22} = \frac{7}{22},$$

$$d = \underline{0{\cdot}565} \text{ inch.}$$

Strain produced by shear.—The kind of strain produced by shearing stress is **change of shape.** This can be clearly shown by means of a thick book. Lay the book on the table and sketch a square on its end edges as shown in Fig. 125. Then

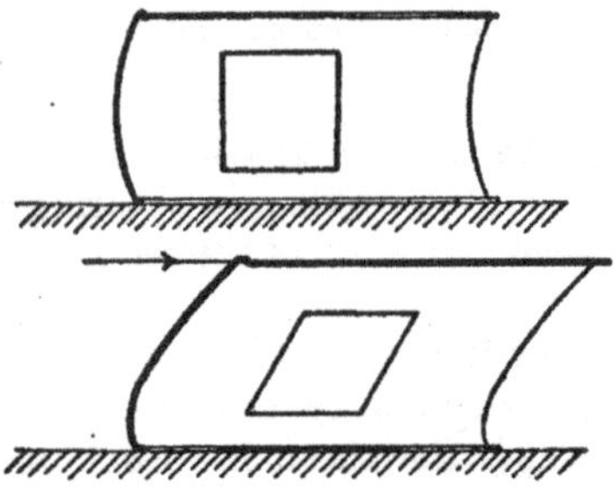

FIG. 125.—Shear strain illustrated by a book.

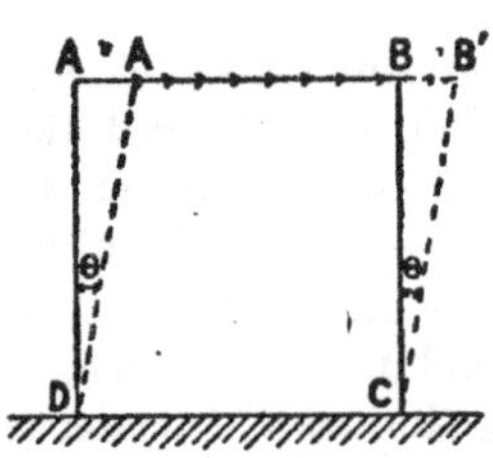

FIG. 126.—Measurement of shear strain.

apply shear stress by holding the bottom cover firmly on the table and pushing the top cover tangentially in the manner indicated. Every page of the book shears past the adjacent page, and the result is to change the shape of the square to that of a parallelogram. The strain is measured by stating the angle θ, in radians, turned through by the vertical sides of the square. Thus, if A (Fig. 126) comes to A' and B to B', then the angle BCB', equal to the angle ADA', measured in *radians*, is the measure of the strain.

In engineering materials, the shear strain is always very small, so that the angle θ is very small. We may therefore measure θ by putting it equal to $\frac{BB'}{BC}$ radians. This assumes that the small distance BB' is a circular arc instead of being straight. The modulus of elasticity for this sort of stress and strain is called the **rigidity modulus**, usually written C. Following the same definition as before

$$C=\frac{\text{stress}}{\text{strain}}=\frac{\text{shear stress applied}}{\theta \text{ radians}}.$$

The average value of C for wrought iron is about 5000 tons per square inch, and about 5400 for steel. The rigidity modulus is of importance when dealing with the stiffness of shafts and other bodies subjected to twisting.

EXERCISES ON CHAP. VII.

1. A tie bar 4″ broad, $\frac{5}{8}$″ thick, is under a tension of 7 tons. Calculate the tensile stress.

2. Find the working load for the bar in Question 1 if the tensile stress is not to exceed 5 tons per square inch.

3. A bar of square section, $\frac{1}{4}$″ edge, is 60 ft. long and is found to stretch 0·6″ when a certain pull is applied. Find the strain. Suppose the pull applied to have been 1562 lbs., and find Young's modulus of elasticity.

4. Suppose the tensile stress is not to exceed 4 tons per square inch, find the diameter of a round tie rod which has to resist a pull of 16 cwts.

5. Taking Young's modulus for wrought iron to be 29,000,000 lbs. per square inch, what decrease in length will take place when a column containing 12 square inches in section and 20 ft. high carries a load of 36 tons?

6. What load, in pounds, must be hung to an iron wire 50 ft. long and 0·1″ diameter to make it stretch $\frac{1}{5000}$ inch.

7. An iron tie bar is 50 ft. long, its section being rectangular 4″ × $\frac{3}{4}$″. Its stretch must not exceed $\frac{1}{16}$″; calculate the maximum load it can carry.

8. Taking the shearing strength of iron to be 20 tons per square inch, calculate the force necessary to punch a $\frac{3}{4}$″ hole in a $\frac{5}{8}$″ plate. Find also the stress on the punch.

9. Suppose it were attempted to punch a $\frac{1}{4}''$ hole in the plate of Question 8, make a similar calculation.

10. A copper wire, previously pulled beyond the elastic limit, was tested again, after 15 hours' rest, under tension. Length of wire, 12′ 3″, diameter of wire 0·036″. The following results were obtained:

Load, (lbs.).	Scale reading (inches).	Extension for differences of 2 lbs. (inches).	Remarks.
0	3·03		Load marked zero in column 1 was actually a load of $4\frac{1}{2}$ lbs., hung on to keep wire taut.
2	3·04	0·01	
4	3·06	0·02	
6	3·08	0·02	
8	3·10	0·02	
10	3·12	0·02	
12	3·14	0·02	Elastic limit reached.
14	3·65	0·51	

Plot columns (1) and (2) on squared paper. Calculate the value of Young's modulus for this sample of copper wire.

11. What do you understand by the terms tensile, compressive, and shearing strength respectively of any material? Define "modulus of elasticity." If a wrought iron bar of 1 square inch sectional area just breaks under a tensile stress of 60,000 lbs., what would be the area of the section of a tie-rod which would just support a load of 20 tons? (1896.)

12. How would you find out for yourself the behaviour of steel wire loaded in tension till it breaks. What occurs in the material? Use the words stress and strain in their exact senses. (1897.)

13. What do we mean by stress, strain, and modulus of elasticity? A wire 10′ long and $\frac{1}{8}$ sq. inch in sectional area is hung vertically, and a load of 450 lbs. is attached to its extremity, when the wire stretches 0·015″ in length. What are the stress and strain respectively? And also the modulus of elasticity? (1899.)

14. An iron wire is loaded with gradually increasing tensile loads till it breaks. We want to know its modulus of elasticity, its elastic limit stress and its breaking stress. What measurements and calculations do we make? (1900.)

CHAPTER VIII.

STRENGTH AND STIFFNESS OF BEAMS.

Bending of a beam.—Beams are parts of a structure, usually supported horizontally, for the purpose of carrying loads applied transversely to their lengths. Suppose we have a beam consisting of a number of planks of equal lengths laid one on the other, and supported at the ends. A load *W*, applied at the centre of the span, will cause all the planks to bend in a similar fashion, and consequently, as the lengths of all the planks will remain the same, they will overlap at the ends as shown (Fig. 127). Strapping the planks firmly together will prevent this occurring and the beam will now bend as a whole, the ends of the planks remaining in one plane (Fig. 128).

FIG. 127.—Bending of a loose plank beam.

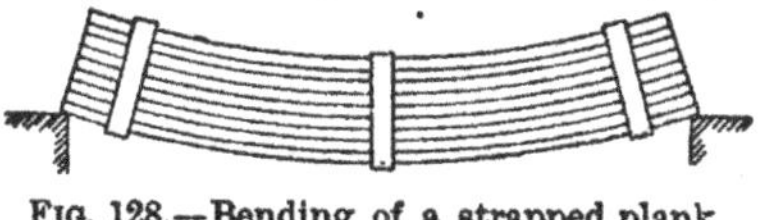

FIG. 128.—Bending of a strapped plank beam.

Assuming the middle plank to remain the same length as at first, it is clear that the top plank, and all those above the middle must have become shorter, and those below the middle one, longer than at first. The planks above the middle must therefore have been subjected to compressive stress in the direction of their length and those below the middle to tensile stress in the same direction. The further we get away from the middle, above or below, the greater will be the change of length of the planks and therefore the greater will be the compressive and tensile stresses producing these changes of length. An ordinary solid metal or timber beam may be looked upon as being built up of a large number of fibres cemented together, corresponding to the planks in our model beam. These fibres will be subjected to stresses in the same manner as the planks. Therefore, in a loaded solid beam, such as that in Fig. 129, the upper fibres will be under compressive stress and the lower ones under tensile stress. These stresses are shown at the section *AB*.

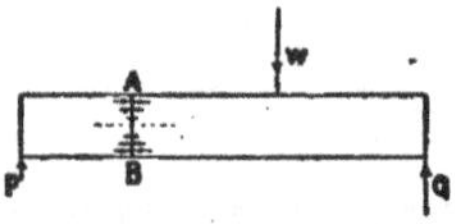

FIG. 129.—Tensile and compressive stresses on the section *AB*.

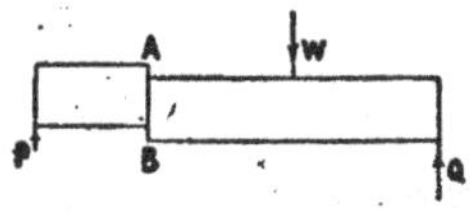

FIG. 130.—Shear at the section *AB*.

Further, the action of *P*, the pressure of the support, acting on the left-hand side of the section *AB*, and of *W* and *Q* acting on the right-hand side of the section, will be usually to give the material a tendency to slide past at the section as in Fig. 130. The material at the section will therefore be under shearing stress as well as the stresses mentioned above.

Models showing the forces in the material.—These stresses may be very well understood by examining models such as those illustrated. Fig. 131 shows a **cantilever**, that is, a beam fixed at one end only and free at the other end. The cantilever has been cut at *AB*. In order to balance the portion outside of *AB* it is necessary to put a cord connection at *A* and a small strut at *B*. The pull and push forces thereby supplied counteract the bending tendency. In addition, an upward force *S* has to be supplied to balance the tendency to shear. In the uncut cantilever these forces would be supplied by the resistance of the material at the section. It will be noticed in the cantilever that the tensile stress occurs above, and the

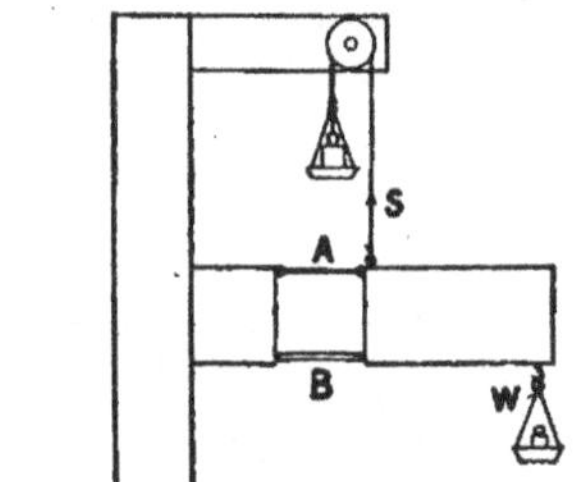

FIG. 131.—Model of a cantilever, cut to show the forces at *AB*.

compressive stress below, the middle. This is just the reverse of what we have seen for a beam supported at both ends. Fig. 132 shows this latter case. This particular model consists of an I section, supported at the ends and cut into two pieces. A compression spring balance at the upper flange of the beam

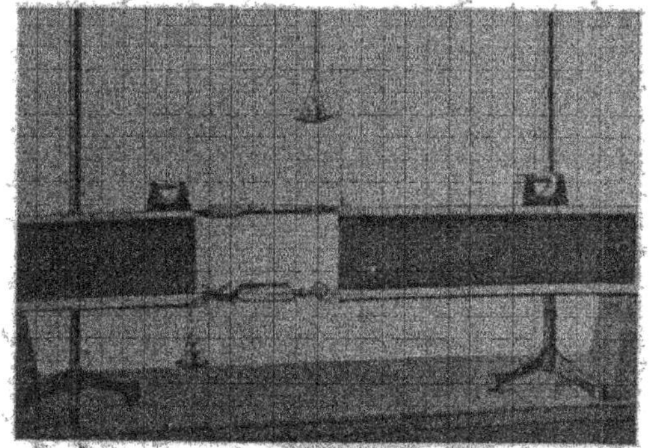

FIG. 132.—Model of a cut beam.

and two ordinary spring balances at the lower flange enable the forces at the section, produced by bending, to be measured. Shearing is balanced by weights applied as indicated, the upward force being supplied by means of a cord passing over a pulley above.

If we again examine the model beam built up of planks, we may notice that when it is bent and the ends of the planks overlap, that they have done so by the planks sliding on one another in the direction of their lengths. The straps obviated this by binding the planks firmly together and so preventing this sliding, or shearing action taking place. So also in a solid beam there are shearing stresses distributed over horizontal longitudinal sections.

The actual distribution and calculation of these stresses is beyond the scope of this book, but sufficient will be said to enable the student to solve many simple practical problems.

Bending moment.—The **bending moment** at any section of a beam is measured by the moment about that section of the applied forces tending to turn the portion of the beam on the right or left hand of the section. For example, in the beam in Fig. 133, carrying a load of 10 tons at the middle of a 12 ft. span, the bending moment at the middle section will be found by considering the forces acting on the right hand or left hand side. If we neglect the weight of the beam itself, there is only one upward force on either side, and its moment about the

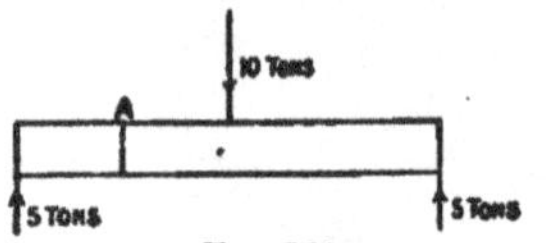

FIG. 133.

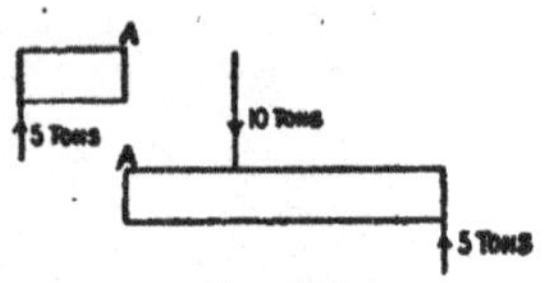

FIG. 134.

middle section is $5 \times 6 = 30$ ton feet. The bending moment at the middle section is therefore 30 ton feet. To find the bending moment at A (Fig. 134), 3 ft. from the left-hand support—consider the left-hand portion of the beam, the only force is an upward one of 5 tons, and its moment about A is $5 \times 3 = 15$ ton feet. If we consider the right-hand portion, there are two forces to be dealt with, an upward one of 5 tons and a downward force of 10 tons. Their moments about A being contrary to one another, the total moment about A will be $(5 \times 9) - (10 \times 3) = 15$ ton feet, as before. The bending moment at A will therefore be 15 ton feet.

In any given case therefore, proceed thus: **first, from the given loads and dimensions, calculate the reactions of the supports; then, to find the bending moment at any section, calculate the algebraic sum of the moments about the section of all the forces acting either on the right-hand or left-hand portion of the beam.**

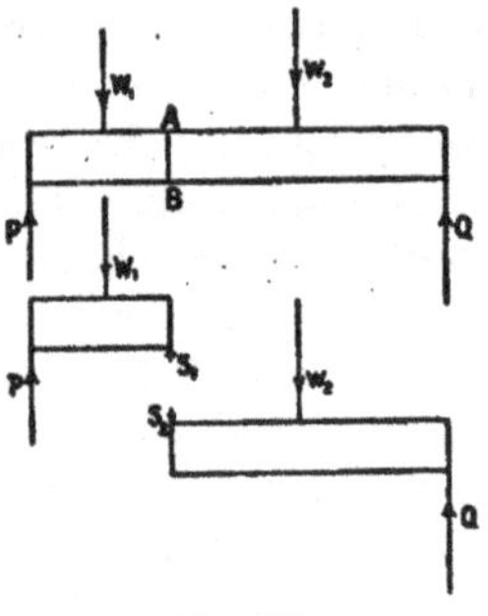

FIG. 135.

Shearing force.—To calculate the total shearing force at any section such as AB in Fig. 135, notice that if the beam were cut at AB we should have to balance the left-hand portion by applying, at AB, a force $S_1 = W_1 - P$; to balance the right-hand portion, a force $S_2 = Q - W_2$ must be applied at AB. These forces S_1 and S_2 in the uncut beam are mutual shearing actions of the two portions of the beam at the section, and therefore must be equal to one another. **In any given case, therefore, we calculate the total shearing force at any section by taking the algebraic sum of all the applied forces acting either on the right-hand or left-hand portion of the beam.**

It is customary to describe bending moments which cause a beam to bend convex in a downward direction as *positive*, and

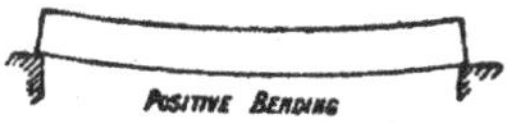

FIG. 136.

FIG. 137.

convex in an upward direction as *negative*. These are shown in Figs. 136 and 137. Shear which causes stresses of the kind

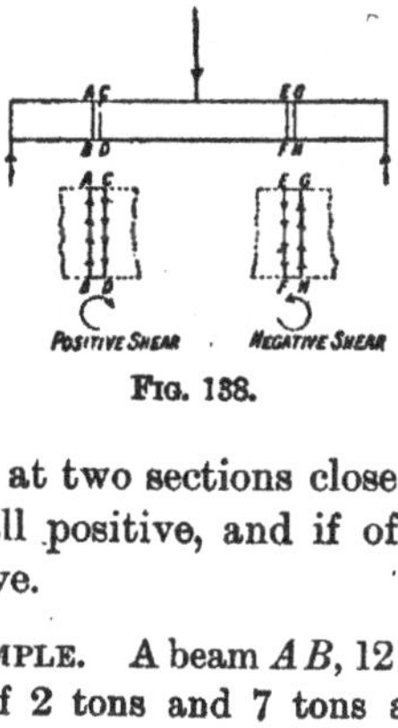

FIG. 138.

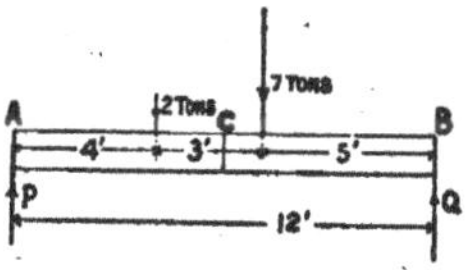

FIG. 139.

shown at two sections close together *AB* and *CD* (Fig. 138), we will call positive, and if of the kind as shown at *EF* and *GH* negative.

EXAMPLE. A beam *AB*, 12 ft. span, supported at the ends, carries loads of 2 tons and 7 tons as shown in Fig. 139. Calculate the Bending Moment and Shearing Force at the centre of the span, neglecting the weight of the beam.

To obtain the reactions of the supports, taking moments about *B*.

$$P \times 12 = (2 \times 8) + (7 \times 5),$$

$$P = \tfrac{51}{12} = 4{\cdot}25 \text{ tons.}$$

Taking moments about *A* (as a check on working),

$$Q \times 12 = (2 \times 4) + (7 \times 7),$$

$$Q = \tfrac{57}{12} = 4{\cdot}75 \text{ tons.}$$

$$P + Q = 4{\cdot}25 + 4{\cdot}75 = 9 \text{ tons} = \text{sum of loads.}$$

Imagine the beam cut at C and consider the left-hand portion.

$$\text{Bending moment at } C=(P\times 6)-(2\times 2)$$
$$=4{\cdot}25\times 6-4$$
$$=\underline{21{\cdot}5} \text{ ton feet; positive.}$$

Or, by considering the right-hand portion,

$$\text{Bending moment at } C=(Q\times 6)-(7\times 1)$$
$$=4{\cdot}75\times 6-7$$
$$=\underline{21{\cdot}5} \text{ ton feet; positive.}$$

These results agree.

For shearing force, considering left-hand portion,

$$\text{Shearing force at } C=P-2$$
$$=4{\cdot}25-2=\underline{2{\cdot}25} \text{ tons.}$$

Or, by considering the right-hand portion,

$$\text{Shearing force at } C=Q-7$$
$$=4{\cdot}75-7=-\underline{2{\cdot}25} \text{ tons.}$$

The effect of these shearing forces is to put two sections close together at C under stresses as indicated in Fig. 140. The shearing force is therefore positive.

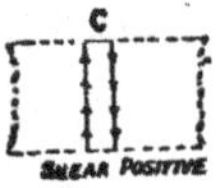

FIG. 140.

Beam, load distributed.—A beam carrying a distributed load must now be studied. Fig. 141 shows a beam supported at its ends, 20 ft. span, with a uniformly distributed load 1 ton per foot length. The reaction of each support will be 10 tons. To calculate the bending moment at any section A, consider, say, the left-hand portion of the beam. There are two forces acting on it, 10 tons upwards from the support, and W, equal to the weight of the distributed load on this portion of the beam, acting downwards

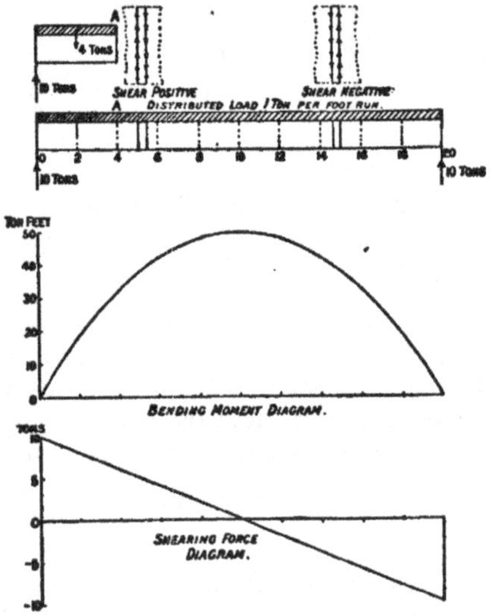

FIG. 141.—Bending moment and shearing force diagrams for a beam carrying a distributed load.

at its centre of gravity. Supposing A is 4 ft. from the left-hand support, then W is equal to 4 tons. Notice also that the turning tendencies about A, of the reaction of the support and of W, are contrary.

$$\begin{aligned} \text{Bending moment at } A &= (10 \times 4) - (W \times 2) \\ &= 40 - (4 \times 2) \\ &= 32 \text{ ton feet; positive.} \end{aligned}$$

$$\begin{aligned} \text{The shearing force at } A \text{ will be} &= 10 - W \\ &= 10 - 4 \\ &= 6 \text{ tons.} \end{aligned}$$

The shearing force puts two adjacent sections at A under stresses as shown and is consequently positive.

It is interesting in this case to calculate the Bending Moments and Shearing Force at several sections and then plot the results. Doing this for sections at intervals of two feet (Fig. 141) we get,

Bending moment at	0	$=0$
" "	2 and 18	$=(10\times 2)-(2\times 1)=18$ ton feet
" "	4 and 16	$=(10\times 4)-(4\times 2)=32$ " "
" "	6 and 14	$=(10\times 6)-(6\times 3)=42$ " "
" "	8 and 12	$=(10\times 8)-(8\times 4)=48$ " "
" "	10	$=(10\times 10)-(10\times 5)=50$ " "

Shearing force at	$0=$	10 tons
" "	$2=10-2=$	8 tons
" "	$4=10-4=$	6 "
" "	$6=10-6=$	4 "
" "	$8=10-8=$	2 "
" "	$10=10-10=$	0 "
" "	$12=10-12=$	-2 "
" "	$14=10-14=$	-4 "
" "	$16=10-16=$	-6 "
" "	$18=10-18=$	-8 "
" "	$20=10-20=$	-10 "

These results plotted as shown in Fig. 141 give **Bending Moment and Shear Diagrams.** It will be noticed that the curve in the Bending Moment Diagram is *parabolic*, and in the Shear Diagram, *a straight line.* Notice that the bending moments increase towards the middle of the span, and that the shearing force diminishes towards the middle, where it is zero.

A common practical section.—The strength of a beam of I section may be calculated on the assumption that the flanges supply all the resistance to bending and the web all resistance to shearing. This assumption leads to results which differ only very slightly from the more correct ones for sections used in practice. Consider a section in which the area of each flange is A square inches and the distance from centre to centre of the flanges D inches (Fig. 142). Let f=stress allowed on the flanges, in tons per square inch. Then $f \times A$=greatest force which may act on one flange.

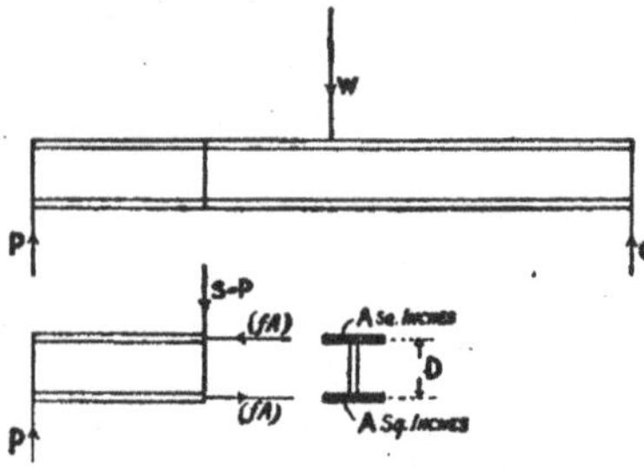

FIG. 142.—Beam of I section.

These forces fA (pull), fA (push), give a couple, the moment of which is $fA \times D$ inch tons, and this is the resisting moment of the section. The Bending Moment at any section of the loaded beam must not exceed this quantity.

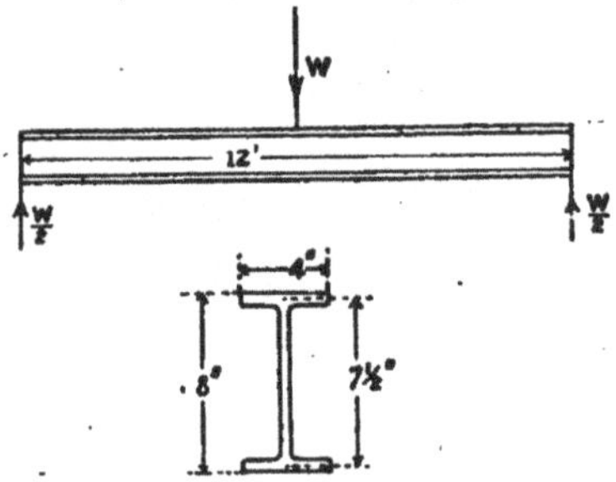

FIG. 143.

EXAMPLE 1. Suppose we have a beam 12 ft. long, supported at its ends, I section, 8″ in. deep, 4″ broad, metal of flanges ½″ thick. Find the maximum load at the centre, if the stress due to bending is not to exceed 5 tons per square inch.

Let W=load at centre in tons.

$$\text{Reaction of each support} = \frac{W}{2}.$$

$$\text{Bending moment at centre of span} = \frac{W}{2} \times 6 \text{ ton feet,}$$
$$= 3W \times 12$$
$$= 36 . W \text{ ton inches.}$$

Resisting moment of section $= fAD$

$$= 5 \times (4 \times \tfrac{1}{2}) \times 7\tfrac{1}{2}$$
$$= 75 \text{ ton inches.}$$

Bending moment = resisting moment

$$36\,W = 75,$$
$$W = \tfrac{75}{36} = \underline{2{\cdot}08} \text{ tons.}$$

EXAMPLE 2. The shear on any section of the beam in the last Example will be 1·04 tons. Suppose 4 tons per square inch to be the shear stress allowed, what thickness of web is required?

Shear = shear stress × area of web section,

$$1{\cdot}04 = 4 \times 7 \times t,$$
$$t = \frac{1{\cdot}04}{28} = 0{\cdot}037''.$$

Webs are not made so thin as this in practice, because there is *buckling* to be guarded against as well as shear. In the case of this beam, the actual thickness would be probably $\tfrac{3}{8}''$ or $\tfrac{7}{16}''$.

Some other practical sections.—Rolled steel beams are usually made of the section considered in Example 1, above, the flanges being made equal as there shown. This form of section gives equal stresses on each flange. Cast iron beams have the tension flange of larger area than the compression flange (Fig. 144), since this material is very much stronger under push than under pull, and the effect is to reduce the magnitude of the tensile stress. A common practice is to make the tension flange of four times the sectional area of the compression flange, which gives a tensile stress approximately equal to one quarter of the compression stress. As the larger flange has to be under tensile stress, this must be arranged for in placing the beam or cantilever. Thus, a cantilever will have the larger flange uppermost; and in a beam supported at the ends, this will be the lower flange.

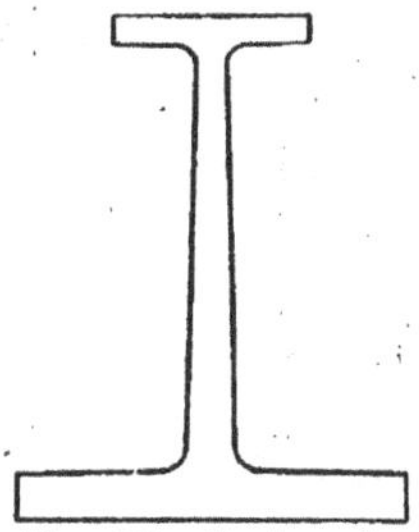

FIG. 144.—Section of a cast iron beam.

The resisting moment of other two sections may be stated.

Let f=maximum stress allowed, in tons per square inch.

Then for a **rectangular section,** b inches broad, d inches deep (Fig. 145),

$$\text{Resisting moment}=\frac{f.b.d^2}{6}\text{ ton inches.}$$

For a **circular section,** r inches radius (Fig. 146),

$$\text{Resisting moment}=\frac{f\pi r^3}{4}\text{ ton inches.}$$

Modulus of the section is the name given to that quantity by which the safe stress must be multiplied in order to give the

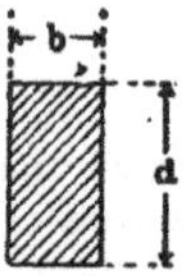

FIG. 145.

FIG. 146.

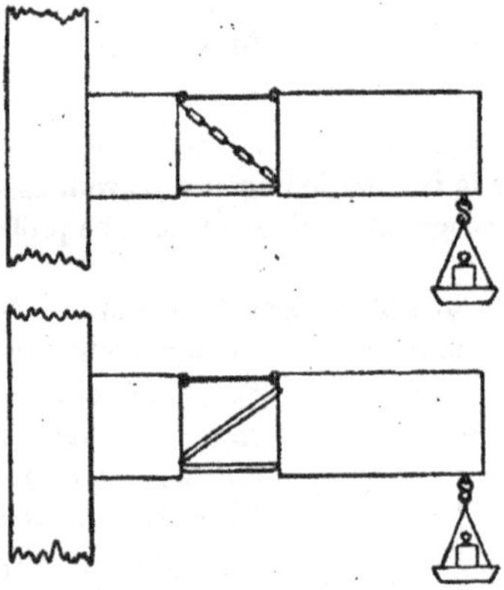

FIG. 147.—The resistance to shear supplied by a diagonal chain or prop.

resisting moment of the section. The letter Z is usually taken to represent the modulus of a section. Thus, in the case of the rectangular section above, $Z=\frac{bd^2}{6}$; and for the circular section, $Z=\frac{\pi r^3}{4}$.

Distribution of material in girders.—Turning again to the model cut cantilever, if we remove the cord and weight which supply the resistance to shear, and substitute a diagonal cord or prop as shown in Fig. 147, we find that the cantilever will be in equilibrium at the cut part. This gives us information as to the functions of the parts in a girder (Fig. 148). The horizontal flanges, or **booms** as they are called, supply the required resistance to bending, the top one in a girder supported at its ends being under push and the bottom one under

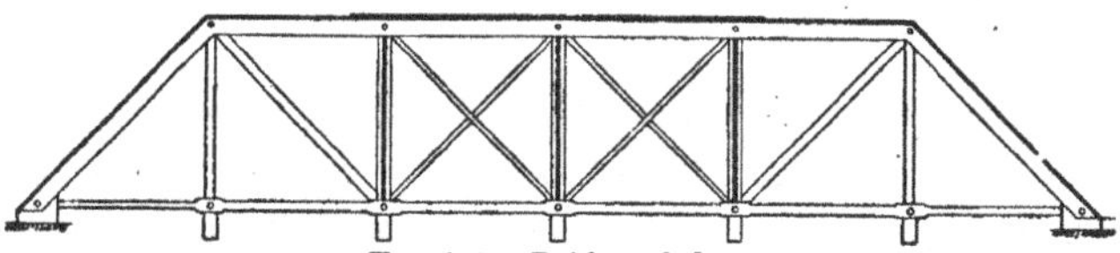

FIG. 148.—Bridge girder.

pull. The diagonal bars supply the required resistance to shearing, being under push or pull depending on the direction in which they are inclined.

As we have already seen in the case of the beam supported at its ends and carrying a distributed load, the bending moment is greatest at the middle of the span and diminishes to zero at the ends, while the shearing force is greatest at the ends and diminishes to zero at the middle. In consequence of this, **the booms of a lattice girder should be thicker towards the middle in order to supply a greater resistance to the bending moment, and the diagonal parts stronger towards the ends in order to cope with the larger shearing forces there.** The same thing is also generally done with large plate girders, that is, girders built up

FIG. 149.—Plate girder.

of plates riveted together with angle irons (Fig. 149). In these, the flanges supply almost all the resistance to bending and the web almost all the resistance to shearing. We generally find that the flanges at the middle consist of several plates riveted together, these being gradually reduced in number as the ends are approached. The webs are often made of thicker plates near the ends and thinner ones near the middle. The large thin plates of which the webs of these girders are constructed are liable to bulge or buckle; to counteract this tendency it is usual to stiffen them at intervals by vertical angle or T sections riveted to them.

Six standard cases of beams.—The student will have no difficulty in understanding and working out for himself the Bending Moments and Shearing Forces for the first four of the six standard cases of beams on pp. 100-101. The process is exactly similar to that used in the example of the beam supported at both ends and loaded with a uniformly distributed load. The Bending Moment and Shear Diagrams should be drawn from the calculations. The last two cases require a more difficult theory for their explanation than can be discussed in this book. For the sake of comparing the strengths and stiffnesses of the beams with different methods of supporting and loading, it is assumed in the following results that the materials, total loads, and lengths are the same, and that the sections are uniform in dimensions and rectangular.

Beams of similar sections.—Beams made of the same material and having similar sections, but with varying dimensions of breadth, depth and span, are found to have strengths and stiffnesses agreeing approximately with the following laws, the loading and manner of support being similar in the compared beams.

Strength *is proportional to the breadth, to the square of the depth and inversely proportional to the length.*

Stiffness *is proportional to the breadth, to the cube of the depth and inversely proportional to the cube of the length.*

Strength is measured by the load which the beam will carry, stiffness by the reciprocal of the deflection of the beam under a given load.

These rules, taken together with the comparative strengths and stiffnesses given in the table (pp. 100-101), enable us to solve many practical problems.

EXAMPLE. Suppose it is found that a beam of cast iron 1″ broad × 1″ deep × 36″ between supports breaks with a load of 6 cwts. at its centre. Calculate the breaking load at the centre of the span for a beam of cast iron 1½″ broad × 3″ deep × 48″ span.

Expressing the above rules for strength in proportional form

$$W_1 : W_2 = \frac{b_1 d_1^2}{l_1} : \frac{b_2 d_2^2}{l_2}.$$

SIX STANDARD CASES OF BEAMS.

Description.	Bending Moment.	Shearing Force.	Diagrams. M, Bending Moment. S, Shearing Force.	Comparative Bending Strength.	Comparative Stiffness.
I. **Cantilever.** Load W at end.	M at $A = Wx$. Maximum $M = WL$, and occurs at support.	$S = W$ at any section.	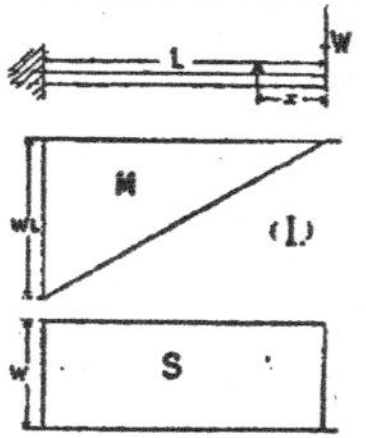	1	$\frac{1}{16}$
II. **Cantilever.** Load W uniformly distributed as w per foot run.	M at $A = wx \times \frac{x}{2}$ $= \frac{1}{2}wx^2$. Maximum $M = \frac{1}{2}wL^2$ $= \frac{1}{2}WL$, and occurs at support.	S at $A = wx$. Maximum $S = wL$ $= W$, and occurs at support.	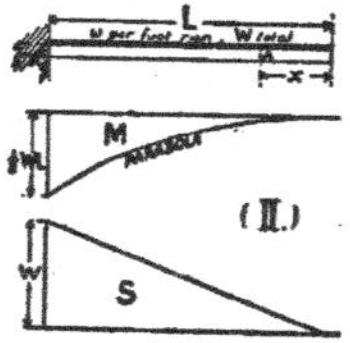	2	$\frac{1}{6}$
III. **Beam supported at ends.** Load W at middle of span.	M at $A = \frac{1}{2}W \times x$. Maximum M $= \frac{1}{2}W \times \frac{1}{2}L$ $= \frac{1}{4}WL$, and occurs at middle of span.	S at $A = \frac{1}{2}W$, uniform over each half span, positive on one half, negative on other.	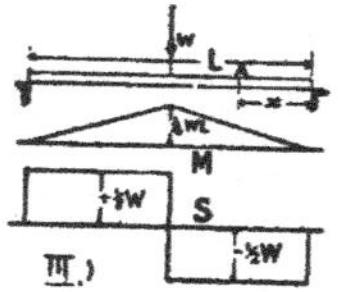	4	1

IV. **Beam supported at ends.** Load W uniformly distributed as w per foot run.	M at A $=\frac{wL}{2}x - wx\frac{x}{2}$ $=\frac{1}{2}wLx - \frac{1}{2}wx^2$. Maximum M $=\frac{1}{4}wL^2 - \frac{1}{8}wL^2$ $=\frac{1}{8}wL^2$ $=\frac{1}{8}WL$, and occurs at middle of span.	S at $A=\frac{wL}{2} - wx$. Maximum S $=\frac{1}{2}wL$ $=\frac{1}{2}W$. Positive on one side, negative on other side of middle, and zero at middle.	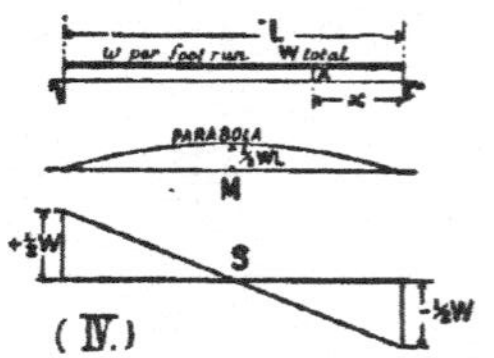 (IV.)	8	$\frac{8}{5}$
V. **Beam fixed at both ends.** Load W at centre of span.	M at supports $=M$ at middle of span $=\frac{1}{8}WL$. M positive at middle, negative at supports.	$S=\frac{1}{2}W$, uniform over each half span, positive on one side, negative on other side of middle.	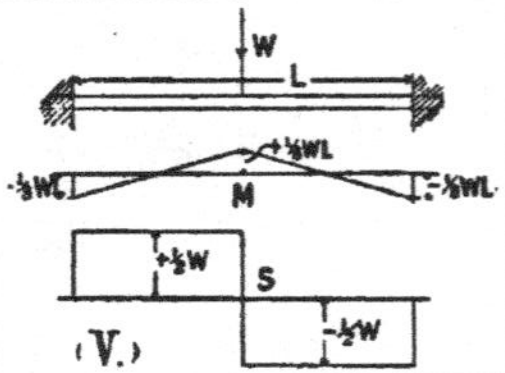 (V.)	8	4
VI. **Beam fixed at both ends.** Load W uniformly distributed as w per foot run.	Maximum $M=\frac{wL^2}{12}$ $=\frac{WL}{12}$, negative, and occurs at supports. M at middle of span $=\frac{wL^2}{24}=\frac{WL}{24}$, positive.	$S=\frac{1}{2}W$ at supports, positive at one support, negative at other, and uniformly diminishes towards middle, where it is zero.	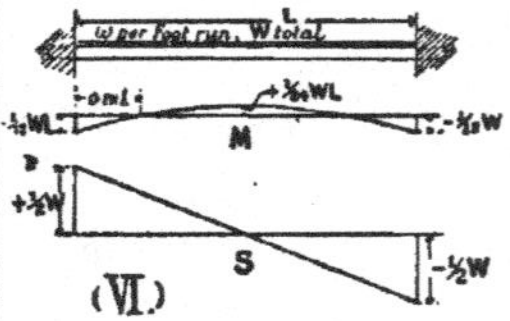 (VI.)	12	8

Suffix 1 refers to the given case and suffix 2 to the one to be worked out; this gives

$$6 : W_2 = \frac{1 \times 1^2}{36} : \frac{1\frac{1}{2} \times 3^2}{48},$$

or

$$W_2 \times \frac{1}{36} = 6 \times \frac{1\frac{1}{2} \times 9}{48},$$

$$W_2 = \frac{6 \times 3 \times 9 \times 36}{2 \times 48}$$

$$= \underline{60\tfrac{3}{4}} \text{ cwts.}$$

Using a factor of safety of 15, about 4 cwts. would be a safe load for this beam.

Commercial tests.—**Cast iron** is generally tested for commercial purposes by supporting pieces at a known span and ascertaining what load at the centre of the span will break them. The test pieces are generally 36″ span, 1″ broad and 1″ deep, or, better, 1″ broad and 2″ deep. The breaking load at the centre of the span for pieces having the first dimensions ranges from 6 to 8 cwts. and for pieces having the second dimensions from 25 to 35 cwts.

Timber is also generally tested by bending. Test pieces of large section are desirable as the effects produced by local flaws are thus minimised.

Let W = the breaking load at centre of span, in tons.

L = span in inches.

b = breadth in inches, d = depth in inches } of rectangular section.

Then $c = \frac{3}{2} \frac{WL}{bd^2}$ is a quantity which is called the **Modulus of Transverse Rupture**, which may be defined to be the result of the calculation from this formula, using experimental data; actually it has no meaning physically, although sometimes an erroneous one is applied to it. Worked out values of c from the results of tests enable us to calculate the breaking load of similar pieces of the same material. Thus, suppose we take the value of c for cast iron to be 16 tons, and it is required to calculate what central load would break a cast iron beam 36″ span 1″ broad and 2″ deep; then

$$16=\frac{3}{2}\,\frac{W.L}{b.d^2}$$

$$=\frac{3}{2}\times\frac{W\times 36}{1\times 4},$$

or

$$W=\frac{16\times 8}{3\times 36}$$

$$=\underline{1{\cdot}18}\text{ tons.}$$

Experiments on beams.—Students should carry out for themselves some experiments on the stiffness and strength of beams. Metal beams are best tested to breaking in a large

FIG. 150.—Apparatus for experiments on the stiffness and strength of beams.

testing machine; wooden ones having a section 1″ × 1″ and 36″ span can easily be broken with apparatus similar to that described below, and the same apparatus will do for experiments on the deflection of both metal and timber beams. The apparatus (Fig. 150) consists of a lathe bed fitted with two or three cast iron brackets which can be clamped anywhere to it. These brackets are arranged to receive either steel knife edge supports resting in **V** slots cut on the tops of the brackets, or cast iron caps held down by studs. The knife edges are used for beams simply supported. The caps are employed for screwing down on to the beam so as to fix it, the knife edges being first removed. A wrought iron stirrup, with a knife edge for resting on the beam, carries a hook for applying a load anywhere to the beam. Deflections may be measured in various ways. If timber is being experimented on, a pointer, fixed to the stirrup and moving over a scale clamped to a support, suffices, as the deflection is usually considerable. Or a light lever may be used, pivoted to a fixed support and attached by a fine wire at its shorter end to the stirrup, the longer end moving over a fixed scale as the beam deflects. This lever has arms having a ratio of 1 to 10, so that the deflection is multiplied 10 times at the fixed scale. Using a scale of inches divided into tenths, the deflection with this apparatus may be easily read to $\frac{1}{200}$ inch. For very fine measurements a micrometer microscope is used, the readings being taken from the movement of the stirrup as shown by a fine silk fibre mounted on it. This instrument is shown in the illustration and need not be described here.

Pieces of tool steel of various breadths, depths and lengths form useful examples for verifying the comparative stiffnesses of beams. These may be used as beams supported at the ends, or as cantilevers, or as beams fixed at both ends. All the foregoing numbers given for comparative stiffnesses may be verified by use of these samples. It must not be expected, however, that they will be arrived at absolutely by experiment, but the results, if the experiments are carefully done, should agree closely with them. Deflection tests also form a very convenient method of approximately determining Young's modulus of elasticity for a given material, as, using simple means, much larger pieces can be tested by bending than by direct pull, the latter test requiring the use of a large machine.

Let W = load applied, in lbs.

L = length of span, in inches.

D = deflection produced by W, in inches.

b = breadth } of beam of rectangular section,
d = depth } both in inches.

Then, if the test piece is used as a cantilever with the load applied at the end,

$$E=4\,.\frac{WL^3}{D\,.\,b\,.\,d^3}\text{ lbs. per square inch.}$$

And, if the test piece is used as a beam simply supported at both ends, with the load applied at the middle of the span,

$$E=\frac{1}{4}\,.\,\frac{WL^3}{D\,.\,b\,.\,d^3}.$$

The theory involved in these equations for E is too complicated to be dealt with here.

Test bars of ductile material.—Good qualities of wrought

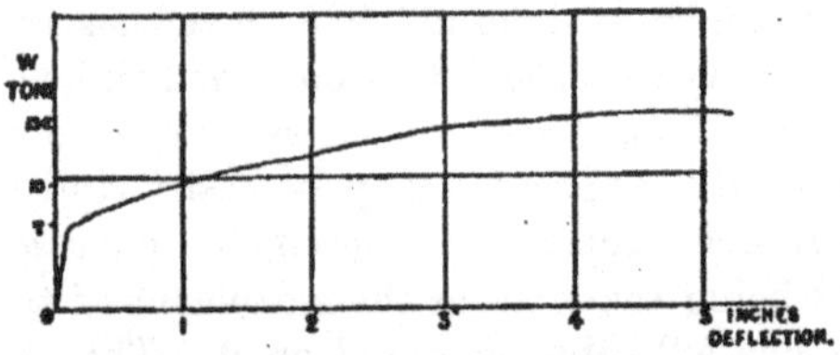

FIG. 151.—Autographic record of a Bessemer steel bar under bending test.

and mild steel bend over double without breaking, and consequently experiments on their bending strength always stop short after the piece has bent through any convenient angle. Fig. 151 shows a copy of an autographic diagram obtained while a bar of Bessemer steel, 24″ span, 2·735″ deep, 0·873″ thick was under test in a 50-ton machine. The bar was tested on edge, simply supported at the ends and the load applied at the centre of the span. Ordinates in this diagram give these loads, abscissae the deflections corresponding to them. The test was stopped when the deflection reached $5\frac{1}{4}$ inches.

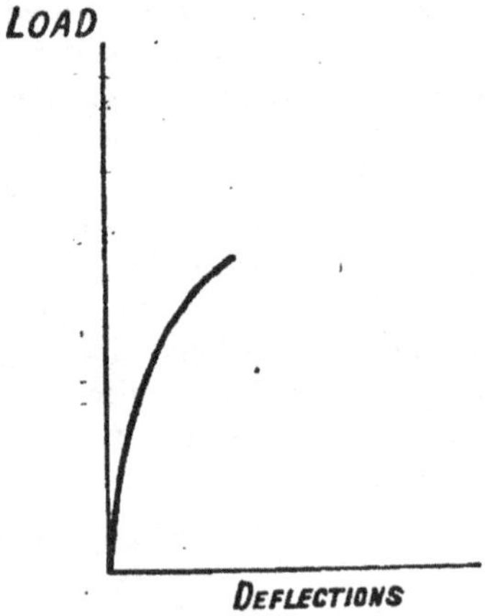

FIG. 152.—Autographic record of a cast iron bar under bending.

Cast iron test bars 36″ span, 1″ broad, 2″ deep usually have a deflection of from 0·35″ to 0·4″

at the middle of the span at breaking. The diagram usually resembles that shown in Fig. 152. It is interesting to note the effect of notching a bar of cast iron. For this purpose two cast iron bars, 1″ × 1″ section everywhere but at the centre, were prepared. One of these, *A* (Fig. 153), had a sharp-bottomed V notch cut in it, the other one, *B*, was thinned down gradually over a length of $3\frac{1}{2}$″ at the middle. The sectional areas of both bars at the middle section were the same. *A* broke with a load of 2·6 cwts. at the centre of the span, the total deflection being 0·23″. *B* broke with a load of 4 cwts. at the middle, the total deflection being 0·44″. The inference from this is, *do not have a sudden change of section in a beam unless it is required to break easily.*

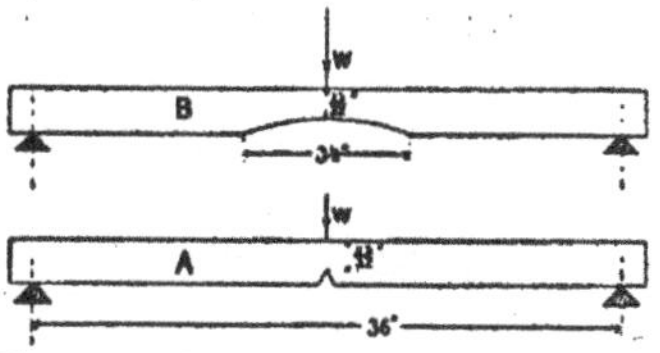

FIG. 153.—Cast iron beams: *A*, having a V notch; *B*, thinned at centre.

Timber test bars.—Samples of various timbers, 36″ span, 1″ × 1″ section should be tested to breaking by the student himself. The deflection corresponding to equal increments of loading should be measured and diagrams should be plotted from these results to show the behaviour of the various pieces. The results for such small samples of the same timbers will be found to vary considerably, owing to the very different qualities found even in pieces cut from the same plank. The pieces should be selected, as far as possible, free from knots, shakes, and flaws of any kind, and should be as straight in the grain as can be obtained.

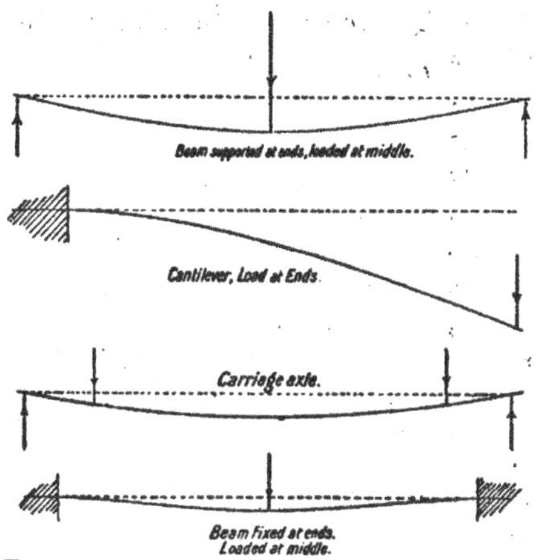

FIG. 154.—Curves of beams loaded in various ways; drawn from a bent knitting needle.

Curve of a bent beam.—The actual shape of the curve in which a given beam will bend may be examined by the student by using a long thin knitting needle. This needle, placed on a sheet of paper secured to a drawing board and "loaded" by means of drawing pins pushed into the board where the supports and applied loads would come in the actual beam, will enable the curve to be drawn. A few cases are shown in Fig. 154. Of course it must be remembered that the deflections in the actual beams will be much smaller in magnitude.

EXERCISES ON CHAP. VIII.

1. A timber beam, 10 ft. span, supported at its ends, carries a load of 600 lbs. at its middle. Calculate the Bending Moment and Shearing Force at places (*a*) very near its ends, (*b*) very near its middle.

2. A cast iron cantilever projects 6 ft. from a wall and carries a load of 500 lbs. at its end. Calculate the Bending Moment and Shearing Force (*a*) at the wall, (*b*) at its middle.

3. A beam 20 ft. span, supported at its ends, carries a load of 2 tons at its middle, also loads of 1 ton at places 5 feet from each end. Neglect the weight of the beam and calculate the Bending Moment at each load, and also the Shearing Force at a place 6 ft. from one end.

4. A cantilever projects 8 ft. from a wall and carries a load of 400 lbs. uniformly distributed. Calculate the Bending Moment and Shearing Force at intervals of 2 ft. throughout its length.

5. A beam 12 ft. span, supported at its ends, carries a uniformly distributed load of 2,400 lbs. Calculate the Bending Moment and Shearing Force at the middle and at 3 ft. from each end.

6. A beam of I-section, 10 ft. span, section 6″ deep, 3″ broad, metal $\frac{1}{2}$″ thick, is supported at its ends and has to carry a load at the centre of the span. Find this load if the maximum stress due to bending is not to exceed 5 tons per square inch.

7. A beam of I-section, 25 ft. long, supported at the ends, has a total depth of 15″, and its top and bottom flanges are each 6″ × $\frac{5}{8}$″. Allowing a stress of 4 tons per square inch, find the load which the beam may carry at its middle. Neglect the weight of the beam.

8. A girder, span 60 ft., weighs 3 tons and is supported at its ends. A load of 1 ton per foot length is distributed over 20 ft. length of the beam from one end. Find the pressures on the supports.

9. Calculate the bending moment at the middle of the span of the girder in Question 8, and find the sectional area of each flange there, if the stress allowed is 4 tons per sq. inch, the depth of the girder being 6 ft.

10. A cast iron cantilever has a mean depth at wall of 12″. The large flange is 10″ × $1\frac{1}{2}$″, and the small flange 3″ × 1″. The cantilever projects 10 ft. and carries a uniformly distributed load of $\frac{1}{2}$ ton per foot length. Find the stresses in the top and bottom flanges at the support.

11. A cast iron cantilever 1″ long, 1″ broad, 1″ deep breaks with a load of 30 cwts. at its end. Calculate the safe load for a cantilever 4″ broad, $1\frac{1}{4}$″ deep, 3″ long, taking a factor of safety = 12.

12. A bar 2 ft. long, 2″ × 2″ square section is supported at its ends and breaks when a load of 3 tons is placed at its middle. Calculate the working load of a bar 5 ft. long, 6″ deep, and 4″ broad, taking a factor of safety = 12. What distributed load would be safe?

13. A wrought iron shaft has its supports 5 ft. apart and carries a load of 2 tons at the middle. If the shaft is 4″ diam., what is the maximum stress due to bending? Neglect the weight of the shaft.

14. A bar of mild steel, section 1″ × 1″, is supported at points 40″ apart. A load of 10 lbs. being applied at the middle of the span, the deflection is observed to be 0·0053″. Calculate the value of E for the material.

15. A beam of wood, rectangular in section, is fixed at one end and loaded at the other. What is occurring at various places in any imaginary cross section? Sketch anything you have seen or used which illustrates your ideas about bending. (1897.)

16. Uniform beams of timber of the same sizes are loaded and supported as follows: 1. Loaded at one end, and fixed at the other. 2. Fixed at one end, and uniformly loaded all over. 3. Supported at the ends, and loaded in the middle. 4. Supported at the ends, and loaded uniformly all over. 5. Fixed at the ends, and loaded in the middle. 6. Fixed at the ends, and loaded uniformly all over. What are their relative strengths? What are their relative stiffnesses? Where is each most likely to break? (1898.)

17. What are the functions of the top and bottom booms and of the diagonal pieces of a railway girder? Why are the booms usually larger in section towards the middle of the girder, and the diagonal pieces usually larger towards the ends of the girder? (1901.)

CHAPTER IX.

RIVETED JOINTS. SHAFTS. SPRINGS.

Riveted joints.—Plates are permanently connected by **riveted joints.** In the simplest form of joint, the edges of the plates *overlap*, and the rivets are closed up in a *single* row of holes. This joint is called a **single riveted lap joint**; if there are *two* rows of rivets—a **double riveted lap joint.** In **butt joints** the plates are brought together, edge to edge, and *cover plates* running along the seam are placed either on one or both sides.

Rivet holes are either *punched* or *drilled.* Punching injures the material of the plate round the hole, and this must be removed by *rymering out* the holes, which, in this case, are punched smaller in diameter than the rivet hole is to be, or else the plate must be *annealed* after punching. Punching must be done with the plates separate, and for this reason the holes will not come exactly opposite one another when the plates are brought together unless a special machine is used for spacing them. The holes produced by punching are slightly *conical*, and the plates are so punched that when they are put together the smaller ends of the holes are on the inside. This produces a sounder job after the rivets are closed.

Drilling does not injure the plate, and is usually done with the plates in position, so that the holes are bound to come fairly opposite one another. The slight burr raised round the edges of the holes by drilling must be removed by separating the plates after drilling and slightly countersinking the holes.

Unless the plates are thin and the rivets small, the rivets are heated before being put into the holes. The head is then formed by hand hammers and finished by a snap, or else machines worked by hydraulic or pneumatic pressure are employed. The plates require to be first drawn tight together by bolts, and the rivet contracts and binds the plates together after the head is formed, as it cools down ; the rivets being thereby put under tensile stress of an indefinite amount.

EXAMPLE. Calculate the force required to punch a $\frac{7}{8}''$ diameter hole in a plate $\frac{1}{2}''$ thick, taking 24 tons per square inch as the ultimate shear stress. Calculate also the compressive stress on the punch.

Area of metal under shear $= \pi d \times t$ (p. 84).

Force required to punch the hole $= \pi d \times t \times 24$

$$= \frac{22}{7} \times \frac{7}{8} \times \frac{1}{2} \times 24$$

$$= \underline{33} \text{ tons.}$$

$$\text{Compressive stress on punch} = \frac{33}{\text{area of punch}}$$

$$= \frac{33}{\frac{\pi d^2}{4}}$$

$$= \frac{33 \times 4 \times 7 \times 8 \times 8}{22 \times 7 \times 7}$$

$$= \underline{55} \text{ tons per square inch.}$$

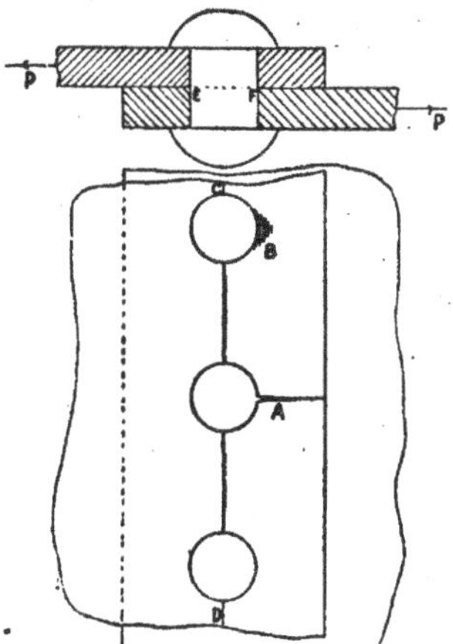

FIG. 155.—Ways in which a riveted joint may fail.

Strength of riveted joints.—Considering the strength of a single riveted lap joint under pull, we see that it may fail in one of *four* ways:

(*a*) By the material between the edge of the rivet hole and the edge of plate opening out (*A*, Fig. 155). This would be due to the holes being too near the edge of the plate. In practice it is found to be sufficient to make the distance from the edge of the hole to the edge of the plate equal to the diameter of the rivet for rivets $\frac{3}{4}''$ and greater in diameter, and slightly more for rivets less than this. Thus, for $\frac{1}{2}''$ rivets, the distance is about $\frac{3}{4}''$.

(*b*) By the material of the plate crushing at *B* (Fig. 155). This would be due to the rivets being too small in diameter, thus providing too small a bearing surface. In practice, a rule such as

$$d = 1{\cdot}2\sqrt{t},$$

t being the thickness of the plate in inches, is used and is found by experience to be sufficient.

(*c*) By one of the plates rupturing under tension along the line *CD* (Fig. 155).

(*d*) By the rivets shearing at *EF* (Fig. 155).

The most economical joint would be obtained by so designing it that the liabilities to rupture in these four ways are equal to one another. There being no exact mathematical information as to the strength against rupture by methods (*a*) and (*b*), it is customary to determine first the diameter of the rivet for the given plates by the rule given in (*b*), and then to decide upon the overlap of the plates as shown in (*a*). Afterwards (*c*) and (*d*) are calculated so as *to make the joint equally strong against failure by tearing along the row and by shearing of the rivets.*

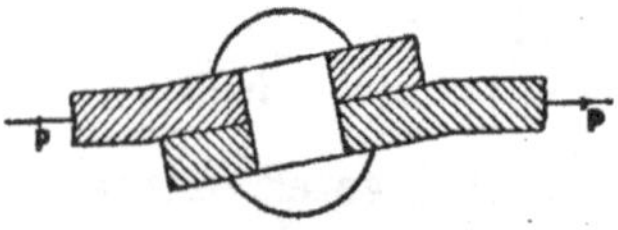

FIG. 156.

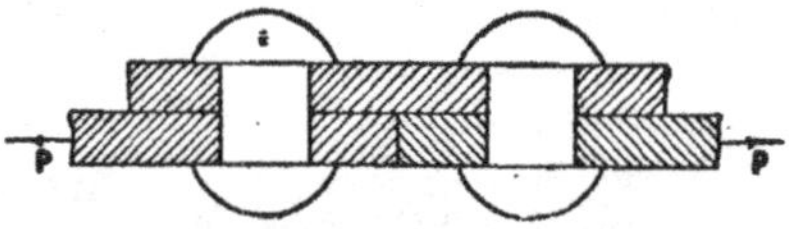

FIG. 157.—Butt joint, single cover strap.

Bending action on joints.—It must be further noticed that when the pulls *P*, *P*, (Fig. 156) are applied to the joint, they produce a couple tending to make the joint assume a form resembling that shown in the illustration, so that *P*, *P*, will act in the same straight line. Joints are sometimes made as shown in Fig. 156 in practice in order to prevent this bending tendency. Butt joints are liable to the same action if there is only one cover strap (Fig. 157), but with a strap on each side (Fig. 158), this is prevented.

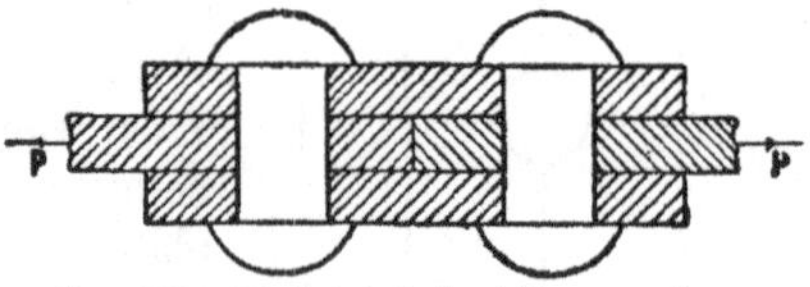

FIG. 158.—Butt joint, double cover strap.

Application to single riveted lap joint.—Consider a strip of the joint equal in breadth to the **pitch** of the rivets, *i.e.* the distance from centre to centre of the rivets, measured along the row. This will be the breadth of joint supported by one rivet and so the conclusions arrived at will be true for the whole joint.

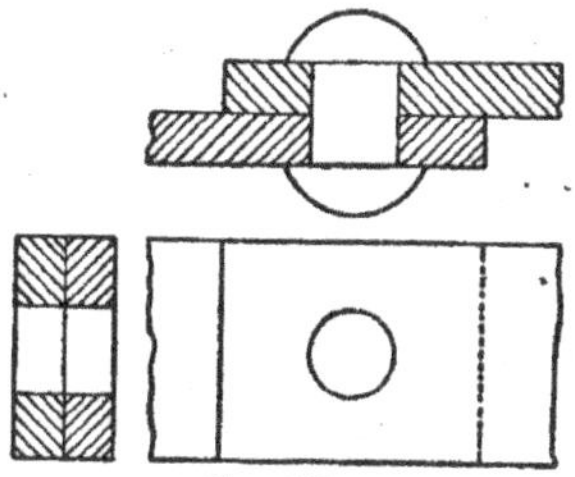

FIG. 159.

In Fig. 159 let

p = pitch of rivets, inches;
d = diameter of rivets, inches;
t = thickness of plate, inches;
f_t = tensile stress permitted;
f_s = shear stress permitted.

Area under tensile stress $=(p-d)t$

Area under shear stress $=\frac{\pi d^2}{4}$.

For equal strength, the following equation must be true:

$$f_t(p-d)t=f_s\frac{\pi d^2}{4},$$

Taking $d=1{\cdot}2\sqrt{t}$, or $t=\frac{d^2}{1{\cdot}44}$, then

$$f_t(p-d)\frac{d^2}{1{\cdot}44}=f_s\frac{\pi d^2}{4}$$

$$f_t(p-d)=f_s\frac{\pi}{4}\times 1{\cdot}44$$

$$=0{\cdot}36.f_s.\pi,$$

or $$(p-d)\frac{f_t}{f_s}=1{\cdot}131$$

f_t ranges in practice from 35,000 to 67,000 pounds per square inch, and f_s from 43,000 to 53,000 pounds per square inch. The values to be taken in any given case depend on the number of rows of rivets, on the material (whether iron or steel), and on whether the holes have been punched or drilled. For iron plates and iron rivets, with drilled holes, the ratio $\frac{f_t}{f_s}$ may be taken as

$$\frac{f_t}{f_s}=0{\cdot}94,$$

which would give for the single riveted lap joint

$$(p-d)0{\cdot}94=1{\cdot}131.$$

EXAMPLE. Calculate the diameter and the pitch of the rivets for plates $\frac{1}{2}''$ thick connected by a single riveted lap joint.

$$d=1{\cdot}2\sqrt{t}$$

$$=1{\cdot}2\sqrt{\frac{1}{2}}=\frac{1{\cdot}2}{1{\cdot}41}=\frac{7''}{8}\text{ nearly.}$$

$$\left(p-\frac{7}{8}\right)0{\cdot}94=1{\cdot}131,$$

$$p=\frac{1{\cdot}131}{0{\cdot}94}+0{\cdot}875$$

$$=\underline{2\tfrac{1}{8}''}\text{ nearly.}$$

Percentage strength of joint.—It will be observed that the sectional area of the plate along the centre line of the row has been diminished after the rivet holes have been punched or drilled, and that therefore the strength of the joint is less than that of the unhurt plate. Taking a width of plate equal to p; its sectional area will be p multiplied by t in the unhurt plate and $(p-d)t$ at the centre line of the row of rivets; therefore

$$\frac{\text{strength of joint}}{\text{strength of unhurt plate}}=\frac{(p-d)t}{p\times t}=\frac{p-d}{p}.$$

In the above example, this will give

$$\frac{2\frac{1}{8}-\frac{7}{8}}{2\frac{1}{8}}=\frac{1{\cdot}25}{2{\cdot}125}=0{\cdot}59\text{ nearly;}$$

or the strength of the joint is about 59 per cent. of the strength of the unhurt plate.

In joints such as that shown in Fig. 158, the rivets are under *double shear*, that is, they would have to shear at two places if the joint gave way by fracture of the rivets. Usually butt joints with double cover straps are double riveted, having four rows of rivets in all. In this case, for a length of joint equal to the pitch of the rivets, there are *four* rivet sections under shear, but as these will probably not all be effective, it is customary to reckon 3·5 rivet sections only. In joints such as this, a percentage strength of 75 can be obtained.

Twisting moment on a shaft.—Shafts are pieces used for the transmission of motion and power from one place to another. They are usually made *round*, and receive a moment tending to rotate them at one place, which moment is transmitted, by stresses in the material of the shaft, to the desired place. Let us consider a shaft AB (Fig. 160) one end of which, A, is fixed in some way, and having an arm BC, mounted at the other end. If a force P is applied at the end of the arm BC, two things will evidently occur: the shaft will tend to *rotate*, or *twist*; also it will tend to *droop*, showing that there is a *bending action*. We may get rid of the tendency to bend by prolonging the arm on the other side of B (as shown dotted in Fig. 160) and applying a force equal and opposite to P at the end E. These forces P, P, now form a couple acting on the shaft, which in consequence is called upon to resist rotation only, and its material will be subjected to pure twist. The moment of the couple is called the **Twisting Moment, or Torque.** Thus

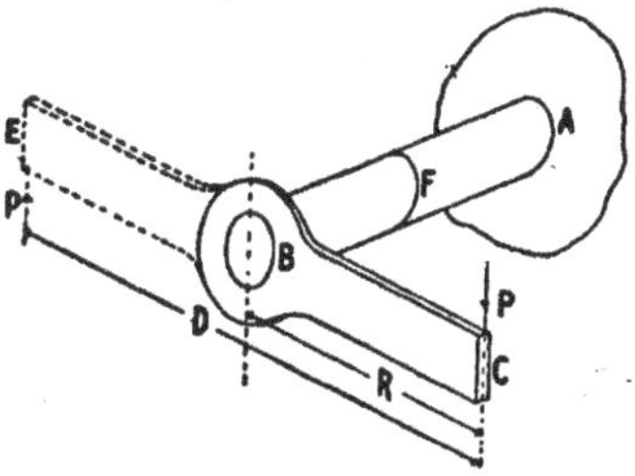

FIG. 160.—Shaft under torque.

$$\text{Torque} = T = P \times D.$$

Shearing stress produced by torque.—Consider now any cross section of the shaft, such as that at F, taken perpendicular to its axis. There will be a tendency, when the torque is applied, for the material on one side of the section at F, by rotation, to slide past the material on the other side of the section. Consequently such sections will be under shear stress. The shear stresses will not be uniformly distributed over the section, and can for round shafts be shown to be in proportion to the distance from the axis of the shaft at any part of the section. Let r be the radius of the shaft section, then we should expect to find at a_1 (Fig. 161), which is at a distance $\frac{1}{2}r$ from the axis, a stress just half that at a_2, which is at a distance r from the axis. The stress at the axis of the shaft would be zero.

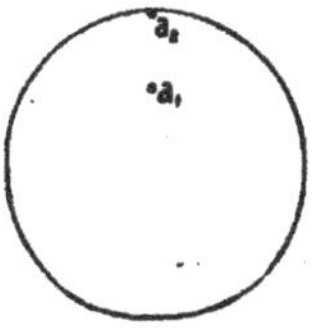

FIG. 161.

Moment of resistance.—The stress on all parts of the section will give forces, each having a moment about the axis of the shaft, and the sum of all these moments will give a resultant moment, called the **Moment of Resistance** of the material. This Moment of Resistance balances the Twisting Moment applied to the shaft and consequently will be equal to it. For a round solid shaft, the Resisting Moment can be shown to be

$$\text{Resisting Moment} = \frac{f\pi r^3}{2} \text{ lb. inches.}$$

Where f = maximum shear stress allowed, lbs. per sq. inch,

r = radius of shaft section, in inches.

Let T = the torque, in lb. inches, applied, then

$$T = \frac{f\pi r^3}{2} \text{ for a round solid shaft.}$$

EXAMPLE. What torque can be applied to a shaft 4″ diameter if the maximum shear stress is not to exceed 10,000 lbs. per square inch?

$$T = \frac{f\pi r^3}{2}$$

$$= \frac{10{,}000 \times 22 \times 2 \times 2 \times 2}{7 \times 2}$$

$$= 125{,}700 \text{ lb. inches}$$

$$= \underline{10{,}470} \text{ lb. feet.}$$

If we measure the strength of a shaft by the torque which may be safely applied to it, the above equation shows us, that if the shaft is under pure twist, without bending action, **its strength will be independent of its length and directly proportional to the cube of its radius.** Thus, a solid shaft 4″ diameter could withstand safely 8 times the torque which could be safely applied to a shaft of the same material, but only 2″ diameter.

Hollow shafts.—Since the material of a shaft near its axis is only under a small shearing stress, and the arm of which, in taking moments, is also small, we could obtain a stronger shaft of the same weight by removing some of this material and putting it instead at the outer circumference of the shaft. This is often done in large shafts for the sake of lightness, and weight for weight, a *hollow shaft will be stronger than a solid one.*

Flanged shaft couplings.—Two pieces of shafting in the same straight line are often connected by having their ends

flanged, or **flanged couplings** *keyed on* and the flanges secured together by *fitted bolts.* These bolts will be under shearing stress when the torque is applied.

Let d be the diameter of the bolts and f_s the shearing stress they may have, then

$$\text{Shearing force on one bolt} = f_s \times \frac{\pi d^2}{4}.$$

If there are N bolts, the total shearing force will be

$$P = N \,.\, f_s \,.\, \frac{\pi d^2}{4}.$$

Let R = radius of the bolt circle, then the Resisting Moment of the bolts to shearing will be P multiplied by R, or

$$\text{Moment of Resistance} = N \,.\, f_s \,.\, \frac{\pi d^2}{4} \,.\, R.$$

And this must be equal to the torque applied; hence, if we know the maximum torque to be applied to the shaft, the diameter of the bolts required may be easily calculated.

FIG. 162.—Apparatus for experiments on the torsion of wires.

Stiffness of wires under torsion.—The elastic properties of a wire under torsion can be examined by a self-contained machine such as is shown in Fig. 162. This particular machine was not specially designed, but made out of some materials which happened to be at hand. It consists of an upper bracket for holding the top end of the wire under test, supported by three mild steel rods, which are tied together near their lower ends by another similar bracket. Two pointers can be clamped to any part of the wire, and move over circular scales divided in degrees. The torque is applied by two cords coiled round a drum 5″ diameter, clamped to the wire, the cords being led over pulleys and having scale pans at their ends. If equal loads are placed in the pans they will produce a couple, giving pure twist to the wire. A permanent weight hung to the end of the wire keeps it taut. This machine can be used for verifying the following elastic properties of wires under torsion.

The angle of twist is proportional to the torque applied, directly proportional to the length of the wire, and for wires of the same material but of different diameters, inversely proportional to the fourth power of their diameters.

The following results were obtained in an experiment for verifying the first two statements.

AN EXPERIMENT ON TWISTING.

Steel wire, annealed, 0·065″ diam.

Load in each pan + weight of pan; W lbs.	Torque $= W \times 5$ lb. inches.	Angle of twist on 4·5″ length. Degrees.	Angle of twist on 20·8″ length. Degrees.
0·0	0·0	0·0	0·0
0·20	1·0	1·0	6·0
0·80	4·0	5·0	23·0
1·05	5·25	7·5	29·0
1·55	7·75	9·0	41·0
2·05	10·25	11·0	53·5
2·80	14·0	15·5	73·0
3·00	15·0	16·0	79·0
3·50	17·5	19·0	92·0
3·60	18·0	20·0	94·0
3·85	19·25	21·5	101·0

Columns 2 and 3 when plotted, and also columns 2 and 4, give very nearly a straight line, showing that the angle of twist is proportional to the torque (Fig. 163).

Selecting two values from the plotted curves, when the torque is 14 lb.-inches, the angle of twist on a length of 4·5″ is 15·5 degrees, and the angle for a length of 20·8″ is 73·5 degrees.

$$\text{Ratio of angles of twist} = \frac{73 \cdot 5}{15 \cdot 5} = 4 \cdot 7.$$

$$\text{Ratio of lengths of wire} = \frac{20 \cdot 8}{4 \cdot 5} = 4 \cdot 6.$$

An approximation to the law that the angle of twist is proportional to the length of the wire.

Angle of twist after passing the elastic limit.—Test pieces of wrought iron or mild steel subjected to torsion, after the elastic limit of the material for shearing stresses is passed, do not twist through angles proportional to the torque. After this, the angle is much larger proportionally, and a piece of great length compared to its diameter will twist through many complete turns before fracture occurs. For this reason, it is convenient to use rather short pieces of the material for such testing purposes.

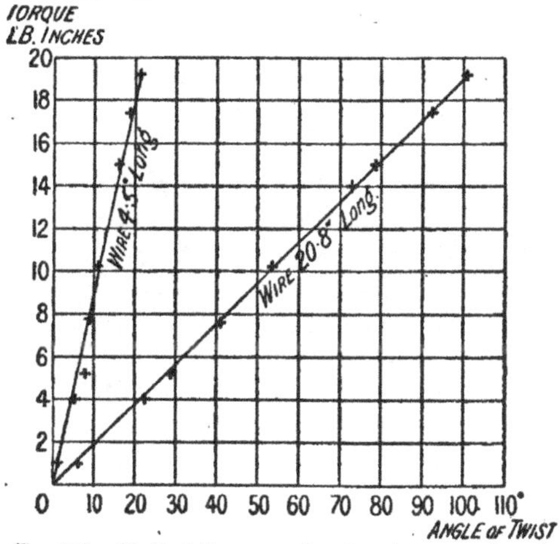

FIG. 163.—Plotted diagram, showing torque and angles of twist for a wire under torsion.

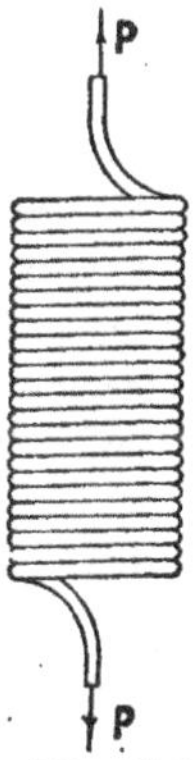

FIG. 164.—Helical spring.

Springs.—Springs are pieces intended to take a relatively large amount of strain, although any piece of material which can be strained and shows a strong tendency to recover its original shape *freely* may be called **a spring**. Thus, a bar of iron or steel, pulled within its elastic limit, may be called a spring. The forms taken by springs depend on the purpose for which they are intended. For spring balances, used to measure forces, **helical springs** are used (Fig. 164). In these the forces are applied in the direction of the axis of the spring and the material of the spring is under torsion. Helical springs are also occasionally subjected to couples, as shown in Fig. 165 ; in this case the material is under bending. A spring resembling this is used in the *Wayne indicator*.

Spiral springs, as in Fig. 166, are used for watches and clocks,

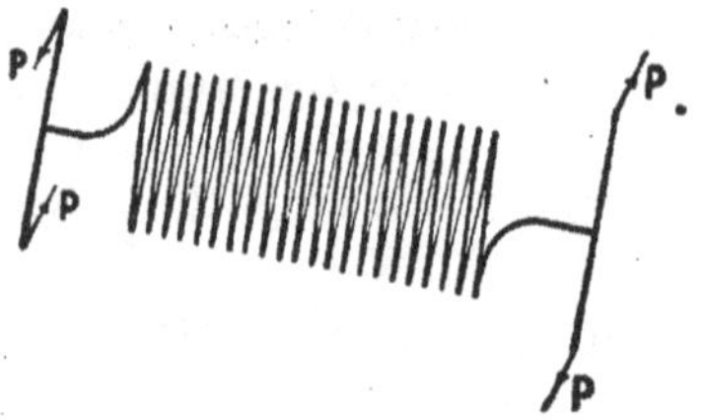

FIG. 165.—Helical spring under torsion.

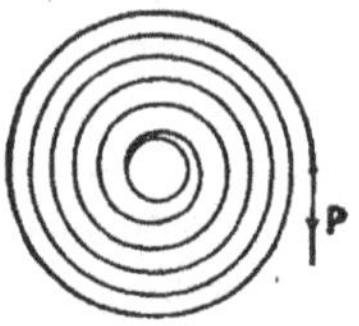

FIG. 166.—Spiral spring.

and have their material subjected to bending. **Volute springs** are used for railway carriage buffers; one is shown in Fig. 167. **Carriage springs** consist of a number of flat strips, of gradually increasing length, secured together and loaded (Fig. 168). Springs consisting of small discs corrugated in a circular direction (Fig. 169), are occasionally employed for measuring, by their deflection, the pressure of a gas or liquid acting on them. Rubber cushions are often used between the body of a carriage and the steel springs. These **rubber springs** are under compression.

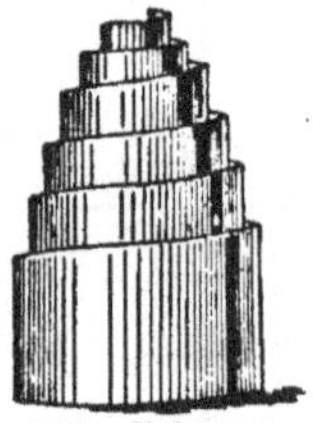
FIG. 167.—Volute spring.

FIG. 168.—Carriage spring.

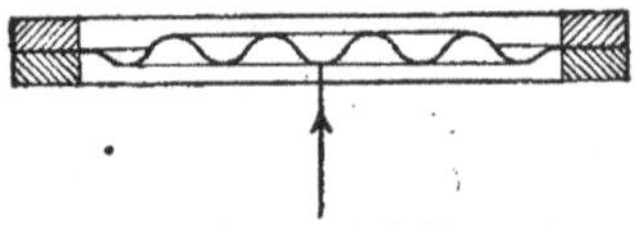
FIG. 169.—Corrugated disc spring.

In employing rubber springs under compression it may be noticed that although rubber can change its shape very easily, it can only change its bulk very slightly even under very great forces. Consequently, if confined so that it cannot swell out laterally when the forces are applied, it will act not as a spring but as a relatively rigid body.

Springs are used for measuring forces, for storing energy, and for minimising the effects of shocks. In all cases, the loading is kept within the elastic limit, and consequently, as a rule, the spring is distorted, or changes its length, by an amount proportional to the applied torque or forces.

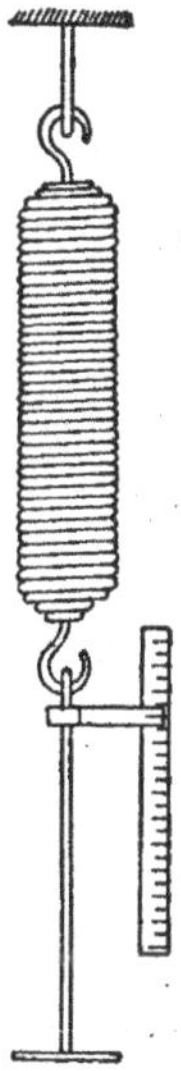

FIG. 170.—Apparatus for measuring extensions of a spring.

Elastic extension of springs.—The student should experiment with a helical spring by loading it with gradually increasing weights, the extensions being given by a pointer attached to the spring and moving over a scale (Fig. 170). It will be found that, if several springs of the same material are available for testing, the following laws are approximately true:

The extensions are proportional to the load, to the cube of the radius of the helix, and to the number of complete turns in the helix; also inversely proportional to the fourth power of the radius of the wire of which the helix is made.

In these experiments, the wire is assumed to be round, and the springs made so that the coils lie close together.

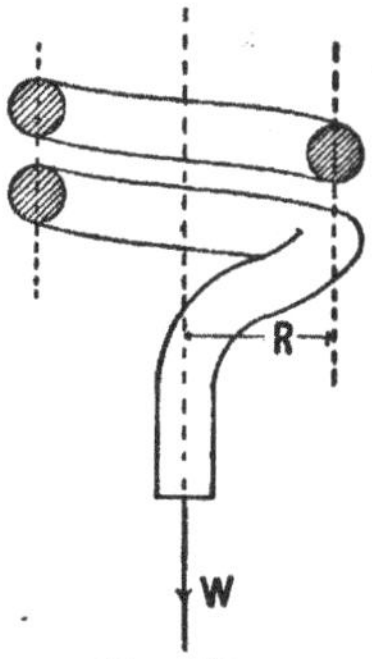

FIG. 171.

The proportional laws may be represented by an equation in this way.

Let

W = load applied, lbs. (Fig. 171).

R = radius of helix to centre of wire, in inches.

N = number of complete turns.

r = radius of wire, in inches.

X = extension produced by W, inches.

Then

$$X = c\frac{WR^3N}{r^4},$$

where c is a constant depending on the elastic qualities of the material. Its value for steel is approximately

$$c = 0{\cdot}00000033.$$

EXERCISES ON CHAP. IX.

1. A hole $\frac{3}{4}''$ diameter has to be punched in a plate $\frac{7}{16}''$ thick. Take the ultimate shearing stress as 24 tons per square inch and calculate the force required.

2. The plates of a boiler are $\frac{7}{16}''$ thick, and the circumferential seams are to be lap joints double riveted. Find the diameter and pitch of the rivets and the percentage strength of the joint, assuming that the sectional area under shear has to be equal to the sectional area under tension.

3. A boiler 7′ 0″ diameter has to work at a pressure of 120 lbs. per square inch. Calculate the thickness of plate required if the maximum tensile stress is not to exceed $4\frac{1}{2}$ tons per square inch (*a*) assuming the strength of the joint to be 100 per cent., (*b*) assuming the strength of the joint to be 75 per cent.

4. Two portions of a tie rod are connected by a knuckle joint. The pull in the rod is 6 tons. Calculate the diameter of pin required if the shearing stress allowed is 4 tons per square inch.

5. Two shafts, one 2″ diameter the other $2\frac{1}{4}''$ diameter, are subjected to equal twisting moments. Compare their maximum shearing stresses.

6. A shaft 2″ diameter is subjected to a torque of 12,000 lb. inches. Calculate the maximum shearing stress produced.

7. Calculate the diameter of shaft required to carry a torque of 5000 lb. feet, if the maximum shearing stress allowed is 10,000 lbs. per square inch.

8. Calculate what twisting moment can be applied to a shaft 10″ diameter if the shearing stress is not to exceed 4 tons per square inch. What force applied at 90° to a crank at an arm of 27″ will produce this torque?

9. Compare the Bending Moment and the Torque which may be applied to a round solid shaft if the safe tensile, compressive, and shearing stresses are assumed equal.

10. A wire of Siemens' steel, 0·1 inch diameter, is to be twisted till it breaks. Sketch the arrangement and show how the angle of twist and the twisting moment are measured, how the results may be plotted on squared paper, and the sort of results that may be expected. In what way may a wire of twice this diameter be expected to behave? (1901.)

11. Two portions of a shaft 2″ diameter are connected by a flanged coupling whose four bolts have their centres on a circle concentric with the shaft centre and $5\frac{1}{2}''$ diameter. Allowing a safe shearing stress of 10,000 lbs. per square inch, calculate (*a*) the twisting moment the shaft may be subjected to; (*b*) the diameter of bolts required.

12. A helical spring is made of round steel $\frac{1}{8}''$ diameter, and has 50 coils $1\frac{1}{2}''$ diameter. Calculate its extension when pulls of 5 lbs. are applied.

CHAPTER X.

WORK. MECHANICAL ADVANTAGE. VELOCITY RATIO OF MACHINES. ENERGY. POWER. EFFICIENCY. DIAGRAMS OF WORK. RESILIENCE.

Definitions of terms.—A force is said to be doing **work** when it acts through a distance, overcoming resistance. The **quantity of work** done is proportional jointly to the magnitude of the force and the distance through which it acts, the distance being always measured along or parallel to the line of action of the force. The **unit of work** used by engineers in this country is the **foot pound, and is that quantity of work which is done when a force of one pound acts through a distance of one foot in its line of action.** The **inch-ton** and **foot-ton** are also sometimes used, these being the work done when a force of one ton acts through a distance of one inch or one foot respectively.

The work done by any force is calculated by taking the product of the magnitude of the force and the distance through which it acts.

EXAMPLE 1. If a weight of 4 tons has to be raised from the bottom of a shaft 100 fathoms deep, find the work done.

$$\begin{aligned} \text{Work done} &= \text{force} \times \text{distance} \\ &= 4 \times (100 \times 6) \\ &= \underline{2400} \text{ foot-tons.} \end{aligned}$$

EXAMPLE 2. A load of 2 cwts. is dragged along a level floor through a distance of 10 feet by means of a rope inclined at 30° to the floor (Fig. 172). The pull P is found to be 80 lbs. Calculate the work done.

First Solution. Notice that P does not act through a distance $AB=10$ feet, but through $AC=5\sqrt{3}$ feet (Fig. 172). Hence,

$$\text{work done} = 80 \times 5 \times 1{\cdot}73$$
$$= \underline{692} \text{ foot-lbs.}$$

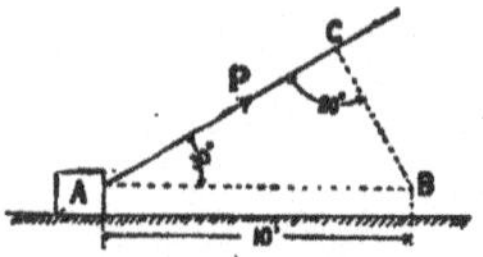

FIG. 172.

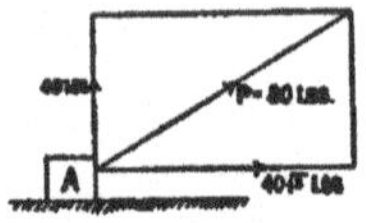

FIG. 173.

Second Solution. Or, we may solve the question in another way: Take horizontal and vertical components of the pull in the rope (Fig. 173) giving $40\sqrt{3}$ lbs. for the horizontal and 40 lbs. for the vertical component.

This vertical component, while the weight is being dragged along, will always be acting in parallel vertical directions, and its point of application will move neither up nor down. Consequently, as it does not act through a distance measured along or parallel to its line of direction, no work is done by it. The horizontal component acts through a distance of 10 feet measured along its line of direction, consequently work done will be

$$40\sqrt{3} \times 10$$
$$= \underline{692} \text{ foot-lbs., as before.}$$

The same quantity of work may be done either by a small force acting through a large distance, or a large force acting through a small distance. Thus, suppose we have to do 1,000 ft.-lbs. of work, we may use a force of 1 lb. acting through 1,000 feet, or 500 lbs. acting through 2 feet, obtaining the desired result in each case.

Machines. — **Machines** are arrangements receiving work from some outside source of supply, which work is modified by the machine and delivered in some form more suitable for the purpose required.

FIG. 174.—Simple winch.

As an example of a machine, we may take a **simple winch** (Fig. 174) used for raising loads. This machine takes in work from the pushes and pulls of two men at the handles, turning the drum. The drum produces a modified pull on the lifting rope by which work is done on the ascending weight.

Usually in machines such as hoisting tackle, machine tools, etc., the force delivering work to the machine is smaller than the resistance which is overcome by the use of the machine. The **Mechanical Advantage** of a machine is the ratio of these two forces. Thus, in the above winch, suppose that each man exerts a constant force of 30 lbs. applied always tangential to the path of his hand (Fig. 175), the load raised being 300 lbs., then the Mechanical Advantage will be $\frac{300}{2 \times 30} = 5$.

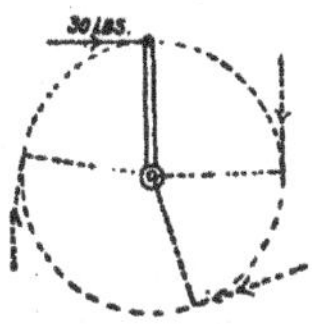

FIG. 175.

The **Velocity Ratio** of a machine is measured by dividing the distance through which the force applied to the machine acts, by the distance through which the resistance is overcome in the same time. Thus, in the above winch, if the men's hands move through a circumferential distance of 6 feet while the weight is being raised a height of one foot, then the velocity ratio is 6.

Energy and its transformation.—If work is imparted to a body so that it stores it up and is capable of giving it out again, the body is said to possess **Energy.**

Energy means capability of doing work. If we raise a one pound weight from the table a height of one foot, we have done one foot-pound of work on the weight, and this work it can give out again if we allow it to descend. The raised weight possesses energy to the amount of one foot-pound. The individual who raised the weight has had to part with one foot-pound of his store of internal energy, and if he goes on for some time raising weights, say for 4 or 5 hours, he will find that there is need to replenish his store of energy by absorbing some food and resting a little. Food possesses energy which, transformed by the organs of our body, enables us to do mechanical work. The energy of food is liberated in our bodies in the form of heat by a process of slow combustion. Coal also possesses heat energy which is utilised by a much more rapid combustion in boiler furnaces or otherwise. Water at an elevation possesses energy which can be transformed into mechanical work while the water is descending, and the atmosphere has energy also when in motion as wind. The principal source of all these stores of energy is the heat of the sun, which (i), in past ages, caused the vegetation to grow from which we at present derive our coal supply, and (ii) now gives us our food supply by giving life to plants. The heat of the sun also raises water by evaporation and so gives it its store of energy, and it also sets the atmosphere in motion as wind.

Conservation of energy.—The principle of the **Conservation of Energy** asserts that *man is not able to create or destroy energy, he can only transform it from one form into another.* This principle is the result of the observation and experiment of many people, including those who have sought in vain for perpetual motion. To the engineer, it is of extreme importance. Thus, if we impart to any machine a certain quantity of energy, and no energy is lost in the machine or used to coil up a spring belonging to the machine or do any other form of work in the machine, then the machine will deliver an exactly equal quantity to that given to it. It cannot deliver more, for then it would create energy; nor can it deliver less, for then energy would be destroyed in the machine.

Actually, it is impossible to construct a machine in which there is no energy lost, whether by the rubbing of surfaces on one another, by churning the atmosphere, or by the development of sound and other causes. But we can assert about all machines:

Energy supplied = Energy given out + energy lost in overcoming resistances in the machine.

If the machine is *running light*, *i.e.* doing no useful work against resistances, then we must supply energy sufficient to make good that lost by resistances in the machine. If the machine is *doing useful work*, then we must in addition supply energy equivalent to this useful work.

Efficiency of machines.—The **efficiency of any machine** is measured by the ratio of energy actually given out as useful work to the energy supplied. Or,

$$\text{Efficiency} = \frac{\text{useful work done}}{\text{energy supplied}}.$$

EXAMPLE. A certain machine is found to give out 125 foot pounds of useful work when 180 foot pounds of energy are supplied to it. Find the efficiency of the machine.

$$\text{Energy lost in machine} = 180 - 125$$
$$= 55 \text{ foot pounds.}$$
$$\text{Efficiency} = \tfrac{125}{180} = \underline{0{\cdot}69.}$$

The efficiency may be stated as a percentage by multiplying this by 100, giving

$$\text{Efficiency} = 0{\cdot}69 \times 100$$
$$= \underline{69} \text{ per cent.}$$

Forms of energy useful to the engineer.—The principal forms of energy that the engineer has to deal with are: **Potential Energy**, such as the energy of a raised weight or a coiled spring; **Kinetic Energy**, which a body possesses when it is in motion, and can give out as mechanical work while it is coming to rest; both these forms of energy are stated in foot pounds. One foot pound of potential energy is exactly equivalent to one foot pound of kinetic energy. For example, a raised weight possesses potential energy, and if it is allowed to fall freely, doing no work against any resistance, the potential energy will be converted into an exactly equal quantity of kinetic energy.

Heat is a form of energy which can be converted into mechanical work. The amount of mechanical work equivalent to a given quantity of heat is known with considerable accuracy from the experiments of Dr. Joule and others. Thus 772, 774, or 778, foot pounds of mechanical work transformed into heat would, if all the heat passed into one pound of water, raise its temperature one degree Fahrenheit. The numbers given are those used by various authorities, and it is immaterial which of them is taken for engineering purposes, as the total difference between the first and the last is less than one per cent., and no engineering problem involving heat quantities can be worked out from practical data to a greater degree of accuracy. The quantity of heat which would raise the temperature of one pound of water through one degree Fahrenheit is the unit of heat used by British engineers, and is called the British Thermal Unit.

Electrical energy is measured in *kilowatts performed per hour.* The **kilowatt** is an electrical *power unit,* and consequently gives the rate of energy production, corresponding to the horse-power (p. 128). A kilowatt is equal to 1000 watts, the **watt** being the rate of working when an electric current of one ampère flows from one point of a conductor to another, the potential difference between which is one volt. The product of ampères and volts is expressed in watts. 746 watts are equivalent to the mechanical horse-power. It will be understood that just as we have to state the time during which a given horse-power has to be maintained in order to produce a certain amount of work, so we must state, as above, the time during which one kilowatt has to be maintained in order to produce a given amount of electrical energy. The Board of Trade unit of electrical energy is that given above as one kilowatt maintained for one hour.

Loss of useful energy.—It should be clearly understood that although energy in one form may be exactly equivalent to a certain quantity of energy in another form, that we never succeed, in any transformation of energy, in obtaining an exactly equal quantity in the new form. There are always losses—sometimes great losses—which are inevitable. As a common example, and one which gives a fair idea of the magnitudes of some of these losses, take the following case of electrical power production. Suppose 100 units of energy to be liberated from some coal in the boiler furnace. About 75 of these will enter the steam and the remaining 25 will be lost by the passage of the smoke and heated gases up the chimney, or by radiation and other causes. Of the 75 units of energy reaching the engine in the steam, about 6 will be converted into mechanical energy and the remaining 69 will be lost. The 6 units of mechanical energy given to the dynamo will produce about 5 units of electrical energy, 1 unit being lost. If these 5 units be reconverted into mechanical energy by an electrical motor, about 4·5 units will be produced. We utilise in this way about 4·5 per cent. of the original energy and lose 95·5 per cent.

Variation of the actual mechanical advantage.—From what has been said it will be seen that the velocity ratio of a machine, which depends solely on the arrangement and nature of its parts, does not change provided the arrangement remains the same. On the other hand, the mechanical advantage, or the ratio of the resistance overcome to the force delivering energy to the machine, depends on the extent of the losses in the machine; these again are variable, depending on the load and on the condition of the machine as regards lubrication and state of the bearing surfaces.

Let P = the force delivering energy to a machine, and W = the resistance overcome, both in same units of force. Let D = distance through which P acts while W is being overcome through a distance = d, D and d being in the same units of distance. Then, if there were no losses in the machine,

$$\text{Energy supplied} = \text{useful work done},$$

$$P \times D = W \times d,$$

$$\frac{D}{d} = \frac{W}{P}.$$

Now $\frac{D}{d}$ is the velocity ratio of the machine and $\frac{W}{P}$ is the mechanical advantage, so that in this hypothetical case, the velocity ratio and mechanical advantage are equal numerically. Actually, however, W will always be less than its value assumed above, and consequently the actual mechanical advantage will always be less than the velocity ratio for any machine. In Chap. XIV. it will be seen how the actual mechanical advantage of a machine can be obtained. The velocity ratio can be calculated from a knowledge of the mechanism, or by direct measurement, at the places where P and W are applied, of the distances through which they act.

Power.—If we state not only the quantity of work done by a force or forces, but also the time in which it is done, this will give us the rate at which work is being performed. **Power, or activity**, is the name given to the rate of performing work. The **unit of power** used generally by engineers in this country is produced when 33,000 foot-pounds of work are done in one minute. This unit was defined by James Watt, who found that the average horse could do about 22,000 foot-pounds of work in a minute. Watt added 50 per cent. to this and took 33,000 foot-pounds per minute as the unit of power to be used in measuring the performance of his steam engines.

The horse-power developed in any given case will be ascertained by first calculating the work done per minute, in foot-pounds, and dividing the result by 33,000.

Thus, suppose we take the previous example (p. 122), in which 4 tons were raised from a depth of 100 fathoms. If the work is performed in 40 seconds, the horse-power required may be calculated thus:

$$\text{Work done in 40 seconds} = (4 \times 2240) \times (100 \times 6) \text{ ft.-lbs.}$$

$$\text{Work done in 60 seconds} = 8960 \times 600 \times \frac{60}{40} \text{ ft.-lbs.}$$

$$= 8{,}064{,}000 \text{ ft.-lbs.}$$

$$\text{Horse-power} = \frac{8{,}064{,}000}{33{,}000}$$

$$= \underline{244 \cdot 4.}$$

In calculating the horse-power required in any given case, the efficiency of the machine employed must be considered, as power will be required for overcoming losses in the machine. Thus, suppose in the above case, that the efficiency of the mechanism of the winding engine employed to raise the 4 tons is 60%. This means that 60% of the energy given to the engine in a given time, say one minute, is turned into useful work. Consequently, the 8,064,000 ft.-lbs. useful work above done per minute is only $\frac{6}{10}$ths of the energy that must be given to the engine per minute.

$$\text{Energy supplied to engine per min.} = \frac{8{,}064{,}000 \times 10}{6} \text{ ft.-lbs.,}$$

$$\text{and, Actual Horse-power required} = \frac{8{,}064{,}000 \times 10}{6 \times 33{,}000}$$

$$= \underline{407.}$$

Graphic representation of work.—Since work is measured by the product of two quantities, force and distance, we may represent it by the area of a diagram. Thus, supposing a uniform force P to act through a distance D, the work done will be $P \times D$. If we set off D to scale in a diagram (Fig. 176), and erect ordinates of constant height equal to P to scale, we obtain a rectangle of area equal to PD, which therefore represents the work done.

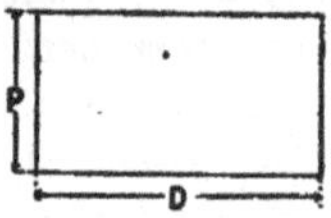

FIG. 176.—Work done by a uniform force.

If the work is done by a varying force, and the amount of it is known at various points in the distance acted through, we may set up ordinates to represent its value at these points, giving a diagram as shown in Fig. 177 bounded by a curve at the top. The work done in this case will be equal to the average value of P multiplied by D. Now the average P will be represented by the average height of the diagram to scale; and as the average height multiplied by D gives the area of the figure, it follows that in this case also, the area of the figure represents the work done.

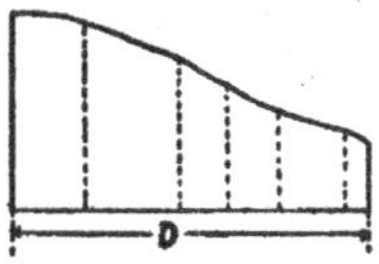

FIG. 177.—Work done by a varying force.

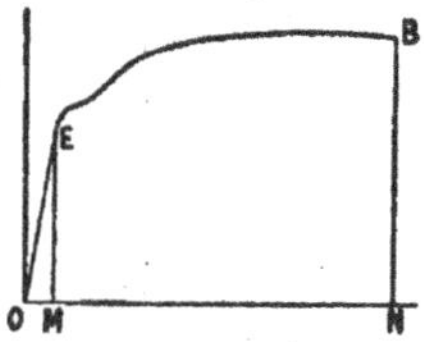

FIG. 178.—Work done in extending a bar.

Work done in pulling a bar.—A useful example of diagrams of work is to be found when the work done against the resistance of a bar to stretching is shown. As we have seen already (p. 72), the resistance of the bar, or the stretching load, is directly proportional to the extension produced, giving, when plotted, a straight line up to the elastic limit of the material. The area under this line will represent the work done in stretching the bar. If the whole diagram is given showing the gradual loads applied and the extensions produced up to breaking, then the whole work done in breaking the bar will be given by its area. Thus, the area EOM (Fig. 178) represents the work done on the bar up to its elastic limit; this work, stored in the bar as energy, and enabling it to spring back to its original length is called the **resilience** of the bar. The whole area $OEBN$ represents the total work done in breaking the bar. The energy which can be stored in a piece of good wrought iron by stretching it up to its elastic limit, that is, its resilience, is about 0·08 inch-ton, and the energy which must be expended while pulling it to rupture is about 55 inch-tons, the bar being one square inch in sectional area and ten inches long.

The resilience of a pulled bar may be easily calculated if the value of E and the elastic limit are known. Thus,

let f = elastic limit in tons per sq. inch.

Then, taking a bar 1 square inch sectional area,

Resilience = average force × extension up to elastic limit.

From the part of the diagram OEM (Fig. 178), we see that the average force will be one half the maximum and is therefore equal to $\frac{1}{2}f$ tons.

The extension may now be calculated from

$$E=\frac{\text{stress}}{\text{strain}}=\frac{f\times L}{e} \text{ tons per square inch.}$$

Taking $L=1''$, so that one cubic inch of material is being considered, this will become

$$E=\frac{f}{e},$$

or,
$$\text{extension}=e=\frac{f}{E} \text{ inches.}$$

Therefore,
$$\text{Resilience}=\tfrac{1}{2}f\times\frac{f}{E}$$
$$=\frac{f^2}{2E} \text{ inch-tons per cubic inch of material.}$$

Effect of suddenly applied loads.—It may now easily be seen what effect a suddenly applied load has on the material of a bar. Suppose a bar AB (Fig. 179) to be suspended vertically and that a load W is just touching the collar on the end B of the bar. If the supports of W are released and it is allowed to rest on the collar, we get a conception of a suddenly applied load. Work will be done by gravity on W, the force acting being equal to W. Set this off as OK (Fig. 180) to scale, and let distances along OM represent extensions of the bar. So long as the bar continues to extend, work will be done by gravity on W. This will be represented by a rectangular area, the force being uniform. But the bar can

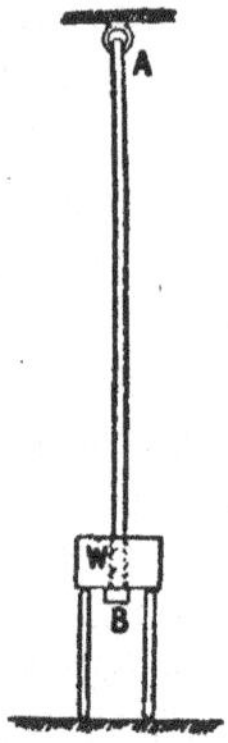

FIG. 179.—Load suddenly applied to a bar.

only resist the load by stresses gradually increasing from zero, so that the work taken up by the bar will be represented by the area of a triangular diagram. At N the resistance of the bar will be equal to W, but at this point more work has been done on W than the bar has taken up. At this point the work done by gravity is represented by the rectangle $OKNP$, while that taken up by the bar is represented by the triangle ONP; the work represented by the shaded area remains to be yet taken up by the bar, which consequently goes on extending. When the work taken up by the bar, represented by OQM, is equal to the work done by gravity, represented by $OKLM$, the extension will cease. This is easily seen to occur when $OM=2OP$, or when $MQ=2ML=2W$. The bar will now spring back and vibrate vertically until the damping action of molecular friction brings it to rest. The whole effect, as regards extending the bar and the stress produced in it, is just as though a load equal to twice W had been gradually applied.

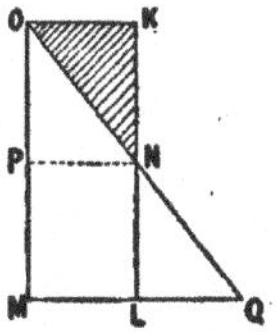

FIG. 180.—Diagram of work.

Horse-power transmitted by shafting.—The horse-power which can be safely transmitted by a given shaft depends on the uniformity or otherwise of the torque applied. Supposing this to be uniform, we proceed thus:

Let $P=$the turning effort in lbs.

$R=$radius in feet at which P is applied.

Then $P\times 2\pi R=$work done in one revolution in foot-pounds.

Let $N=$revolutions per minute,

Work per minute $=P\times 2\pi R\times N$ foot-pounds,

and
$$\text{H.P.}=\frac{P\times 2\pi R\times N}{33{,}000},$$

assuming, as we stated above, the torque $T=PR$ to be uniform. In the case of factory line shafts it is often fairly so, but in other cases it varies considerably.

Take, for example, the crank shaft of an engine having one double-acting cylinder only. The effort of the steam on the piston does not remain constant during the stroke, also the effect of the connecting rod acting at different angles to the centre line while transmitting the varying force from the piston rod to the crank, cause the torque to vary widely during the stroke. The maximum torque will occur usually when the crank and connecting rod are at right angles to one another. The torque will be zero when the crank is on the *dead centres*, as the crank and connecting rod will then be in the same straight line.

If the steam pressure and dimensions of the engine are given, we can easily obtain the torque when the crank and connecting rod are at 90° to one another, by setting out an outline diagram of the engine to scale.

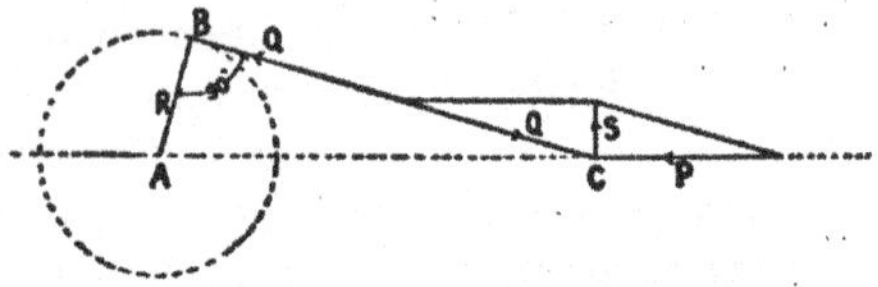

FIG. 181.—Maximum torque on the crank shaft.

Let AB (Fig. 181) represent the crank, and BC the connecting rod. Let P=total force along the piston rod, Q=thrust along the connecting rod, S=pressure of guide. P, Q, and S will balance one another, and Q may be found by an application of the parallelogram of forces. The connecting rod pushes the crosshead at one end and the crank pin at the other with equal forces Q, consequently the torque is equal to $Q \times R$. This value must be taken in estimating the diameter of crank shaft required.

EXERCISES ON CHAP. X.

1. A load of 2 tons is raised from the bottom of a shaft 300 ft. deep. How much work is done? Draw a diagram to represent this work.

2. In Question (1) the load is raised by a wire rope weighing 3 lbs. per foot length. Calculate the work done in raising the rope alone, and draw a diagram of work done.

3. A loaded truck, weight 12 tons, is pulled along a level track. If the resistances to motion are 11 lbs. per ton weight, calculate the work done in pulling the truck a distance of one mile.

4. A bridge girder weighs 15 tons, and is to rest on supports 25 ft. above the level of the ground. Calculate the work done in raising the girder into position.

5. A man weighs 140 lbs. Calculate the total work he has to do in carrying his bicycle, weight 30 lbs., upstairs to a room 20 ft. above the street level.

6. A man exerts a constant force of 30 lbs. in turning a handle of 14″ radius; calculate the work done per revolution if (*a*) the force is always exerted in a horizontal direction, (*b*) the force is always exerted tangential to the circle described by the handle.

7. A tank measures 10 ft. long × 6 ft. wide × 3 ft. deep, and is at a height of 200 ft. above the level of the pump used for filling it with water. Calculate the work done in filling it, taking one cubic foot of water to weigh 62·5 lbs.

8. What work is done in raising a bucket, weight 2 lbs., containing 25 lbs. of water from a well the surface level of which is 12 feet below ground level?

9. A shaft, 10 ft. diam., 100 feet deep, is full of water. Calculate the work done in emptying it.

10. The weight of a pile driver is 1250 lbs., and it is raised 6 ft. above the pile head before delivering a blow. Calculate its potential energy when raised.

11. One cubic foot of a gas contains 600 British thermal units. To how much mechanical work is this equivalent?

12. Find the mechanical work equivalent to the heat contained in a pound of petroleum of heating value 20,000 British thermal units.

13. A horse exerts a constant pull of 80 lbs. in dragging a cart along a level road. If he walks at the rate of 3 miles an hour, what horse-power is he developing?

14. A locomotive exerts a steady pull of 2500 lbs. in hauling a train along a level track. If the speed is 4200 feet per minute, calculate the horse-power.

15. Calculate the horse-power required to pump 5000 gallons of water per minute from a well 40 feet deep to the surface of the water if the efficiency of the machinery employed is 60 per cent.

16. 15,000,000 ft.-lbs. of energy are given to an engine per hour, and the horse-power developed is $1\frac{1}{2}$. What is the efficiency of the engine?

17. A shaft running at 120 revolutions per minute is subjected to a torque of 7000 ft.-lbs. Calculate the horse-power transmitted.

18. A bar of mild steel, 10 feet long, has a sectional area of 3 square inches. Calculate the work done in stretching it when a load of 12 tons is applied gradually. Take $E=30,000,000$.

19. What work in foot-pounds is done in raising the materials for building a brick wall 50′ high, 12′ long, and 2′ 3″ in thickness, if one cubic foot of brickwork weighs 112 lbs.? (1897.)

20. A man of 150 lbs. climbs a hill regularly 1200′ vertically per hour; at another time he climbs a staircase at $2\frac{1}{2}$′ per second; in each case find the useful horse-power in lifting himself. (1897.)

21. A chain weighing 10 lbs. per foot of its length is 240 ft. long and hangs vertically, what work is done in winding up the chain on to a drum? (1899.)

22. A body weighing 1610 lbs. was lifted vertically by a rope, there being a damped spring balance to indicate the pulling force F lb. of the rope. When the body had been lifted x feet from its position of rest, the pulling force was automatically recorded as follows:

x	0	11	20	34	45	55	66	76
F	4010	3915	3763	3532	3366	3208	3100	3007

Find approximately the work done on the body when it has risen 70 feet. How much of this is stored as potential energy, and how much as kinetic energy? (1901.)

CHAPTER XI.

FRICTION OF DRY AND LUBRICATED SURFACES.

Definitions.—When two bodies are pressed together, resistance has to be overcome before they can be made to slide on one another. This resistance is called the **force of friction.** The force which friction gives to a body always acts contrary to the direction of motion of the body and tends to arrest the motion, or, if the body is at rest, to prevent motion taking place. In many engineering problems, frictional forces have to be considered with a view to their reduction; in others, friction is useful by preventing slipping taking place.

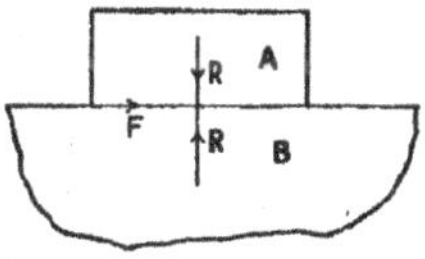

FIG. 182.—Frictional resistance to sliding.

If two bodies A and B (Fig. 182) are pressed together so that the mutual pressure perpendicular to the surfaces in contact is R, and if F is the force of friction which has to be overcome before sliding will take place, then $\frac{F}{R}$ is called the **coefficient of friction of rest,** or the **static coefficient of friction** of the bodies. If the bodies are sliding steadily on one another, and F' is the steady resistance to sliding, then $\frac{F'}{R}$ is called the **coefficient of friction of motion**, or the **kinetic coefficient of friction** of the bodies.

Conditions influencing friction.—The value of the coefficient of friction for two given bodies depends on the nature of the materials of which they are made, especially *on their hardness and ability to take on a smooth regular surface, and on the state of the rubbing surfaces as regards cleanliness and lubrication.* With dry, clean surfaces, the force of friction is produced largely by roughnesses on the surface of one body interlocking with roughnesses on the surface of the other. The surfaces which have to be dealt with in engineering work are usually of fair shape and satisfactorily fitted to one another, but even these do not bear on one another all over, but only in places, and when sliding takes place, the projections on one body have to get over, or, if the forces pressing them together are large enough, to cut away, or abrade, the projections on the other. That body which is the more easily replaced is generally made of softer material, in order to confine the wear principally to it.

When clean, dry surfaces, well fitted to one another, are brought together, a film of air may exist between them and thus prevent the bodies from actual contact. This is very noticeable when one surface plate is laid on another. If the cast iron surfaces are perfectly clean, the upper plate seems to float on the lower one. By pressure and working a little, the air film may be eliminated. The plates then adhere strongly together, or *seize*, partly on account of the vacuum between them, but more, since the effect takes place even in a good vacuum, on account of molecular forces of attraction being brought into play. Seizing takes place more readily with bodies of the same than with those of different materials. In practice a film of lubricant is used to keep the rubbing bodies as far as possible, from contact with one another, and the working load is such that there is no danger of the film being squeezed out.

Laws of friction for dry surfaces.—For bodies with rubbing surfaces dry, and perfectly clean or only slightly contaminated by films of foreign matter, the following laws of friction have been deduced from the results of experiments: *The static coefficient of friction is greater than the kinetic coefficient of friction;* in other words, *the resistance offered to sliding when the bodies are at rest is greater than that after steady rubbing has been attained.*

The force of friction is practically proportional to the perpendicular pressure between the surfaces in contact and is independent of the extent of such surfaces and of the speed of rubbing, if moderate. From this we infer that for two given bodies, *the coefficient of friction is practically constant for moderate pressures and speeds.*

Considerable increase in the speed lowers the value of the

coefficient of friction, and heating of the bodies produces the same effect. It has also been found that the coefficient of friction is a little greater for light pressures on large areas than for heavy pressures on small areas.

Experiments on friction.—Students should verify experimentally as many as possible of the above statements.

Expt.—Set up a board *AB* (Fig. 183) as nearly horizontal as possible, and arrange a slider *C* (which can be loaded to any

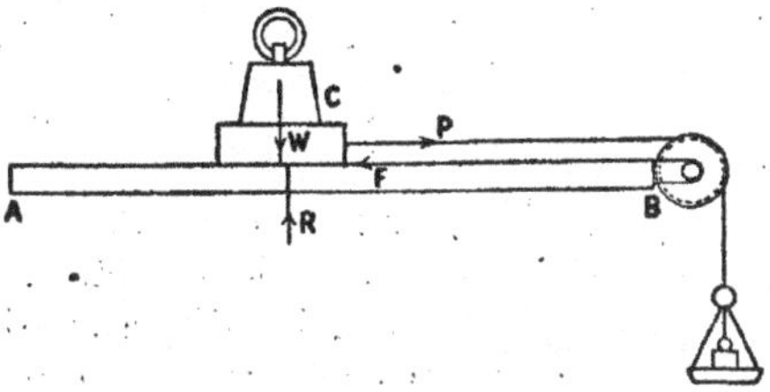

Fig. 183.—Friction of a slider.

desired amount) with a cord, pulley, and scale pan, so that the horizontal force *P* required to overcome frictional resistances to sliding may be measured. Under these conditions, the perpendicular pressure between the surfaces in contact will be equal to the weight of the slider and loads placed on it, *W* say, and its actual distribution over the surfaces in contact need not concern us at present. The force of friction *F* will be equal to *P*, and this will very nearly equal the weight of the scale pan and loads placed in it, provided the pulley used is finely mounted on pivot bearings and oiled so as to run very freely.

The coefficient of friction will be $\frac{P}{W}$.

First, make a number of experiments on the static coefficient of friction, using different loads on the slider. In each experiment place loads carefully into the scale pan so as to avoid jerks until the slider starts off. From the observations the static coefficient will be found.

Using the same loads on the slider, perform the same processes, only this time help the slider to start by jerking it. Adjust the load in the scale pan until steady uniform motion, as nearly as you can judge, has been obtained. From these observations the kinetic coefficient will be found.

Some results are given in order to show the method of recording.

AN EXPERIMENT TO DETERMINE COEFFICIENTS OF FRICTION.

Material of slider—mahogany.

Material of board—teak.

Rubbing surfaces—slightly contaminated with dust and finger marks.

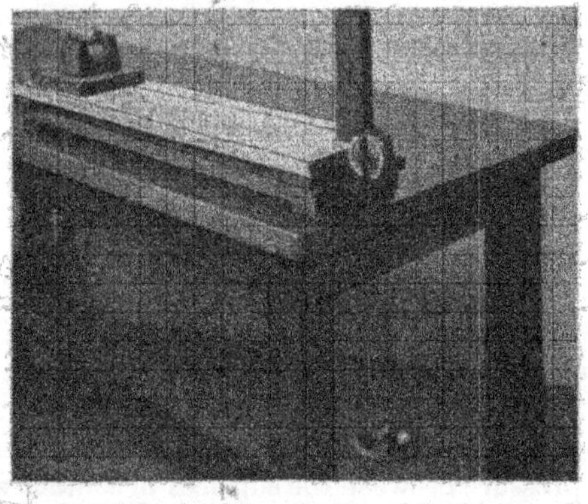

FIG. 184.—Apparatus for determination of the friction of a slider.

Area of sliding surface of slider $= 6'' \times 6'' = 36$ square inches.

Weight of slider $= 0{\cdot}701$ lb.

Weight of scale pan and hook for applying $P = 0{\cdot}701$ lb.

W lbs.	Grain of slider parallel to direction of motion.				Grain of slider perpendicular to direction of motion.	
	Static Values.		Kinetic Values.		Kinetic Values.	
	P_1 lbs.	$\frac{P_1}{W}$	P_2 lbs.	$\frac{P_2}{W}$	P_3 lbs.	$\frac{P_3}{W}$
2·701	0·74	0·337	0·49	0·223	0·51	0·232
4·701	1·43	0·306	0·901	0·191	0·901	0·191
6·701	2·69	0·401	1·301	0·194	1·301	0·194
8·701	3·04	0·35	1·701	0·195	1·65	0·19
10·701	3·54	0·335	2·201	0·206	2·09	0·195
12·701	4·501	0·355	2·501	0·197	2·401	0·189
14·701	4·701	0·32	2·901	0·197	2·90	0·197
16·701	5·04	0·32	3·501	0·209	3·40	0·203

Fig. 185 shows the plotted results for the experiments in which the grain of the slider was parallel to the direction of motion. The plotted points, especially those for the kinetic values, fall approximately on the straight lines which have been drawn to lie fairly among the plotted observations. The plotted points would all lie exactly in a straight line had the friction been proportional to the load and the experiment been perfectly performed.

The coefficient of friction may also be found by another method.

EXPT.—Raise one end of the board XZ (Fig. 186), until the block A, if started off, will slide down with steady speed. Consider the forces acting on the block when sliding occurs. Its

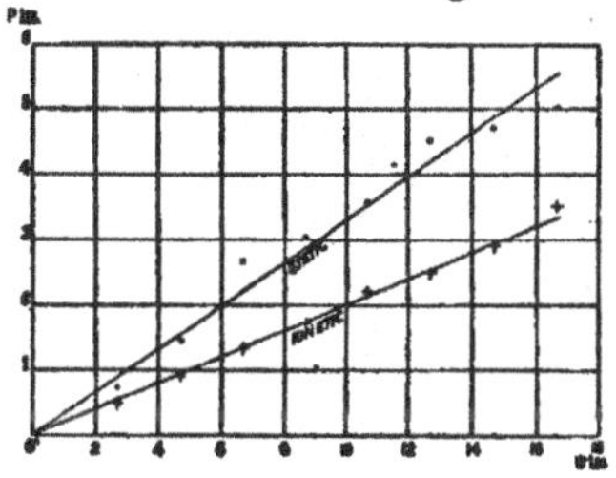

FIG. 185.—Plotted results of an experiment on the friction of a slider.

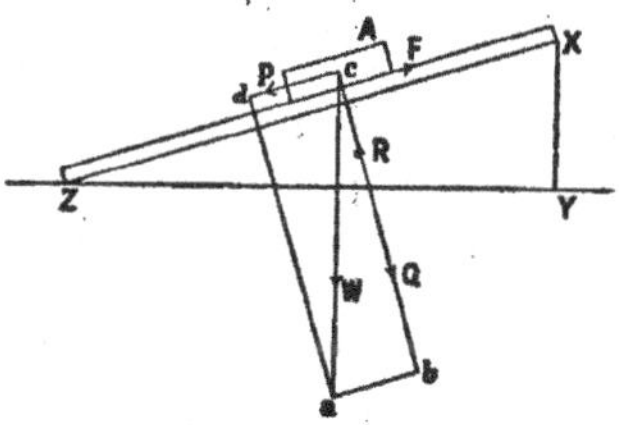

FIG. 186.—Coefficient of friction determined by inclining the board.

weight W may be resolved into two forces, P parallel to the board and Q perpendicular to it. R is the reaction of the board perpendicular to its surface and will be equal to Q. F is the force of friction acting up the plane, and will be equal to P, $abcd$ is the parallelogram of forces for W, P, and Q; therefore

$$P : Q = cd : bc$$
$$= ab : bc.$$

Now the triangles abc and XYZ are, from the construction, similar to one another; therefore

$$ab : bc = XY : YZ;$$
$$\therefore\ P : Q = XY : YZ,$$

or

$$F : R = XY : YZ;$$

$$\therefore\ \text{Coefficient of friction} = \frac{F}{R} = \frac{XY}{XZ}.$$

It therefore follows that, when the board is so adjusted that the block will slide steadily down, the kinetic coefficient of friction may be found by dividing the height of the plane by its base.

Repeat the experiment by raising one end of the board until the block starts unaided. A calculation, similar to that given above, will determine the static coefficient of friction.

Limiting angle of resistance.—Consider again the horizontal board and slider. Let $AB=R$, and $AC=F=$the force of friction when the block is about to move. Suppose a force P_1 less than F to be applied, as in Fig. 187, the force of friction $F_1=AH$ will be equal to P_1, and as this is less than F, the block will not slide. Under these conditions the table gives two forces to the block, R and F_1. Find the resultant of these R_1 by the parallelogram of forces. R_1 is the resultant reaction of the table on the block, and acts at an angle $BAG=\phi_1$, to the perpendicular. As we increase P_1, ϕ_1 will also increase, until a maximum value ϕ will occur when $P=F$, and the block will then slide. This angle $DAB=\phi$, is called the **limiting angle of resistance**, or sometimes the **angle of friction**. Notice from the diagram that $\frac{F}{R}=\frac{DB}{AB}=$the tangent of ϕ. Now $\frac{F}{R}$ is the coefficient of friction, so that

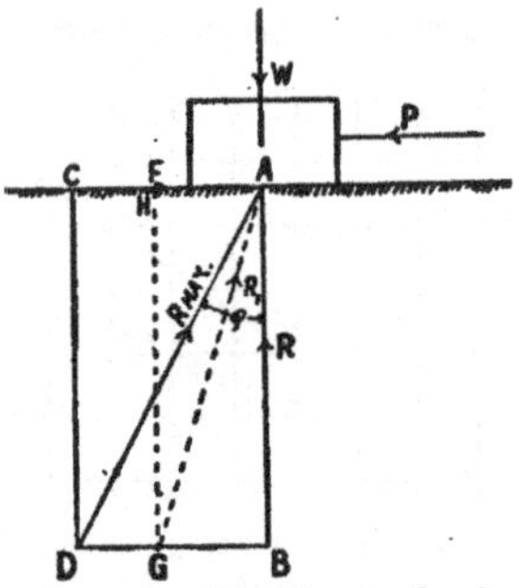

FIG. 187.—Limiting angle of resistance.

$$\text{Coefficient of friction}=\tan\phi.$$

Referring to the previous diagram (Fig. 186) of the slider on a board with one end raised; when the block is just beginning to slide, $\phi=$angle $acb=$angle XZY, so that

$$\text{Coefficient of friction}=\tan\phi=\tan XZY$$
$$=\frac{XY}{YZ}\text{ as stated before.}$$

The angle of resistance will be greater when the block is just on the point of moving than after motion has occurred.

Call ϕ_s the angle in the first case and ϕ_k the angle in the second case, then

$$\text{Static coefficient of friction} = \tan \phi_s,$$
$$\text{Kinetic coefficient of friction} = \tan \phi_k.$$

If the coefficient of friction is known, ϕ may be obtained from the table of natural tangents at the end of the book.

EXPT.—For showing any difference in the coefficient of friction produced by changing the *extent of the surfaces in contact*, four sliders cut from the same teak plank may be used. Each slider measures 6″×6″. One has the full surface, the others are cut away (Fig. 188) on the under side so as to have rubbing surfaces respectively of 27, 18 and 9 square inches. The rubbing surfaces are thus in the proportion 1 : 0·75 : 0·5 : 0·25.

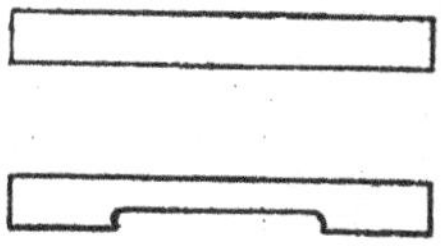

FIG. 188.—Sliders having different areas of rubbing surfaces.

The average results of some experiments are given, the sliders having their grain parallel to the direction of motion and sliding on a teak board.

EFFECT OF EXTENT OF SURFACE ON THE COEFFICIENT OF FRICTION.

Slider.	Proportional Area of Rubbing Surface.	Weight of Slider, lb.	Average Coefficient of Friction.
A	1·0	0·88	0·225
B	0·75	0·83	0·213
C	0·5	0·78	0·206
D	0·25	0·718	0·200

It will be observed from these figures that the coefficient of friction is rather less for the smaller rubbing surfaces, that is, it diminishes as the pressure per square inch rises. The law that *friction is independent of the extent of the surfaces in contact* is, however, shown to be approximately true.

The influence of speed on the coefficient of friction for unlubricated surfaces may be tested by a machine such as that shown in Fig. 189. In this machine a cast-iron pulley mounted on a shaft can be driven at any desired speed. Blocks of different materials have their under surfaces made concave to fit the rim

FIG. 189.—Apparatus for experiments on the influence of speed on the coefficient of friction.

of the wheel and rest on its top. These can be loaded by weights placed in a scale pan underneath. The force of friction is measured by the pull on the block of a horizontal cord, led over a pulley and having a scale pan at its end. A few results obtained with this machine are given.

EXPERIMENT ON INFLUENCE OF SPEED ON THE COEFFICIENT OF FRICTION.

Friction of brass block on cast-iron pulley at different speeds. Surfaces dry and clean.

Diameter of pulley, 6″.
Circumference of pulley, 1·571 feet.
N=revolutions per minute of pulley.

Rubbing speed $=\frac{N}{60}\times 1{\cdot}571$ feet per second.

Weight of block, links and pan, 1·73 lbs.
Load in pan, 1 lb.
Constant load on block = 2·73 lbs.
Frictional resistances $= P$ lbs.

RESULTS OF EXPERIMENT.

Revolutions per min. N.	Rubbing speed, feet per second.	W lbs.	P lbs.	Coefficient of friction. $\frac{P}{W}$.
1000	26·2	2·73	0·41	0·15
850	22·2	2·73	0·56	0·205
740	19·4	2·73	0·83	0·304
650	17·0	2·73	0·856	0·317
560	14·7	2·73	1·042	0·382
450	11·8	2·73	1·10	0·404
360	9·45	2·73	1·08 ?	0·396 ?

At 360 revolutions per minute there was a tendency to seize, which made it difficult to obtain a reading of P. The increase in the coefficient of friction as the speed falls is fairly regular.

Friction of a rope on a drum.—When a rope is coiled round a fixed drum, **slipping** will not occur until the pull on one end is considerably greater than that on the other end. This is due to the friction between the rope and the drum having to be overcome. Suppose $ABCD$ (Fig. 190) is a drum, AC and BD being at 90° to one another, and that a rope is coiled round 90° of arc from A to B. We might find that P_1, when W is just sliding down, is equal to $\frac{3}{4}W$.

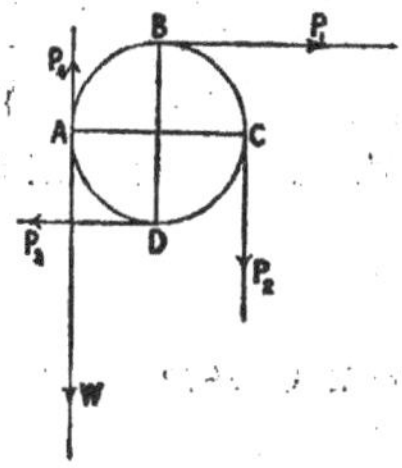

FIG. 190.

If we give the rope other 90° of lap from B to C, and apply P_2, then, as the rope from B to C is under the same conditions as between A and B, we should expect to find

$$P_2 = \frac{3}{4}P_1 = \frac{3}{4} \times \frac{3}{4}W = 0·562\,W.$$

Other 90° lap, to D, should make

$$P_3 = \frac{3}{4}P_2 = \frac{3}{4} \times \frac{3}{4} \times \frac{3}{4}W = 0·422\,W,$$

and for one complete turn round the drum

$$P_4 = \frac{3}{4}P_3 = \frac{3}{4} \times \frac{3}{4} \times \frac{3}{4} \times \frac{3}{4}W = 0·315\,W.$$

We see, then, that P is greatly reduced for the same load W by coiling the rope further round the drum.

In an actual experiment a student obtained, using a silk cord coiled on a pine drum, the following results :

AN EXPERIMENT ON SLIPPING.

Angle of lap.	W lbs. descending.	P lbs. ascending	Experimental Ratio $\frac{P}{W}$.	Calculated Ratio $\frac{P}{W}$.
90°	0·397	0·29	0·73	0·73
180°	0·56	0·29	0·518	0·533
270°	0·79	0·29	0·367	0·39
360°	1·1	0·29	0·263	0·28

Taking the first ratio of $\frac{P}{W}=0{\cdot}73$ for 90° lap,

$\therefore$ the ratio for 180° should be $0{\cdot}73 \times 0{\cdot}73 = 0{\cdot}533$;
for 270°, $0{\cdot}533 \times 0{\cdot}73 = 0{\cdot}39$;
for 360°, $0{\cdot}39 \times 0{\cdot}73 = 0{\cdot}28$.

FIG. 191.—Apparatus for experiments on the friction of a cord coiled on a drum.

The experimental results agree as closely with the calculated ones as we have any right to expect when we remember the assumption we have made with regard to the condition of the rope at different parts of the circumference, and also consider that the coefficient of friction will certainly not be the same at different parts of the drum and cord. The effect of want of perfect flexibility of the cord also tends to make the actual results differ from the calculated ones. But the law simply stated as above enables us to arrive at results, in questions concerning the slipping of ropes on drums or belts on pulleys, which do not differ very greatly from the actual ones. Suppose, for example that we had only performed the first experiment above, $W=0{\cdot}397$ lb., $P=0{\cdot}29$ lb. for 90° and we wished to predict what W would be for the same P with 360° lap.

For

$$90° \text{ lap}, \ P=0{\cdot}73\,W;$$

$$360° \text{ lap}, \ P=(0{\cdot}73)^4\,W;$$

$$0{\cdot}29=0{\cdot}28\,W,$$

$$W=\frac{0{\cdot}29}{0{\cdot}28}=1{\cdot}03 \text{ lbs.},$$

which does not differ greatly from the actual result 1·1 lbs.

For *leather belts on iron pulleys* the ratio of the pulls, when the angle of lap is 180°, is about 0·385 when the belt is just slipping. It varies considerably with different cases, and the number given is only of service to us when we want to calculate the strength of belt required for a given drive, and therefore require to calculate the greatest pull in it. It should be noticed that the diameter of the pulley or drum does not affect the ratio of the pulls.

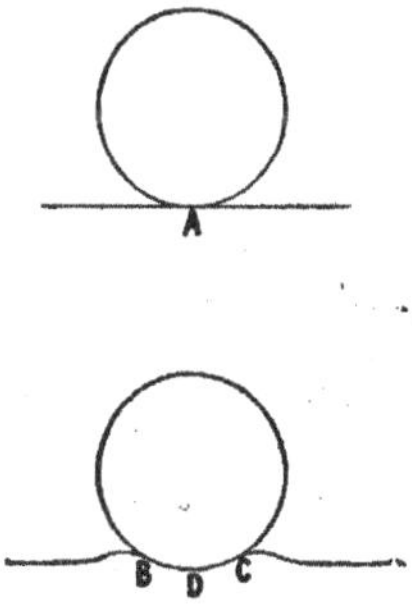

FIG. 192.

Friction of a rolling wheel.—In **rolling friction**, experiments show that the *frictional resistances are roughly proportional to the load and inversely proportional to the radius of the wheel or roller.* In the case of a wheel or roller rolling on a plane surface, the surface is indented by the wheel, so that the rim is in contact not at a line only as at A (Fig. 192), but over a portion of the circumference BC. This introduces a certain amount of sliding from B to C instead of pure rolling.

In **roller and ball bearings** the surfaces are usually lined with hardened steel, and the rollers or balls are also of hard steel. This reduces indentation as far as possible, and consequently lowers the frictional resistances. In roller bearings, a cage is necessary to keep the rollers in their proper relative positions, and this is also often done in the case of ball bearings. An example of a ball bearing for taking the thrust of a lathe spindle is shown in Fig. 193, and a roller bearing for a truck axle in Fig. 194.

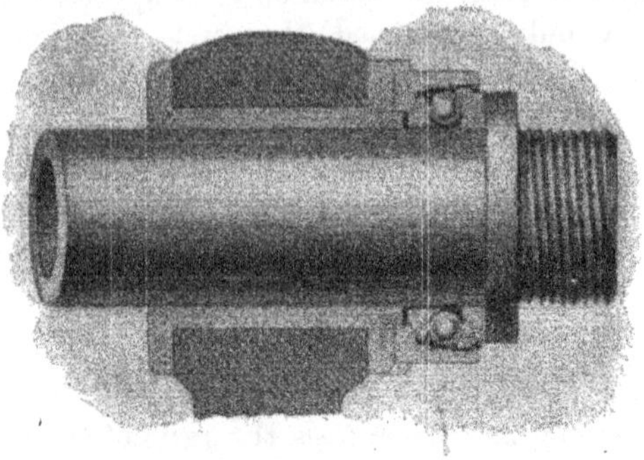

FIG. 193.—Ball bearing for a lathe spindle.

Experiment on rolling friction.—In order to obtain some idea of the *resistance to rolling of a carriage along a level road of different material*, a small three-wheeled carriage may be used, the wheels being of gun-metal and fixed to steel axles carried on fine pivot bearings so as to be as frictionless as possible. The roads consist of slabs of different material, and are levelled as accurately as possible. The horizontal force required to maintain steady, slow motion is measured in the usual way by a horizontal cord passing over a finely mounted pulley and having a scale pan at its end.

FIG. 194.—Roller bearing for a truck axle.

Weight of carriage, 3·09 lbs.
Weight of scale pan, 0·04 lb.
W= weight of carriage + weight placed on it.
P= weight of scale pan + weights placed in it.
Diameter of wheels, $1\frac{31}{32}''$.
Breadth of each of the pair on the same axle, $\frac{1}{4}''$.
Breadth of the single one, $\frac{5}{8}''$.

The following are some of the results obtained :

AN EXPERIMENT ON ROLLING FRICTION.

W lbs.	Resistance P lbs. to rolling on roads of		
	Smooth cast iron.	Teak.	India-rubber.
3·09	0·03	0·047	0·047
10·09	0·125	0·16	0·20
17·09	0·20	0·25	0·36
24·09	0·28	0·36	0·51
31·09	0·36	0·47	0·70
38·09	0·44	0·61	0·95

The resistance to be overcome includes not only the rolling friction, but also the frictional resistances of the pivot bearings of the wheels and of the pulley used for the cord to run over. The effect of the change of roads is, however, clearly shown by plotting the results (Fig. 195). Notice that for the cast iron road, the resistance is practically proportional to the load for all loads, and that for the teak and rubber roads, especially the latter, the resistance is proportional to the load up to a limit and then increases much more rapidly. This is due to the wheels being forced into the material of the road, and thereby introducing a considerable amount of slipping. We may deduce from these experiments, that to minimise frictional resistances in wheels rolling on roads, in roller bearings, or in ball bearings, that *the materials should be hard and not liable to be indented by the loads applied.*

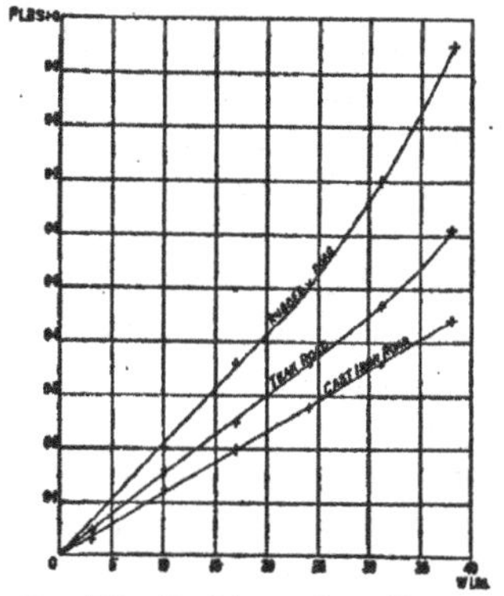

FIG. 195.—Resistance to rolling on roads of cast iron, teak, and rubber.

Fluid friction.—The laws of friction for fluids differ considerably from those stated for dry surfaces. For liquids such as water and oils they have been shown experimentally to be as follows:

The resistance is proportional to the extent of the surface wetted by the liquid.

The resistance is independent of the material of which the boundary is made and of its surface, provided it is not too rough.

The resistance is independent of the pressure to which the liquid is subjected.

Rise of temperature of the liquid diminishes the resistance.

At slow speeds the resistance is very small.

Below a certain critical speed, the resistance is proportional to the speed; at speeds above this, the resistance is proportional to some power, approximately the square, of the speed.

The critical speed depends on the liquid used and its temperature. Below this speed the motion of the liquid is steady; above it, the liquid breaks up into eddies.

Liquids which can flow, or change their shapes more easily than others, are said to be less **viscous**, or to possess less **viscosity**. All liquids have one property in common—*they are unable to resist shearing forces and yet remain at rest.* Now friction is always manifest as a tangential or shearing force, and it therefore follows that if a liquid is at rest there can be no frictional resistances of any kind present.

Laws of friction for ordinary bearings.—The laws of friction for ordinary bearings are intermediate between those for liquids and for dry surfaces. In bearings running in a bath of oil, the laws have been shown to be approximately the same as those of fluid friction, and in other bearings the resistances experienced depend on the success which is achieved in getting the oil into the bearing and in preserving the oil film. It is not usual, in investigating the losses due to frictional resistances of a number of bearings such as we find in any machine, to attempt to ascertain how much is lost at each bearing, but to find simply how much is lost at all the bearings collectively. *Almost always it is found that the frictional losses, when the machine is loaded, are equal to those of the unloaded machine together with a small fraction of the load.* If we were to find that the frictional losses in a machine were constant for all loads, then we might deduce that the frictional resistances have been altogether due to fluid friction, this being independent of the load.

Effect of friction in machines.—It is useful, in dealing with simple machines such as hoisting tackle, to deduce an equation connecting the **effect of friction** in the machine with the actual load applied.

Let P = force applied to work the machine;
v = velocity ratio of machine;
W = actual load raised;

P and W being in the same units.

Suppose W to be raised one foot, then P will act through v feet.

Energy supplied to machine $= Pv$;
Useful energy obtained from machine $= W \times 1 = W$.

Imagine the actual frictional resistances of the machine to be removed, and an equivalent addition to the load W to be made, so that P is unaltered. Call this additional load F = effect of friction, and measure it in the same units as P and W.

Energy lost in overcoming frictional resistances $= F \times 1 = F$.

By the principle of the conservation of energy, energy supplied = useful energy obtained + energy lost.

$$\therefore \quad Pv = W + F,$$

or,

$$F = Pv - W \quad \ldots\ldots\ldots\ldots (1)$$

If experiments have been carried out on a given machine, a series of values of P and W will have been obtained. From (1), corresponding values of F can be calculated. On plotting the values of F and W so found, it will generally be found that the plotted points lie approximately on a straight line, showing that the connection between F and W can be represented by the equation

$$F = aW + b, \quad \ldots\ldots\ldots\ldots (2)$$

where a and b are constants for the machine.

Heating of bearings.—Almost the whole of the work done in overcoming frictional resistances is transformed into heat. The rubbing bodies therefore rise in temperature, until the loss of heat by conduction, etc., balances the heat produced by the continued rubbing. The temperature of the bodies then remains constant.

Let F=frictional resistance of a bearing, in lbs.,

V=distance in feet through which F is overcome in one minute;

then FV=work expended per minute, in foot-lbs.

$$\therefore \text{ Horse-power absorbed in overcoming friction} = \frac{FV}{33{,}000}.$$

Taking 772 foot-lbs. as equivalent to one British thermal unit (p. 126),

$$\text{Heat generated per minute} = \frac{FV}{33{,}000} \text{ B.T.U.}$$

EXAMPLE. A shaft 6″ diameter makes 90 revolutions per minute. The load on it is 4 tons, and the coefficient of friction 0·02. Calculate the H.P. absorbed in overcoming the friction of the bearings and also the heat generated per minute.

$$W = 4 \times 2240 = 8960 \text{ lbs.}$$

$$\text{Force of friction} = 0{\cdot}02 \times 8960 = 179{\cdot}2 \text{ lbs.}$$

$$V = \pi d \times 90$$

$$= \frac{22}{7} \times \frac{6}{12} \times 90 = 141 \text{ feet in one minute.}$$

$$\text{Work absorbed} = 179{\cdot}2 \times 141$$

$$= 25{,}300 \text{ ft.-lbs. per minute.}$$

$$\text{H.P.} = \frac{25{,}300}{33{,}000} = \underline{0{\cdot}768.}$$

$$\text{Heat generated} = \frac{25{,}300}{772} = \underline{32{\cdot}8} \text{ B.T.U. per minute.}$$

EXERCISES ON CHAP. XI.

1. It is found that a horizontal force of 8 lbs. can keep a load whose weight is 30 lbs. in steady motion along a horizontal surface. What is the coefficient of friction?

2. A block whose weight is 10 lbs. rests on a horizontal table; it is found that a pull of 4 lbs., applied at 30° to the table, just starts it off. What is the static coefficient of friction?

3. Answer Question 2 supposing the force applied had been a push.

4. The crank of an engine is 1 foot long and the connecting rod 5 feet. When the crank is at 90° to the centre line of the engine, the push of the piston rod is 12,000 lbs. Taking the coefficient of friction as 0·06, find the frictional force opposing the sliding of the slipper on the guide in this position.

5. The working face of a slide valve of a steam engine measures $8\frac{1}{2}'' \times 15''$; its travel is 4″. Steam pressure on the back of the valve, 120 lbs. per square inch. If the coefficient of friction is $\frac{1}{10}$, calculate the force required to move the valve and the horse-power absorbed when the engine is running at 60 revolutions per minute.

6. An oak plank, 8 feet long, has a block of oak resting on it. If the coefficient of friction is 0·45, how high must one end of the plank be raised before slipping down will occur?

7. In a belt driving a pulley, the ratio of the pulls in the two parts of the belt is 0·4. A difference in the pulls of 250 lbs. is required for driving. Calculate the actual pulls.

8. A train whose speed is $\frac{1}{2}$ mile per minute has frictional resistances amounting to 12 lbs. per ton weight of train. If this weight is 150 tons, calculate the pulling force required and the horse-power of the engine.

9. A shaft 4″ diameter rotates 300 times per minute. If the load on it is $1\frac{1}{2}$ tons, and the coefficient of friction 0·025, calculate the H.P. absorbed in driving it and also the heat generated per minute.

10. What is friction? What is meant by limiting friction, by sliding friction, and by the coefficient of friction? A weight of 5 cwts. resting on a horizontal plane requires a horizontal force of 100 lbs. to move it against friction. What in that case is the value of the coefficient of friction? (1896.)

11. Define "force," "work," "foot-pound," and "horse-power." A small metal planing machine, the table of which weighs 1 cwt., makes 6 backward and 6 forward strokes each of $4\frac{1}{2}$ feet in a minute, and the coefficient of friction between the sliding surfaces is 0·07. What is the work performed in foot-pounds per minute in moving the table? (1896.)

12. How would you experimentally determine the nature of the friction between clean smooth surfaces, say of oak, and what sort of law would you expect to find? (1897.)

13. An express train going at 40 miles per hour weighs 150 tons; the average pull on it is 12 lbs. per ton, what is the horse-power exerted? This power is only 40 per cent. of the total indicated power of the engine; find the indicated power. (1898.)

14. Describe any experiments you have made or seen for finding the laws of solid friction. What are the laws so found? Are they quite true? How do they differ from the laws of fluid friction? (1899.)

CHAPTER XII.

VELOCITY. ACCELERATION. INERTIA. KINETIC ENERGY. RELATIVE VELOCITY. CHANGE OF VELOCITY.

Velocity.—The term **Velocity** has been used before when considering machines. A commoner term with the same meaning is **speed.** Both terms refer to **the rate at which a body is changing its position relative to other bodies.** Velocity has magnitude, direction, and sense, and, like force, may be represented by a straight line.

Velocity is measured by stating the distance travelled by the body in a given time. Thus, a velocity of 15 feet per second means that in one second the body will travel a distance of 15 feet. When we say that the speed or velocity of a train is 55 miles an hour, we mean that if that velocity were kept up constantly, the train would travel a distance of 55 miles in one hour.

Velocities are seldom constant, but the velocity a body has at any instant may be stated by considering what space the body would travel in unit time if the velocity it has at the given instant were kept constant. Thus, it is found that if a body falls freely from a height, at the end of the first second of the fall its velocity is 32·2 feet per second. This does not mean that the body has fallen during the first second a distance of 32·2 feet, or that it is going to travel that distance during the next second. The meaning is that if the velocity possessed by the body at the end of the first second were kept unaltered, it would travel a distance of 32·2 feet during the next second. Actually it does not remain unaltered, for at the end of the next second its velocity will be found to be 64·4 feet per second, and at the end of the third second 96·6 feet per second and so on, the speed continually increasing.

Uniform velocity will occur if the body travels over equal distances in equal intervals of time. Velocity which is not uniform is said to be **accelerating**. If the speed is becoming greater the acceleration is said to be *positive*, and if the speed is diminishing, *negative*.

Parallelogram of velocities.—Although it is impossible for a rigid body to be moving in two directions at one and the same time, yet it is often convenient to think of a velocity as made up of two component velocities. Thus, supposing a point A to have component velocities represented by $AB = V_1$ and $AC = V_2$, its resultant velocity may be found by the parallelogram of velocities, which is similar to the parallelogram of forces. On completing the parallelogram $ABDC$ (Fig. 196), $AD = V$ will be the resultant velocity of the point A. Notice that here, as with the parallelogram of forces, both velocities must be either towards A or away from A before applying the parallelogram.

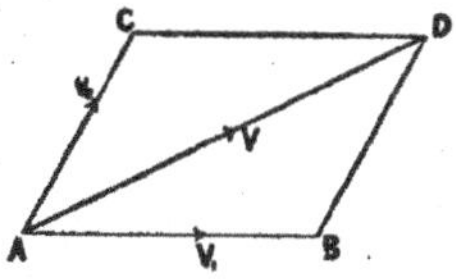

FIG. 196.—Parallelogram of velocities.

Velocity-time diagrams.—The velocity which a body has at any instant can be very conveniently shown in a diagram. Take the following case:

A train leaves Liverpool Street Station at 6.0 p.m., its speed gradually increases and at 6.2 p.m. is 15 miles an hour, and keeps uniform till 6.4 p.m., when brakes are applied, and at 6.4¼ p.m. the train stops at Bethnal Green Station. After 1 minute stop, the train starts again, and in 1½ minutes its speed is 30 miles an hour and keeps uniform for 9 minutes. Brakes are again applied, and it comes gradually to rest at Stratford Station ½ minute later.

Time being taken for abscissae and velocities for ordinates, the diagram (Fig. 197) shows at a glance all that has occurred to the train's speed.

The average speed of a body during a given journey can be calculated by dividing the total distance by the time taken to perform the journey. In the time taken should be included the time lost in any stops. Thus, suppose a train takes 12 minutes

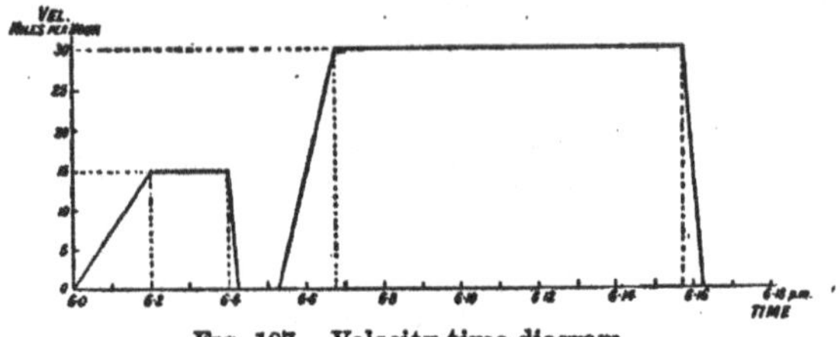

FIG. 197.—Velocity-time diagram.

to perform a journey of 5 miles, its average velocity will be $\frac{5}{12}=0{\cdot}4166$ mile per minute, including stops.

Or, the average velocity may be found from its velocity-time diagram by any of the well-known mensuration rules. In the given diagram (Fig. 198), if the base be divided into 10 equal parts and the sum of the velocities measured by the height of the diagram at the centre of each part be taken, this sum divided by 10 will give the average velocity.

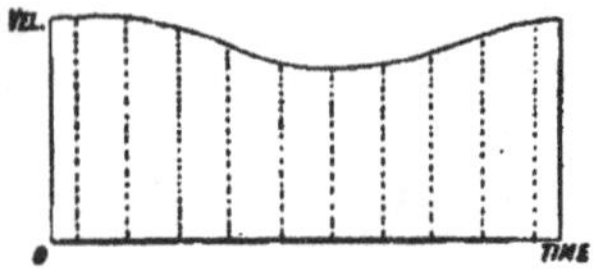

FIG. 198.

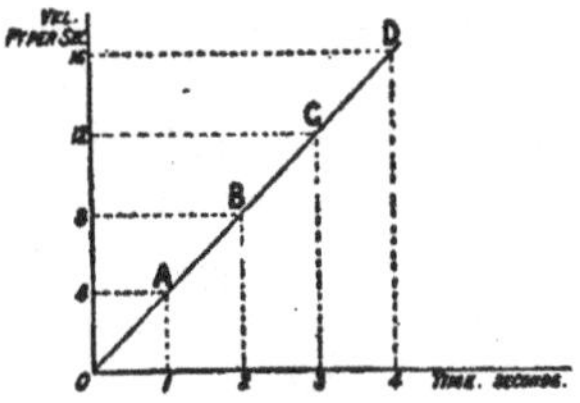

FIG. 199.

Acceleration.—The case of a body gaining speed must now be studied in more detail. Supposing it starts from rest at time 0, and gains speed gradually throughout. Let its velocity at the end of the first second be 4 ft. per second. Its velocity at any instant during the first second will be shown by the height of the diagram at that instant (Fig. 199). During the next second, its velocity will gradually increase again, and at the end of this second will be 8 feet per second, represented by $2B$ in the figure. At the end of the third second its velocity will be

12 feet per second and at the end of the fourth second 16 feet per second, represented by $3C$ and $4D$ respectively. The gain of velocity in any particular second, or positive acceleration, as it is called, will be 4 feet per second. We state this by saying,

Acceleration = 4 feet per second, every second, or,
Acceleration = 4 (feet per second) (per second),

the part of the units in the first bracket referring to the gain of velocity, and in the second bracket to the time in which that change of velocity took place.

Notice, in the diagram, that the total change in velocity in 4 seconds was 16 feet per second. So that we may find the change per second in velocity by dividing the total change by the time in which that change took place. Thus,

Acceleration $= \frac{16}{4} = 4$ (feet per second) (per second).

We must be careful always to state not only the *change in velocity*, but the *time* in which that change took place. Had the body in the above case gone on moving for 10 seconds, its velocity at the end would have been 40 feet per second.

Its acceleration = 40 (feet per second) (per 10 seconds),
or " = 16 (feet per second) (per 4 seconds),
or " = 4 (feet per second) (per second).

Referring to Fig. 199 again, if the body moves for 4 seconds, its velocity at the end is 16 feet per second, its velocity at the start is 0 and the change was gradual and uniform. Its average velocity will therefore be 8 feet per second. The distance travelled may now be calculated from,

$$\begin{aligned} \text{Distance} &= \text{average velocity} \times \text{time} \\ &= 8 \times 4 \\ &= 32 \text{ feet.} \end{aligned}$$

This result can be represented by the area of the triangular diagram $D04$ (Fig. 199), for its area will be the base 04 multiplied by half the height $D4$. Measuring the base in units of time and the height in units of velocity, we get

Distance = area of $D04 = 4 \times \frac{16}{2} = 32$ feet, as before.

Equations of motion.—Taking an example with general terms, let a body start from rest and gain every second a velocity a feet per second. Let this go on for a time t seconds, and call its velocity, at the end of the time, v feet per second. Let S be the distance travelled in feet. The diagram shown in Fig. 200 represents these conditions.

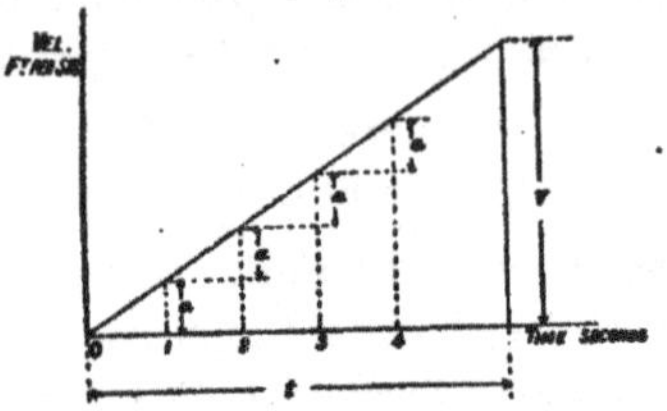

FIG. 200.

The acceleration will be a (feet per second) (per second), and since a feet per second of velocity are gained each second, in t seconds the gain will be $a \times t$, so that

$$v = at. \qquad (1)$$

The distance travelled, S, will be represented by the area of the diagram, so that

$$S = \tfrac{1}{2}v \, . \, t. \qquad (2)$$

Or, substituting $(a \, . \, t)$ for v in (2),

$$S = (\tfrac{1}{2}at) \times t = \tfrac{1}{2}at^2. \qquad (3)$$

Again, from (1) $\quad t = \frac{v}{a}.$

Substitute this in (3) giving

$$S = \tfrac{1}{2}a \times \left(\frac{v}{a}\right)^2,$$

$$= \frac{v^2}{2a},$$

or,

$$v^2 = 2aS. \qquad (4)$$

Body falling freely.—We have already seen that when a body falls freely, its velocity increases by 32·2 ft. per second every second. A special symbol, g, is used for the acceleration in this case, so that

$$g = 32{\cdot}2 \text{ ft. per second per second.}$$

It should be remembered that the acceleration of a falling body is not always the same. Its magnitude depends on the latitude of the place. Applying the above equations to the case of a falling body, let h=height fallen in time t seconds, and v=velocity at the end of this time, then

$$v=gt. \qquad (5)$$

$$h=\tfrac{1}{2}vt. \qquad (6)$$

$$h=\tfrac{1}{2}gt^2. \qquad (7)$$

$$v^2=2gh. \qquad (8)$$

In these equations the body is supposed to be falling freely, that is, no atmospheric or other resistances oppose it.

Inertia.—The importance of the study of acceleration from the engineer's point of view lies in the fact that force must act on a body to produce acceleration. If no resultant force acts on a body free to move when it is at rest, it will remain at rest, or, if the body is moving, it will continue to move with uniform velocity in a straight line. This laziness, or **inertia**, as it is called, of matter, must be overcome by a force or forces, if any change in a body's velocity is to be made. Thus, if a machine is at rest and is to be started, the pulls of the belt must not only overcome the frictional resistances, but must also overcome the resistances due to the inertia of the parts which move when the machine is running. When the belt is thrown on the loose pulley to stop the machine, the machine would go on moving uniformly but for the frictional resistances applying forces to the moving parts, thereby producing negative acceleration, and so bringing the machine to rest.

We may estimate how much force is required to produce a given acceleration in a body, by considering again the case of a falling body. If the body has a mass of one pound, there will be a force of one pound weight acting on it. This produces, in all parts, practically, of the British Isles, an acceleration of 32·2 feet per second per second. If we could reduce the weight of the body to $\frac{1}{2}$ lb. without altering its mass, we should find that the acceleration produced would be 16·1 feet per second per second; and if we could reduce the body's weight to $\frac{1}{32 \cdot 2}$ lb. the acceleration produced would be 1 foot per second per second. In each of these cases the resistance due to the inertia of the body will be an upward force of 1 lb., $\frac{1}{2}$ lb., and $\frac{1}{32 \cdot 2}$ lb. respectively. Again if we could increase the mass of the body to 32·2 lbs., still keeping its weight 1 lb., we should find that its acceleration would be one foot per second per second. In fact, the law is, **the force required to produce a given acceleration in a given body is proportional to the product of the body's mass and the required acceleration**, and as we know that a force of 1 lb. weight acting on a mass of 1 lb. gives an acceleration of 32·2 feet per second per second, we may calculate the force P lbs. required to give an acceleration a feet per second per second to a mass m lbs. from

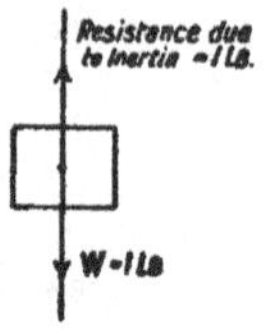

FIG. 201.

$$P = \frac{ma}{32 \cdot 2} \text{ lbs.}$$

If, instead of the pound weight as the unit of force, we were to use a unit of force equal to $\frac{1}{32 \cdot 2}$ lb. weight, or $\frac{1}{g}$ lb. weight, then this unit, acting on the one pound mass, would give an acceleration of one foot per second per second. This unit of force is called an **absolute unit of force**, and in our system of units—the **poundal**. The absolute unit of force for the metric system is the **dyne**, and is of such magnitude that it gives an acceleration of one centimetre per second per second when it acts on a mass of one gram. Using the poundal as the unit of force, the equation stated above may be written

$$P = ma \text{ poundals.}$$

EXAMPLE. A body has a mass of 150 lbs. and we have to give it an acceleration of 100 feet per sec. per sec. Find force required.

$$P = \frac{ma}{g} \text{ lbs.}$$
$$= \frac{150 \times 100}{32 \cdot 2}$$
$$= \underline{456} \text{ lbs.}$$

Or in poundals,

$$P = 456 \times 32 \cdot 2 = \underline{15{,}000} \text{ poundals.}$$

Since the acceleration of a body, falling freely under the action of its weight, is g feet per second per second, it follows that we may express its weight, W, in poundals, from

$$W = mg \text{ poundals.}$$

EXPT.—To verify the law $P=\frac{ma}{g}$, arrange an apparatus as shown in Fig. 202. This consists of two light pulleys attached to a support as high as possible, and having a light cord passing over them, with scale pans at its ends.

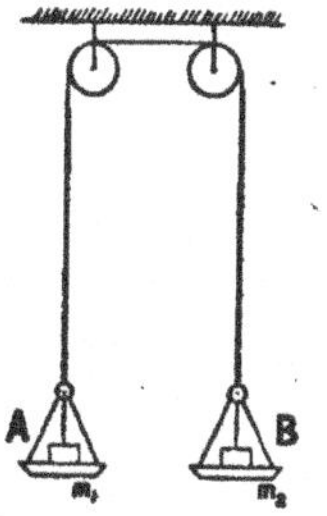

FIG 202.—Apparatus for verifying the law $P=\frac{ma}{g}$.

1. Place equal masses in the pans. It will be found that the pans remain at rest, and that if motion is started by the hand, the frictional resistances in the apparatus rapidly bring it to rest.

2. Increase the mass in one of the pans, until its excess weight enables steady motion to be maintained.

3. An additional mass placed in the same pan will now, by its weight, produce acceleration in the whole of the moving parts of the apparatus. Place a known additional mass in this pan, and elevate it a measured height. Allow it to descend, and note the time of its descent, using a stop-watch for this purpose. From these data, calculate the acceleration of the masses, and also the force required to produce this acceleration, using the equation $P=\frac{ma}{g}$. The result should agree closely with the weight of the additional mass used to produce acceleration in the experiment if the law is true.

In one experiment the following results were obtained:
Mass of each scale pan = 0·72 lb.
Mass placed in *A* (Fig. 202) = 1 lb.
Mass placed in *B* to produce *steady* speed of descent = 1·15 lbs.
Mass placed in *B* in acceleration experiment = 1·2 lbs.
Excess weight = force producing acceleration = 1·2 − 1·15
= 0·05 lb. weight.

In the acceleration experiment, *B* was allowed to descend 9 feet, and was found to do so in an average time of 6·3 seconds.

$$s=\tfrac{1}{2}at^2.$$
$$9=\tfrac{1}{2}a(6{\cdot}3)^2=\tfrac{1}{2}a\times 39{\cdot}7$$
$$a=\frac{18}{39{\cdot}7}=0{\cdot}45 \text{ feet per sec. per sec.}$$

Let m = the whole mass set in motion, neglecting the masses of the pulleys and cord.

$$m = 0{\cdot}72 + 0{\cdot}72 + 1 + 1 + 0{\cdot}2 = 3{\cdot}64 \text{ lbs.}$$

$$P = \frac{ma}{g}$$

$$= \frac{3{\cdot}64 \times 0{\cdot}45}{32{\cdot}2} = 0{\cdot}0509 \text{ lb. weight,}$$

which agrees closely with the actual P used in the experiment, viz. 0·05 lb. weight.

Measurement of kinetic energy.—When a body falls freely from a given height, its potential energy is gradually changed into kinetic energy of an equal amount. Let W be its weight in lbs., and h the height in feet from which it falls. Then its potential energy $= Wh$ foot-lbs. At the end of its fall, its velocity will be

$$v = \sqrt{2gh},$$

or $$v^2 = 2gh,$$

giving $$h = \frac{v^2}{2g}.$$

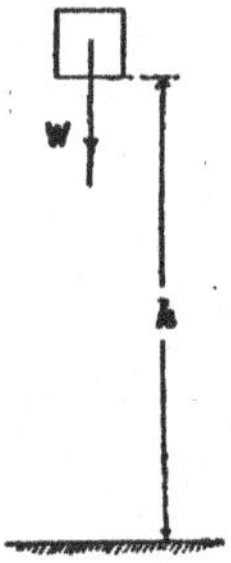

FIG. 203.

We may therefore write,

$$\text{Potential energy lost} = Wh$$

$$= W\frac{v^2}{2g}$$

$$= \text{kinetic energy gained.}$$

Consequently, the kinetic energy of a body is given by this equation in terms of its weight and velocity. Notice that it makes no difference in the kinetic energy possessed by a body moving with a given velocity, whether it is moving vertically or in any other direction, so that in general,

$$\text{Kinetic energy} = \frac{Wv^2}{2g} \text{ foot-pounds,}$$

or foot-tons if W is in tons.

Relative velocity.—When we speak of a body being at rest, what meaning do we attach to the statement? Thus, a house appears to be at rest, that is, it is not shifting its position on the earth; yet being attached to the earth, it possesses the complicated motion of the earth at that place and is therefore not actually at rest. In fact, no body is absolutely at rest. When we speak of rest, we usually mean *at rest relative to the earth*, that is, an observer standing on the earth perceives no motion. If we say that a train has a velocity of 60 miles an hour, we do not mean that this is the absolute velocity of the train, but only its velocity relative to the earth. If two trains are moving side by side with equal speeds, an observer in one of them perceives no motion in the other and therefore says that the relative velocity of the trains is zero. If the train carrying the observer has a velocity of 30 miles an hour, and the other, one of 35 miles an hour, he will see the other train moving past him at a rate of 5 miles an hour, which velocity he would call the relative velocities of the trains. If the second train were going in the opposite direction at 35 miles an hour the relative velocity would be 65 miles an hour. **Relative velocity of two bodies may be defined as the velocity which an observer on one of them would perceive in the other.** Thus, if a stream of water moving at 8 feet per second reaches a water wheel the buckets of which are moving at 6 feet per second, the water will enter the buckets with a relative velocity of 2 feet per second.

If two bodies A and B have velocities as shown at V_1 and V_2 in Fig. 204, their relative velocity can easily be obtained in the following manner. Stop one of the bodies by giving each of them a velocity equal and opposite to the velocity of that body. Then the resultant velocity of the other one will be the relative velocities of the two bodies; thus, giving both A and B velocities equal and opposite to V_1 as shown in Fig. 204, the result will be to bring A to rest, and B's velocity will now be V_R, the resultant of V_1 and V_2 at B. This velocity V_R will be the relative velocity of A and B.

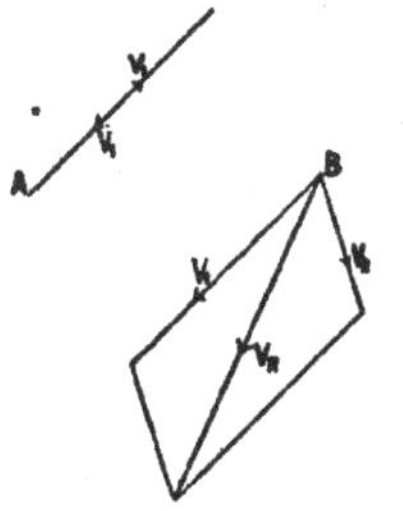

FIG. 204.—Relative velocity of *A* and *B*.

As an example of relative velocity, take the case of a person entering a compartment of a railway carriage when in motion.

Suppose that the velocity of the carriage is 6 feet per second. If the person desires to enter without being thrown against the seats, he will contrive matters so that his velocity relative to the carriage is along the line *BA* (Fig. 205), at 90° to the direction of motion of the carriage. Suppose that this relative velocity is to be 2 feet per second. Stop the carriage by giving both it and the person at *P* velocities, towards the right, of 6 feet per second. This is shown at *P* by the line $PD=6$ feet per second. Now $V_R=2$ feet per second, represented by *PE*, has to be the relative velocity of person and carriage, and hence must be the resultant of his actual velocity along the platform and *PD*. Completing the parallelogram *PDEF*, we get the velocity of the person along the platform represented by $PF=6{\cdot}32$ feet per second, and if he runs along the platform in the direction *PF* with this speed, he will enter the compartment without shock along the line *BA*.

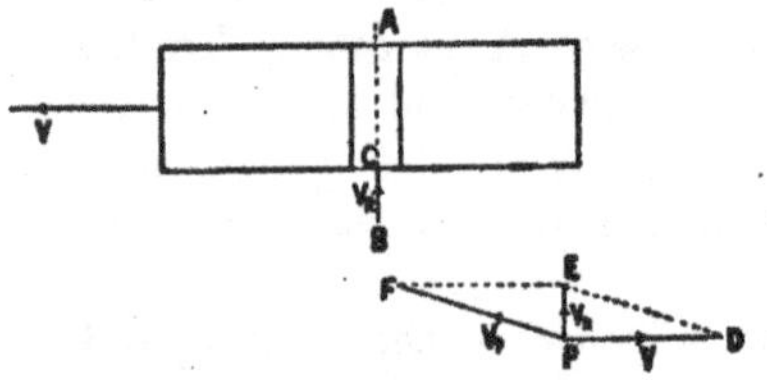

FIG. 205.—Velocity of a person entering a railway carriage.

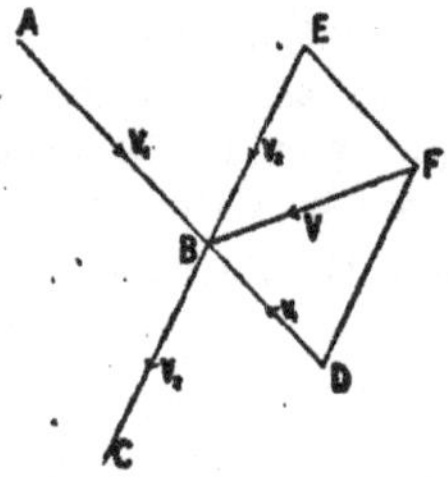

FIG. 206.—Velocity changed from V_1 along *AB* to V_2 along *BC*.

Velocity changed in direction.—Supposing a point is moving along a line *AB* (Fig. 206), with a velocity V_1, and that when it arrives at *B*, something is done to it, in consequence of which it moves off along *BC* with a velocity V_2. Let us examine what change must be effected in its velocity to produce this result. First stop the point at *B* by giving it a velocity equal and opposite to V_1. This is shown by *DB* in the figure. Then give it a velocity V_2 in the direction *BC*, this will fulfil the required conditions; V_2 being represented by *EB*. To find the resultant change in velocity, find the resultant of V_1 and V_2 by the parallelogram of velocities *DBEF*. Then $FB=V$ is the resultant change in velocity. A moving *point* has been considered instead of a *body*, in order to avoid drawing on the imagination in following out the process of reasoning, for we have to think of velocities applied suddenly at B, and a sudden change in velocity implies an infinitely great acceleration, and therefore an infinitely large force to be applied to the body at B. This, of course, is impossible, but what actually occurs is a gradual change in velocity causing the body to turn gently into the direction BC along a curve (Fig. 207). The line FB in Fig. 206 shows, however, the total change in the velocity of the body.

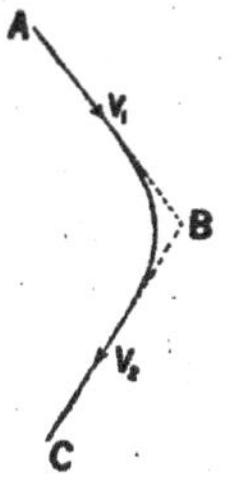

FIG. 207.—The change in velocity must be gradual.

EXERCISES ON CHAP. XII.

1. What distance will be travelled in 5 seconds by a train running at 60 miles per hour?

2. A train is observed to pass two points 480 ft. apart in 10 seconds. What is its speed in miles per hour?

3. A ship is moving due north with a speed of 10 knots. A person crosses the deck from port to starboard, a distance of 40 feet, in 10 seconds. What is his actual velocity relative to the earth? Give a diagram.

4. What is the average speed of a train which travels from London to Edinburgh, a distance of 400 miles, in $8\frac{1}{2}$ hours?

5. A ship steadily acquires a speed of 15 knots, from rest, in 5 minutes. What has been its acceleration in foot and second units?

6. A body falling freely passes two points, the vertical distance between which is 120 feet, in two seconds. From what height above the higher point was it dropped?

7. A train the mass of which is 300 tons is started from rest and gains a speed of 30 miles an hour in 4 minutes. Calculate the force required, additional to that utilised in overcoming frictional resistances, to overcome the inertia of the train.

8. An engine, of stroke 8″, makes 300 revolutions per minute. What is the average speed of the piston in feet per minute?

9. At 300 revolutions per minute, it is found that the piston of an engine, when leaving the inner dead point, has an acceleration of 400 feet per second per second. The mass of piston, rod, and cross-head is 150 lbs. Calculate the force required to overcome the inertia of these parts.

10. The moving parts of a steam hammer have a mass of 500 lbs., and are raised a height of 3 feet above the work before each blow. What is the kinetic energy of these parts when the hammer head is just reaching the work, assuming no frictional losses, and that steam is used for lifting the hammer only?

11. What is the kinetic energy possessed by a hammer head, mass 2 lbs., moving with a velocity of 40 feet per second?

12. What kinetic energy has a ship of 15,000 tons mass when its speed is 20 knots?

13. What exactly does a man mean when he says "this train is going at 30 miles an hour"? Suppose you have a watch with a seconds hand, and know that the telegraph posts are 200′ apart, how can you approximately find the speed of the train? (1897.)

14. A body is moving towards the north at 40 feet per second. In two seconds later, we find it moving towards the north at 50 feet per second. What velocity has been added in these two seconds? (1898.)

15. A body is moving towards the north at 50 feet per second. In two seconds afterwards we find that it is moving towards the north-east at 60 feet per second. Find by drawing what is the added velocity. State the magnitude and direction of the added velocity. (1898.)

16. Suppose a body to have fallen h feet in t seconds from rest according to the law $h=16{\cdot}1\,t^2$. Find how far it falls between the times $t=3$ and $t=3{\cdot}1$; between $t=3$ and $t=3{\cdot}01$; between $t=3$ and $t=3{\cdot}001$. Find the average velocity in each of these intervals of time. What do we mean by the actual velocity when t is 3 seconds? (1898.)

17. A bullet weighing 1 oz. leaves the muzzle of a rifle with a velocity of 1350 feet per sec. What is the kinetic energy of the bullet in foot-lbs.? (1899.)

CHAPTER XIII.

MECHANISM. TRANSMISSION OF MOTION AND POWER.

Driving by belt.—Motion may be transmitted from one shaft to another in many different ways. If the shafts are parallel to one another and a considerable distance apart, the most convenient way is by the use of a belt, or rope, lapping round pulleys fixed to the two shafts. Both shafts will rotate in the same direction if the belt is open, as in Fig. 208, and in opposite directions if the belt is crossed, as in Fig. 209. *A*, the shaft supplying motion, is called the **driver**, *B* is the **driven shaft**. The action of driving is possible by reason of the frictional resistance to slipping of the belt on the pulleys, but there will always be a certain amount of slipping, introducing some loss of motion. Neglecting this, the velocity ratio of the shafts may be found thus.

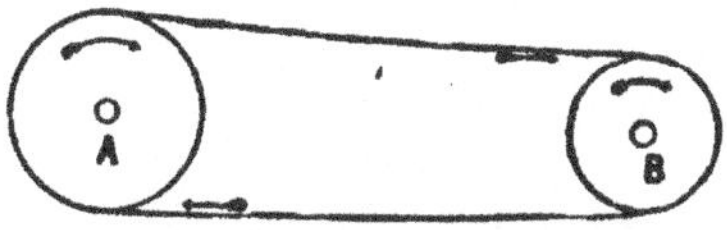

FIG. 208.—Driving by open belt.

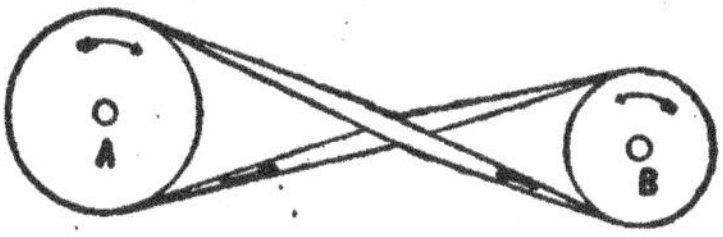

FIG. 209.—Driving by crossed belt.

Let R_A = radius of the pulley on *A*.
R_B = " " " *B*.

Then, if there is no slipping, the circumferences of both pulleys will move through the same distance in a given time, for each will have the same speed as the belt. Suppose, then, that *A* turns once; its circumference will travel a distance $= 2\pi R_A$.

The circumference of B will move through an equal distance, and consequently B will turn through $\frac{2\pi R_A}{\text{circumference of } B}$ revolutions, or

$$\text{Revolutions of } B \text{ for one of } A = \frac{2\pi R_A}{2\pi R_B} = \frac{R_A}{R_B}.$$

If then, A rotates N_A times in a minute, and B rotates N_B times also in a minute,

$$\frac{N_B}{N_A} = \frac{R_A}{R_B},$$

or **the revolutions of the shafts are inversely proportional to the radii of the pulleys mounted on them.**

The power transmitted in any given case can easily be calculated if we know the tensions in the two parts of the driving belt. This has been seen in Chap. XI. to be about 0·385 for a leather belt on a cast iron pulley.

Let T_1 = pull in the tight part of the belt, lbs.
T_2 = „ slack „ lbs.
V = distance travelled in one minute by a point on the belt, in feet.

Then, considering the driver A (Fig. 210), T_2 is assisting the pulley to turn and T_1 is retarding it, so that the driver delivers a net pull $(T_1 - T_2)$ pounds, by means of the belt, to the driven pulley. The work done in one minute will be W foot-pounds.

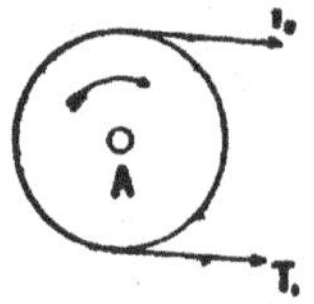

FIG. 210.—Pulls transmitted by the belt.

$W = (T_1 - T_2) \times V$ ft.-lbs. per minute, and

$$\text{Horse-power transmitted} = \frac{(T_1 - T_2)V}{33{,}000}.$$

Let us take roughly $T_2 = 0{\cdot}4 \,.\, T_1$, then

$$T_1 - T_2 = 0{\cdot}6 \,.\, T_1$$

and $$\text{Horse-power transmitted} = \frac{0{\cdot}6\, T_1}{33{,}000} \,.\, V.$$

EXAMPLE. **A belt running at 900 ft. per minute has a pull in its tight part of 400 lbs. Calculate the horse-power transmitted.**

$$\text{H.P.} = \frac{0{\cdot}6 \times 400 \times 900}{33{,}000}$$
$$= \underline{6{\cdot}5}.$$

If instead of the speed of the belt, we are given the diameter, D ft., of the pulley, and N, the number of revolutions per minute of the shaft on which it is mounted, then

$$V = \pi D \times N,$$

and $$\text{Horse-power transmitted} = \frac{(T_1 - T_2)\pi . D . N}{33{,}000}.$$

Losses by slip.—Slipping of the belt introduces a loss of energy in overcoming frictional resistances between the belt and pulley. The amount of slipping is variable, and depends on the power transmitted and the tightness of the belt. Its amount may be found experimentally by first calculating what revolutions the driven shaft should make, using the expression for the velocity ratio

$$\frac{N_B}{N_A} = \frac{R_A}{R_B}.$$

Then, actually count the revolutions of the two shafts for, say, one minute, repeating two or three times and taking the average. The actual velocity ratio will be found by dividing the actual revolutions of A by those of B.

Slipping can then be expressed as a percentage. Thus, suppose a pulley of 3 ft. diameter is driving one of 1 ft. diameter and that the driver rotates 120 times in one minute.

The calculated velocity ratio $= \frac{3}{1} = 3$, so that B should rotate 360 times in one minute. Suppose it is found on trial that B rotates 340 times per minute. Then

$$\text{Actual velocity ratio} = \frac{340}{120}$$
$$= 2{\cdot}83.$$

The lost revolutions of the driven shaft will be 20 in 360, or $\frac{20}{360} \times 100 = 5{\cdot}5$ per cent. We are therefore losing 5·5 per cent. of the motion, and also the same percentage of the power which might have been transmitted if no slip.

Use of guide pulleys.—The belt must always be delivered to the pulley while moving in the same plane as the pulley is rotating, but may leave the pulley in a different plane. There is no difficulty in arranging this condition when the shafts are parallel—they must be simply placed opposite one another, but in cases where the shafts are not parallel, **guide pulleys** may be required in order to direct the belt into the proper plane. A case of this is shown in Fig. 211, where the belt leaving the top of the driving pulley is directed into the plane of rotation of the driven pulley by the guide pulley *C*. The lower side of the belt would be similarly guided. Belt pulleys are generally rounded on the face. The tendency is for the belt to climb to the highest part of the rim and consequently this helps it to stay on the pulley.

FIG. 211.—Belt guided by pulleys at *C*.

FIG. 212.—Section of rim of a rope pulley.

Ropes used for driving are generally of round white cotton. The rims of the pulleys are grooved in this case (Fig. 212) to receive the ropes. The effect of the V-shaped groove is to give better adhesion and thereby minimise slipping.

Belt pulley arrangements.—Belt driving can be conveniently arranged for machines which have to be often stopped or started. This may be managed as shown in Fig. 213, by putting a pulley *F* on the driven shaft, keyed to the shaft, and also one *L*, arranged to run loose on the driven shaft, *F* and *L* being close together. The belt runs on a pulley fixed at *A* on the driving shaft, this pulley being as broad as *F* and *L* together. The belt is guided on to either *F* or *L* by means of **forks** secured to a sliding bar *C*, which may be moved by hand. These forks must engage with the belt on that part which is advancing towards the pulleys on *B*. When the belt is running on *F*, *B* will be driven, and when on *L*, *B* will stop, the pulley *L* then simply running on *B* without driving it.

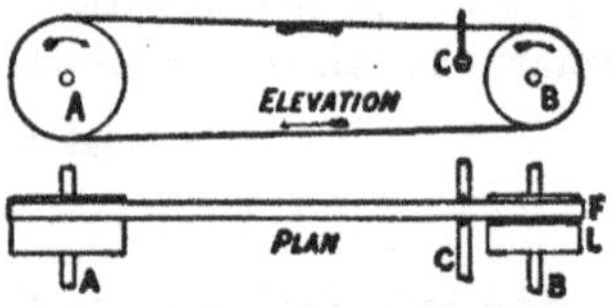

FIG. 213.—Arrangement for starting and stopping a machine.

An arrangement of crossed and open belts may also be used for giving a **reversing motion** to a machine. Thus, in Fig. 214 B is the shaft from which the machine is driven, and it carries two loose pulleys, L_1 and L_2, with a fast pulley F between them. D is an open belt and E a crossed one, both guided by forks on a bar C. In the present position, the open belt D is on the fast pulley F and the machine will be driven in one direction. If the bar C is moved downwards in the plan, D will run on L_2 and E on L_1, the machine will therefore be at rest. Further motion of C downwards will bring E on to the fast pulley F and the machine will consequently run in the reverse direction to the former. The motion of C may be automatically effected, as in an ordinary planing machine, by rods and levers worked by the machine itself. The machine will then be **self-acting**.

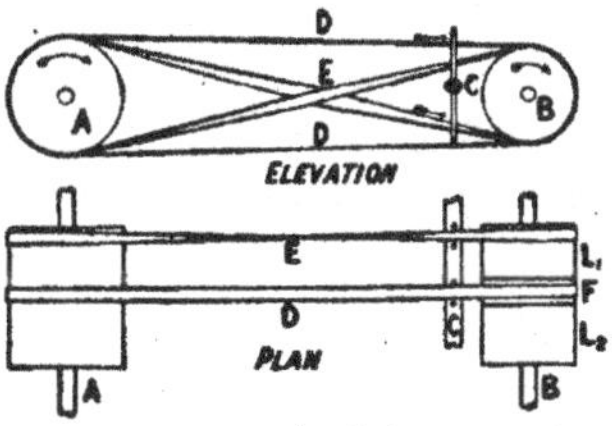

FIG. 214.—Arrangement for reversing a machine.

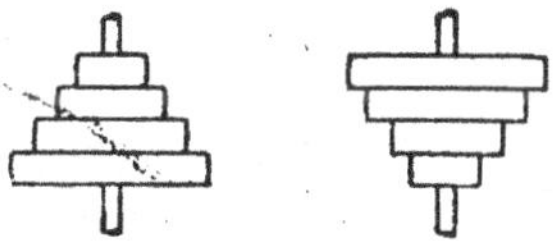
FIG. 215.—Speed cones.

Speed cones are belt pulleys having several steps on which the belt may run (Fig. 215). Their object is to secure a velocity ratio which may be varied to suit the particular work on which the machine is engaged. The velocity ratio when the belt is on any particular pair may be calculated as before, from the radii of the steps.

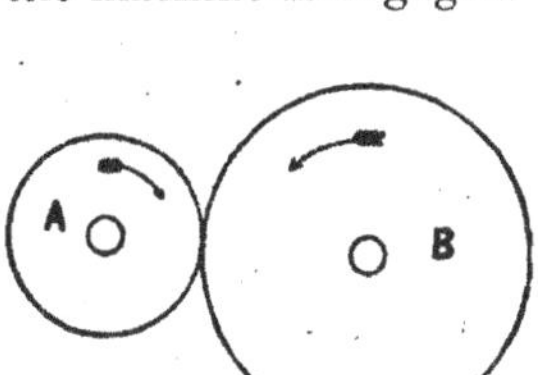

FIG. 216.—Friction wheels.

Friction gearing.—If the driving shaft and the driven shaft are close enough together, the rims of the pulleys may touch one another, and if pressed together, enough friction will be produced to enable the one to drive the other, provided the power being transmitted is not too great. In Fig. 216, A is the driver and B the driven pulley. It will be noticed that the shafts now rotate in opposite directions. There will always be a certain amount of slipping with friction gears, and to minimise this, one wheel may have its rim made of compressed paper or leather, thereby giving a larger coefficient of friction than can be obtained with metal surfaces. If we neglect slipping, and as before, call

Radius of $A = R_A$,
Radius of $B = R_B$,
Revolutions of $A = N_A$ per minute,
Revolutions of $B = N_B$ per minute.

Then, since both circumferences will have the same speed if there is no slip, we may obtain in the same manner as for belt pulleys,

$$\frac{N_A}{N_B} = \frac{R_B}{R_A}.$$

The pressure on the shaft bearings due to the forces pressing the wheels together is objectionable, especially if the shaft is running at high speed. This may be got rid of by using two pulleys B and C (Fig. 217), considerably larger than the driver A, and with the bearings mounted so that each has a limited horizontal travel. A belt passed round B and C will by its tensions cause the driving pulley A to be nipped between B and C without thereby giving pressures to the shaft bearings. B and C will each be driven in the opposite direction to A, and power may be taken from another pulley mounted on the same shaft as either B or C.

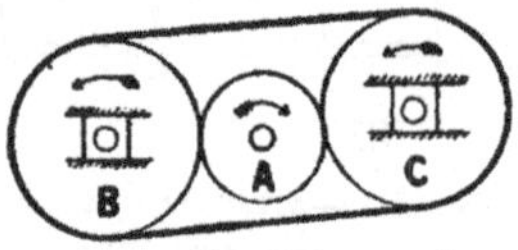

FIG. 217.

If the shafts are not parallel, but have their axes inclined and meeting at a point C (Fig. 218), then two cones, one mounted on each, may be used for driving. These must be pressed together also. The radius of each cone at any place will be proportional to the distance from C, and consequently the circumferences will also be proportional

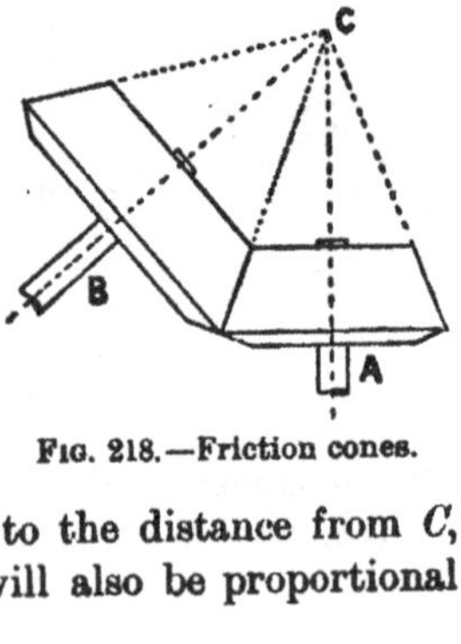

FIG. 218.—Friction cones.

to the same distance. This being the case, when motion occurs, the cones will roll on one another without slip at any of the parts in contact, or if slip does occur, the amount will be proportional anywhere to the distance from C.

Toothed wheels.—To prevent slipping, the discs composing friction wheels may have **teeth** cast or cut on the rims (Fig. 219), the teeth on each disc projecting beyond the edge of the rim, and also coming inside the edge. The teeth on one wheel engage with the teeth on the other, and so prevent any slip. The edges of the original discs have now disappeared, but we may imagine them still to be there, and the wheels will rotate just as though these discs were rolling on one another. This imaginary circle showing the original disc is called the **pitch circle**, and the distance from centre to centre of two teeth measured along the pitch circle is called the **pitch of the teeth.**

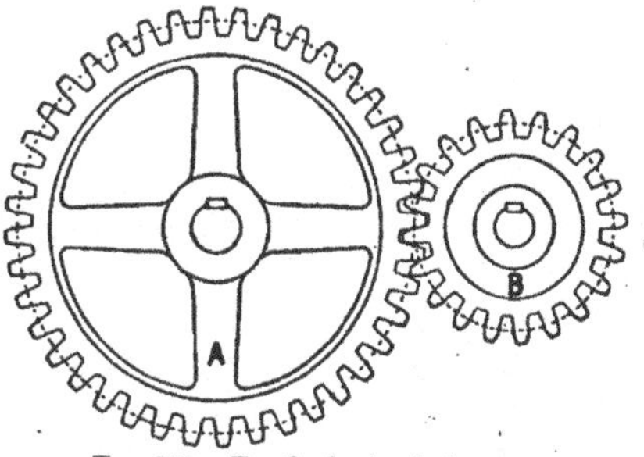

FIG. 219.—Toothed wheels in gear.

Let n_A = number of teeth on A's rim.
n_B = " " B's "

Then, calling the pitch p, which must obviously be the same for each wheel,

$$\text{circumference of } A = 2\pi R_A = p \times n_A,$$
$$\text{circumference of } B = 2\pi R_B = p \times n_B,$$

and from this,

$$\frac{R_A}{R_B} = \frac{n_A}{n_B}.$$

Also
$$\frac{\text{revolutions of } A}{\text{revolutions of } B} = \frac{R_A}{R_B};$$

$$\therefore \frac{N_A}{N_B} = \frac{n_B}{n_A},$$

or **the revolutions per minute of the two wheels are inversely proportional to their numbers of teeth.**

Use of idle wheels.—Two toothed wheels in gear with one another must rotate in opposite directions. If both are required to rotate in the same direction, then another wheel, mounted on an intermediate shaft, and gearing with both driver and driven wheels, is required. This is shown in Fig. 220. *A* and *B* will now rotate in the same direction, and since the speed of the circumferences of all three pitch circles will be the same, it follows that

$$\frac{N_A}{N_C}=\frac{R_C}{R_A};$$

also

$$\frac{N_C}{N_B}=\frac{R_B}{R_C}.$$

Multiplying the left-hand sides of these equations together and also the right-hand sides, we get

$$\frac{N_A}{N_C}\times\frac{N_C}{N_B}=\frac{R_C}{R_A}\times\frac{R_B}{R_C},$$

or

$$\frac{N_A}{N_B}=\frac{R_B}{R_A}.$$

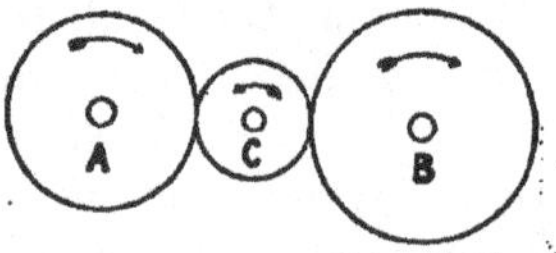

FIG. 220.—*A* drives *B* through the idle wheel *C*.

FIG. 221.—*C*, *D*, and *E* are idle wheels.

The relative speeds of rotation of the driver and driven wheels is therefore the same as if they geared direct. The only object of *C* is to change the direction of rotation, and as it does not alter the velocity ratio it is generally called an **idle wheel.** In the same way we may show that any number of idle wheels, as in Fig. 221, may be interposed without affecting the velocity ratio.

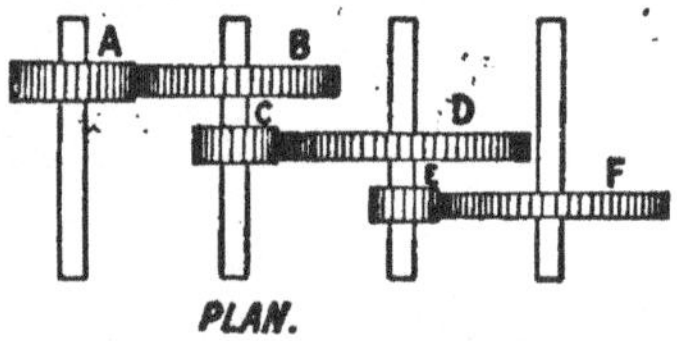

FIG. 222.—Train of wheels

Trains of wheels.—Where a considerable velocity ratio is required, trains of wheels arranged as in Fig. 222 may be employed. In this case, the velocity ratios of the various pairs in gear will be respectively,

$$\frac{R_B}{R_A}, \frac{R_D}{R_C}, \frac{R_F}{R_E};$$

or

$$\frac{n_B}{n_A}, \frac{n_D}{n_C}, \frac{n_F}{n_E}.$$

Supposing F to rotate once, the revolutions of E will be $\frac{n_F}{n_E}$, and D will have the same number of revolutions. For one revolution of D, C will have $\frac{n_D}{n_C}$ revolutions, and consequently for one of F, C will have $\frac{n_D}{n_C} \times \frac{n_F}{n_E}$ revolutions, and B will have the same number. For one revolution of B, A will have $\frac{n_B}{n_A}$ revolutions, and therefore for one revolution of F, A will have

$$\frac{n_B}{n_A} \times \frac{n_D}{n_C} \times \frac{n_F}{n_E} \text{ revolutions.}$$

FIG. 223.—Driving arrangements in a self-acting lathe.

We see therefore, that **the velocity ratio of the first and last wheels in the train is found by taking the product of the numbers of teeth on all the drivers, *F*, *D* and *B*, and dividing this by the product of the numbers of teeth on all the driven wheels, *E*, *C* and *A*.**

Trains of wheels such as these are much used in machinery, and the engineering student will have no difficulty in observing many examples of them for himself. Fig. 223 shows one which can be examined on an ordinary self-acting lathe.

Epicyclic trains of wheels contain usually one fixed wheel, such as *A* (Fig. 224), secured to a bracket so that it may not rotate, and several others in gear, such as *B* and *C*, mounted on pins secured to an arm *D* which can rotate about the axis of *A*. Each wheel *B* and *C* may rotate on its own axis at the same time that it is carried round *A* by the rotating arm. The simplest way of finding the revolutions of any wheel of the set is to proceed thus :— Imagine the whole of the wheels to be locked to the arm, and the bracket carrying *A* to be loosened from its supports. Looking at the plan (Fig. 224), give the whole arrangement one clockwise revolution. Then we have given revolutions to each wheel and the arm as shown :

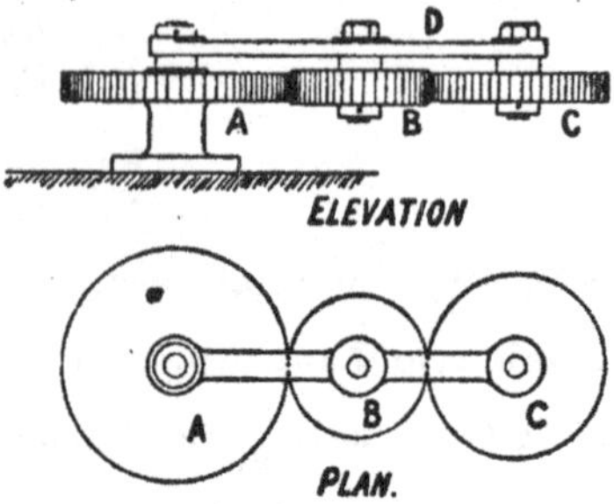

FIG. 224.—Epicyclic train of wheels.

A	*B*	*C*	*D*
One revolution clockwise	One revolution clockwise	One revolution clockwise	One revolution clockwise

Now in the actual arrangement, *A* should not rotate, being a fixed wheel, and as we have given it above a clockwise revolution, we correct this by keeping the arm fixed and rotating the wheel *A* once anticlockwise. This will give revolutions to *B* and *C* thus :

A	*B*	*C*	*D*
One revolution anticlockwise	$\frac{n_A}{n_B}$ revolutions clockwise	$\frac{n_A}{n_C}$ revolutions anticlockwise	0

We have now fulfilled in this manner the conditions of the gear, and have given the arm *D* one revolution, clockwise, the fixed wheel A being where it was at first. The effect on the other wheels will be given by the algebraic sum of the corresponding columns of the tables, thus,

$$B \text{ will have } \left(1+\frac{n_A}{n_B}\right) \text{ revolutions, clockwise.}$$

$$C \text{ will have } \left(1-\frac{n_A}{n_C}\right) \text{ revolutions, clockwise}$$

if positive, anticlockwise if negative.

It will be noticed that if A and C have each the same number of teeth,

$$\frac{n_A}{n_C}=1,$$

and the revolutions of C will be

$$(1-1)=0.$$

In this case C does not rotate on its axis while the arm rotates.

A model epicyclic train such as shown above is easily arranged and will be found very useful in explaining the action and what has been discussed. Epicyclic gears have been used in various machines, such as rope-spinning machines, and also for reducing from a higher to a lower speed of rotation.

Shape of teeth.—If the velocity ratio of a pair of wheels in gear is not to alter at any time, and this is essential to smooth running, then the condition which must be attended to in giving the teeth their **proper shape**, is that the common perpendicular to the outline of two teeth, at any place where they are in contact, must pass through the point where the pitch circles of the two wheels touch (Fig. 225).

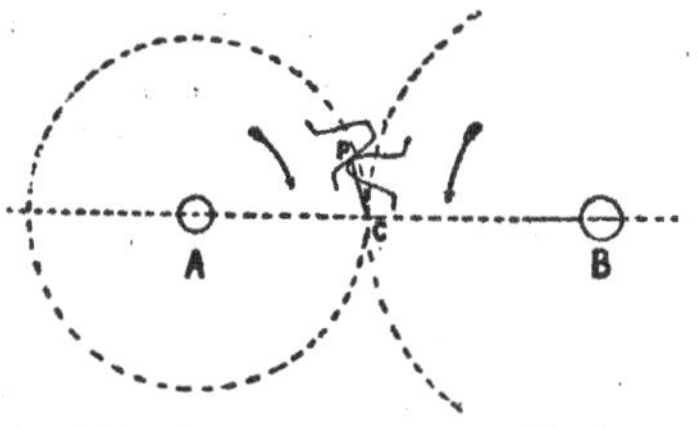

FIG. 225.—The common perpendicular at P must pass through C.

This condition is fulfilled if portions of *cycloidal curves* are used for the teeth outlines, although other curves may be used also. The student is referred to books on Machine Design for the methods of drawing these curves and designing the teeth.

Power transmitted by toothed wheels.—The resultant tangential driving force between two toothed wheels may be easily found if we know the power being transmitted by one of the wheels and also its radius. Thus, let A (Fig. 226) be a wheel the teeth of which encounter a resultant resistance of P lbs. from another wheel which it is driving. Let R = radius of A in feet.

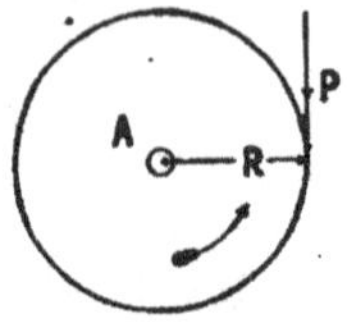

FIG. 226.

Then in one revolution of A, P is overcome through a distance $= 2\pi R$ feet, and therefore

$$\text{the work done} = P \times 2\pi R \text{ ft.-lbs.}$$

If the revolutions per minute $= N$, then

$$\text{Horse-power} = \frac{P \times 2\pi R \times N}{33{,}000},$$

or,

$$P = \frac{33{,}000 \times \text{Horse-power}}{2\pi R \times N}.$$

Having obtained P, the thickness of tooth required for strength may be found by the ordinary proportional rules for cantilevers as described in Chap. VIII.

Bevel wheels are evolved from coned friction wheels by giving teeth to the cones in the same way as for ordinary toothed wheels.

The screw consists of two portions, one A (Fig. 227) cylindrical, and free to rotate but not to slide axially, and the other B,

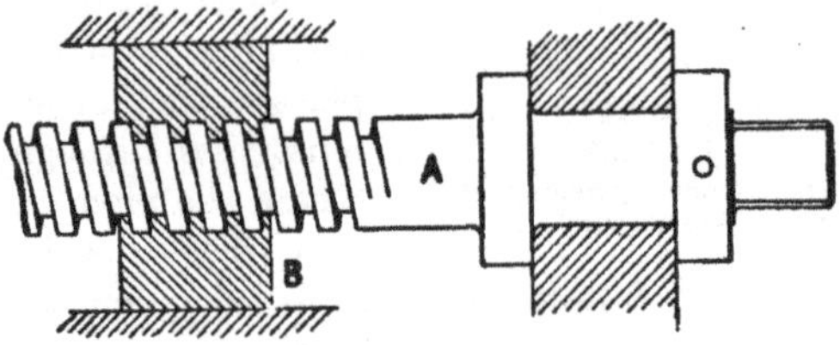

FIG. 227.—Section through a nut, *B*, showing screw, *A*.

called the nut, free to slide axially but not to rotate. A helical thread is cut on the outside of A and a corresponding one on the inside of B, so that A may fit in B. If A is rotated, B will slide axially. Combinations of the screw and nut are very often used and take many forms. The threads also take many different shapes. As regards the relative motions of A and B. Let B have N threads per inch, then the distance from thread to thread, measured from corresponding places on the threads, will be $\frac{1}{N}$. This distance is called the **pitch of the thread**, and may be written p inches. For one revolution of A, B will slide a distance $=p$ inches.

EXAMPLE. The ordinary bolt and nut is an example of a screw. Supposing a $\frac{1}{2}''$ bolt, 12 threads per inch, to be screwed down by a spanner, the turning force being applied 7″ from the axis of the bolt, and equal to 20 lbs., what pull will be produced on the bolt, neglecting friction?

Let P = force on spanner, lbs.,
R = radius of P, inches,
Q = pull on bolt, lbs.,
p = pitch of screw, inches.

Then, if there is no friction,

Work done by P in one revolution = Work done in overcoming Q through a distance equal to the pitch.

$$P \times 2\pi R = Q \times p,$$

$$Q = \frac{20 \times 2 \times 22 \times 7 \times 12}{7}$$

$$= \underline{10,560} \text{ lbs.}$$

If 75 per cent. is lost in overcoming frictional resistances, then, pull on the bolt will be

$$\frac{25}{100} \times 10,560 = \underline{2640} \text{ lbs.}$$

Worm-wheel gearing.—A useful application of the screw is to be found in the **worm and worm wheel**. A short screw of comparatively large diameter is mounted on the driving shaft, and this gears with specially shaped teeth on the rim of a wheel mounted on a shaft, the axis of which is perpendicular to the driving shaft. On the driving shaft rotating, the worm mounted on it will drive the worm wheel, one tooth being advanced for every revolution of the worm. The relative velocities of rotation will therefore be simply equal to the number of teeth on the worm wheel. The driving shaft must be prevented from axial movement under the thrust produced by driving the worm wheel. In Fig. 228 the driving shaft is fitted with ball thrust bearings so as to minimise as far as possible frictional losses

FIG. 228.—Worm and worm wheel.

due to collar friction. The worm and worm wheel is often employed for reducing from high to low speeds of rotation, and is specially adapted for this on account of the large velocity ratio easily obtainable.

Link mechanisms.—**Links** are pieces used for transmitting motion from one point to another, the motion being usually modified. Thus, in the ordinary **crank and connecting rod,** the reciprocating motion of the crosshead is transmitted by a connecting link—the *connecting rod*—to the *crank pin*, which has a motion of rotation. In link work, the motion of each piece is usually quite definite, that is, it does not move at random in any manner, but has a limited degree of freedom, being constrained by the other pieces to which it is connected to move always over the same path in the same manner. The principal problems we require to solve in connection with link work mechanisms are :

(*a*) The complete path traversed by any given point of the mechanism ;

(*b*) The velocity of any point at any instant ;

(*c*) The acceleration of any point at any instant.

The best way of answering (*a*) for any given mechanism is to draw the arrangement to scale in several different positions of the motion, the point under consideration being marked on each. A curve showing the complete path can then be drawn through these points. Such a curve is shown for a point D on a connecting rod in Fig. 229.

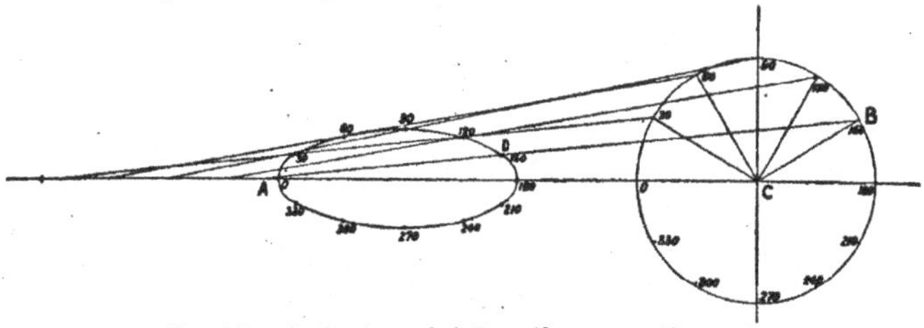

FIG. 229.—Path of a point D on the connecting rod.

Velocity of any part of a rotating body.—If a body is uniformly rotating, then the velocity of any point in it can easily be found if we know the velocity of a given point. Thus, given any body rotating about C in the plane of the paper (Fig. 230), the velocity of a given point A will be always directed tangentially to the circular path of A, and is shown so in the given position, perpendicular to the radius AC. Let the velocity of A be V_1. The velocity of any other point in the body, such as B, when the body is in this position, will be V_2, perpendicular to the radius BC, that is, tangential to the path of B. In one revolution of the body, A will move a distance equal to $2\pi \times AC$, and B will move a distance $2\pi \times BC$, so that

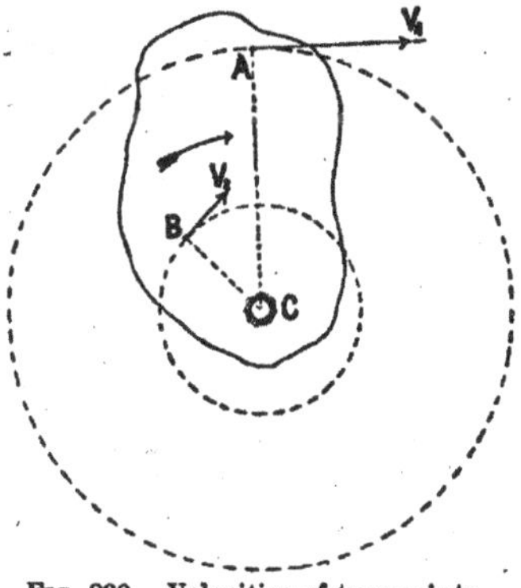

FIG. 230.—Velocities of two points, A and B, in a rotating body.

$$V_1 : V_2 = 2\pi \times AC : 2\pi \times BC$$
$$= AC : BC$$

or *the velocities of any two points are proportional to their radii.*

The velocities of any point in a link may be easily found after the following facts have been studied. Consider a point A

(Fig. 231), moving with uniform velocity v in the circumference of a circle the centre of which is C. For an instant the point A may be considered to be moving with uniform velocity v in the tangent to the circle, AB, and in imagining this, we need not consider the magnitude of the radius of the circle. In fact, A will be moving in the direction AB if it is turning about any centre whatever in the line AC, or AC produced. Thus, D might be taken for an instant as the centre of rotation without thereby altering the direction of the motion of A.

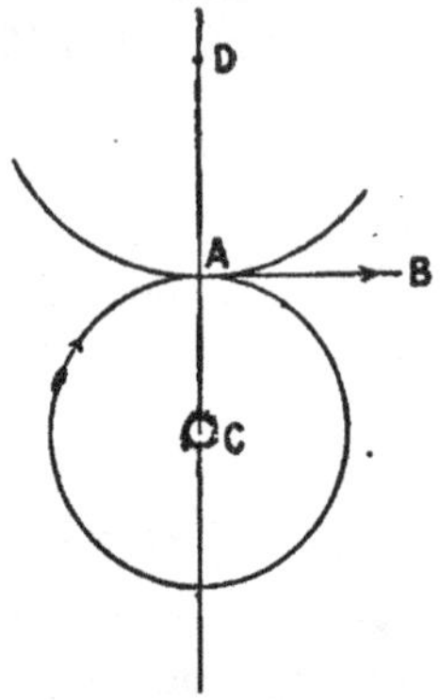

FIG. 231.

Crank and connecting-rod mechanism. —Let us now examine the connecting rod of a crank and connecting-rod mechanism. In Fig. 232, AB is the connecting rod and BC the crank. A always moves in the straight line AC and B always

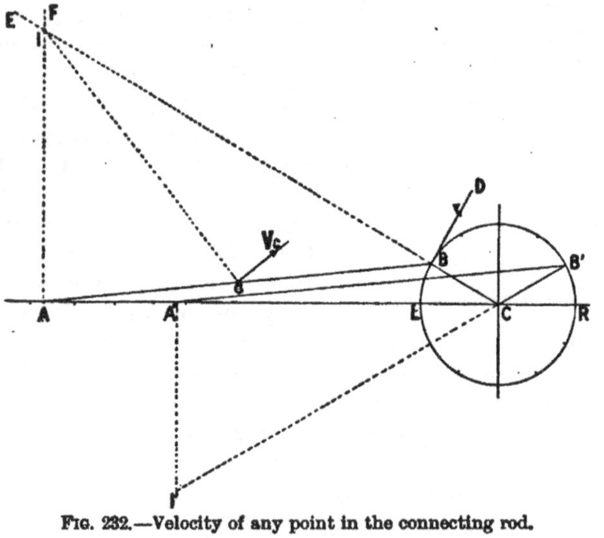

FIG. 232.—Velocity of any point in the connecting rod.

moves in the circumference of the circle LBR. Let the velocity of B be uniform and equal to v. In the given position,

B is actually turning about *C*, and the direction of its motion is in the tangent *BD* to the circle at *B*. This direction will not be changed if we consider *B* to be moving about any point in *CB* or *CB* produced; thus, we might imagine it to be turning about *E* for an instant. *A* is actually moving in *AC*, but we may imagine it to be turning about any centre in *AF*, perpendicular to *AC*; thus, if turning about *F* for an instant, the motion of *A* will be along *AC*. Now *B* may turn about any point in *CE*, and *A* may turn about any point in *AF*, so that if we choose *I*, where *CE* and *AF* intersect, both *A* and *B* may be considered as rotating about the same point *I* for a very brief interval of time. The velocity of *A* may now be found, for the velocities of *A* and *B* will be directly proportional to their radii *AI* and *BI*.

Let V=velocity of *A*, then

$$v : V = BI : AI$$

or

$$V = \frac{v \times AI}{BI}.$$

Since two points, *A* and *B* in the rod, are turning for an instant about *I*, the whole rod is also turning about *I* for an instant, so that the velocity of any other point in it, such as *G*, will be found by joining *G* to *I*. The velocity of *G* will be directed at 90° to *GI* and can be found from

$$V_G : v = IG : IB$$

or

$$V_G = \frac{v \times IG}{IB}.$$

I is called the **instantaneous centre** of the rod for the given position. In order to find the velocity of any point in the rod, the mechanism should be drawn to scale in the required position, the instantaneous centre found, and the radii measured for substitution in the above equation. If this is done for, say, twelve crank positions differing by 30°, values may be found for the velocity of *A*, which, when plotted, will give the velocity of *A* for any crank position. This is shown in Fig. 233.

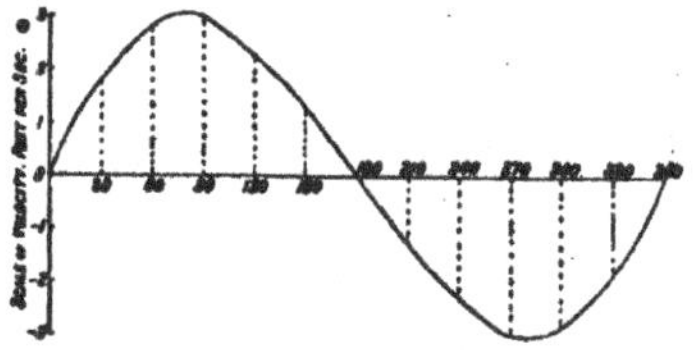

FIG. 233.—Velocity curve for end *A* of the connecting rod.

As another example, take two cranks CA and DB (Fig. 234), connected by a link AB. Let the velocity of A be uniform and equal to v; it is required to find V, the velocity of B in the given position. The mechanism having been drawn to scale, I will be at the intersection of CA and DB, and

$$v : V = AI : BI$$

or

$$V = \frac{v \times BI}{AI},$$

which may be found numerically by measuring BI and AI and inserting the values in this equation.

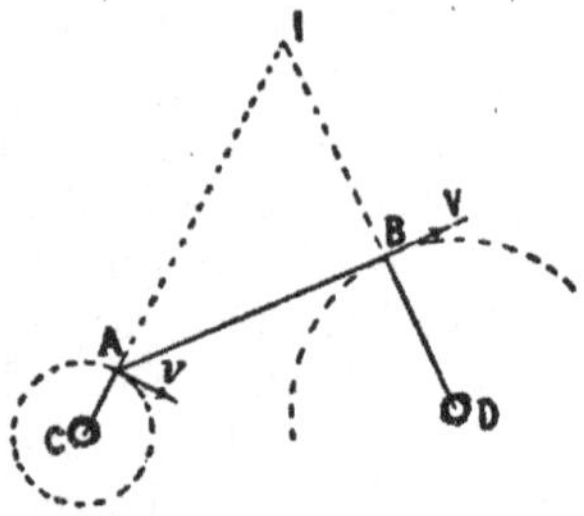

FIG. 234.—Double crank and connecting rod.

Use of models.—In the crank and connecting-rod mechanism, and, indeed, in any linkwork mechanism, a good deal can be learned by studying simple models. Fig. 235 shows a crank and connecting-rod model in which the connecting rod may be

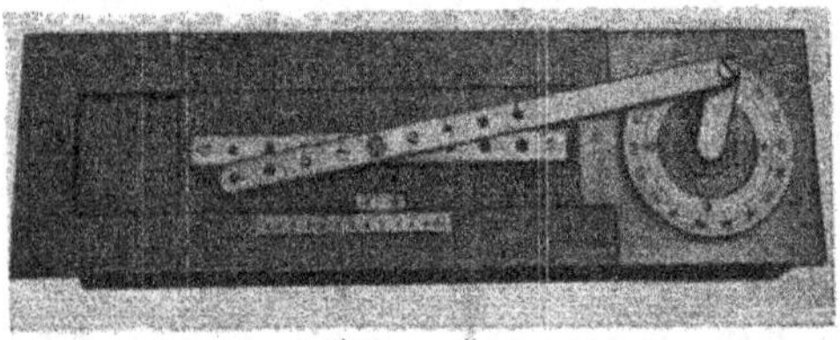

FIG. 235.—Adjustable crank and connecting-rod model.

varied in length. The piston position for any crank angle can be read on a scale against which a pointer attached to the model piston slides. The disturbing effect of the connecting rod, and the influence of its length may be conveniently studied with such a model.

Acceleration of a point in a mechanism.—Until the student has acquired a more extensive knowledge of the subject, the following approximation to the acceleration of a point in a link may be used. Considering the velocity diagram shown in Fig. 233 for the velocity of the crosshead at any crank position. The diagram is drawn on a base line showing equal angles turned through by the crank, and as the crank is supposed to be moving uniformly, we may think of this as being a base line showing equal times.

EXAMPLE. Suppose the crank is $\frac{1}{2}$ ft. radius, and that the velocity of the crank pin is 3 ft. per second. Find the time in which the crank rotates 30°.

$$\text{Circumference of crank pin circle} = 2\pi \,.\, r$$
$$= 2 \times \frac{22}{7} \times \frac{1}{2}$$
$$= \frac{22}{7} \text{ feet.}$$

$$\text{Time in which crank pin revolves once} = \frac{\text{circumference}}{\text{velocity}}$$
$$= \frac{22}{7} \times \frac{1}{3}$$
$$= \frac{22}{21} \text{ seconds.}$$

$$\text{And time in which it rotates } 30° = \frac{22}{21} \times \frac{30}{360}$$
$$= \frac{22}{252}$$
$$= \underline{0{\cdot}0873 \text{ sec.}}$$

Each interval in the base line of the velocity diagram may therefore be taken as 0·0873 second. Now in the interval from 0° to 30° the crosshead changes its velocity from 0 to 1·84 feet per second, so that its average acceleration during this interval will be $\frac{1{\cdot}84}{0{\cdot}0873} = 21{\cdot}1$ feet per sec. per sec.

Repeating this process for all the intervals, we obtain a series of values showing the average acceleration of the crosshead during each interval, and if these values are set up as ordinates from the centres of the intervals, a curve may be drawn showing approximately the acceleration of the crosshead throughout the crank revolution. The values are best set down in a table thus :

ACCELERATION OF CROSSHEAD.

Crank Angle.	Velocity of Crosshead, feet per second.	Difference in Velocity of Crosshead.	Acceleration = $\frac{\text{difference}}{0{\cdot}0873}$ feet-sec. sec.
0	0		
		+1·84	+21·1
30	+1·84		
		+1·00	+11·4
60	+2·84		
		+0·16	+ 1·83
90	+3·00		
		−0·70	− 8·02
120	+2·30		
		−1·05	−11·00
150	+1·25		
		−1·25	−14·3
180	0		
		−1·25	−14·3
210	−1·25		
		−1·05	−11·00
240	−2·30		
		−0·70	− 8·02
270	−3·00		
		+0·16	+ 1·83
300	−2·84		
		+1·00	+11·4
330	−1·84		
		+1·84	+21·1
360	0		

This curve is shown in Fig. 236, positive accelerations being shown above, and negative accelerations below, the base line. The crosses show the points plotted from the above table and the dots are plotted from exact data, both being given in order to show how little the actual acceleration differs from that found approximately above.

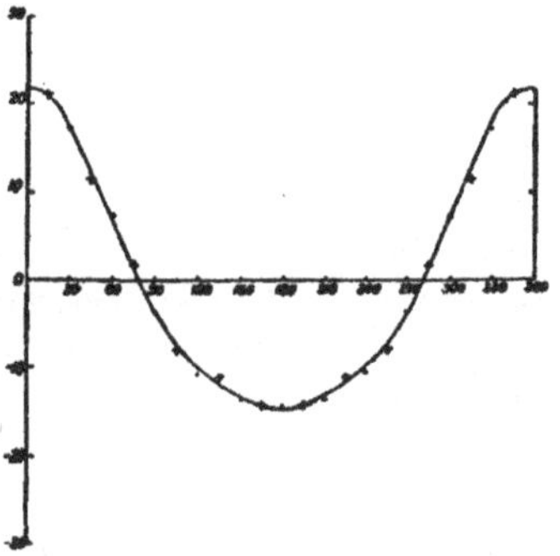

FIG. 236.—Acceleration curve for end *A* of the connecting rod.

Infinite connecting-rod mechanism.—The disturbing effects produced by the varying inclination of the connecting rod in a crank and connecting-rod mechanism, can be got rid of, if, for the connecting rod, a slotted bar is substituted (Fig. 237), the bar being guided to move in a straight line. The crank pin works in the slot, and, as the slot is perpendicular to the direction of motion of the bar, horizontal movements of the crank pin are cancelled by it and its vertical ones alone are copied by the slotted bar. A model of two of these bars, both worked from the same crank pin, and furnished with scales for observing positions of the bar corresponding to various crank positions,

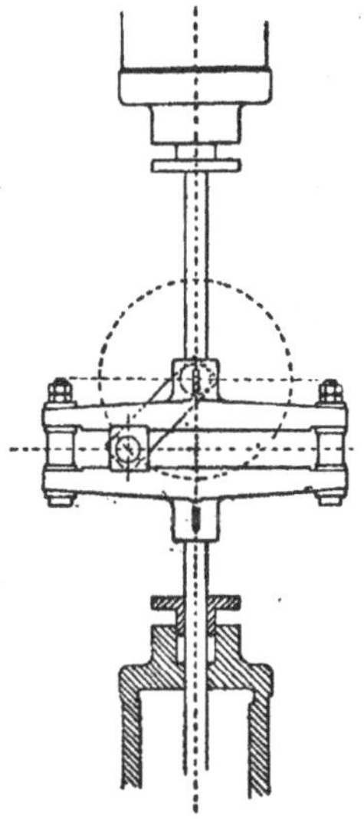

FIG. 237.—Infinite connecting rod applied to a donkey pump.

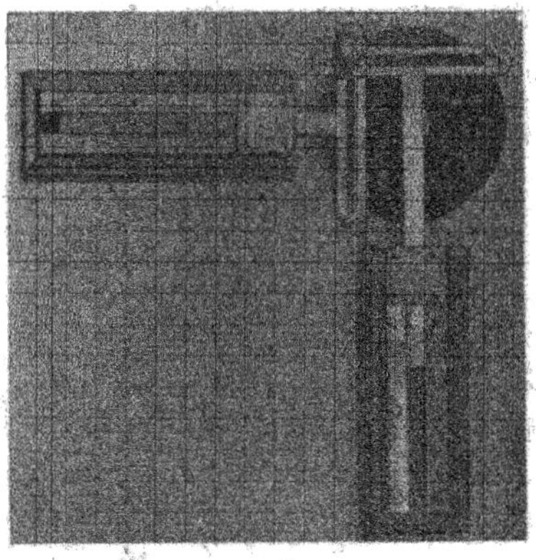

FIG. 238.—Model of infinite connecting rod.

is shown in Fig. 238. The mechanism is usually called the **infinite connecting rod**, as the motion of the sliding bar is the same as would be produced by a connecting rod of infinite length.

Oscillating engine mechanism.—In the mechanism of the oscillating engine, the connecting rod is dispensed with, and the piston rod end is connected direct to the crank pin, the cylinder being mounted on *trunnions*, so that it may oscillate and follow the motion of the crank pin. Fig. 239 shows a model of the arrangement. Scales are attached for showing crank angles, piston position and cylinder angles.

Parallel motion mechanisms.—Linkwork mechanisms for producing straight line motion are commonly called **parallel motions**. In the Scott-Russell parallel motion *AP* (Fig. 240) is a rod, the end *A* being constrained to travel always in the straight line *AB*; *C* is the centre of *AP* and the parts *CB*, *AC* and *CP* are equal. The link *CB* is connected to *C* and to a fixed centre *B*.

Since *CA*, *CB* and *CP* are equal, a semicircle drawn with centre *C* will pass through *A*, *B* and *P*. So that *ABP*, being

Fig. 239.—Oscillating engine model.

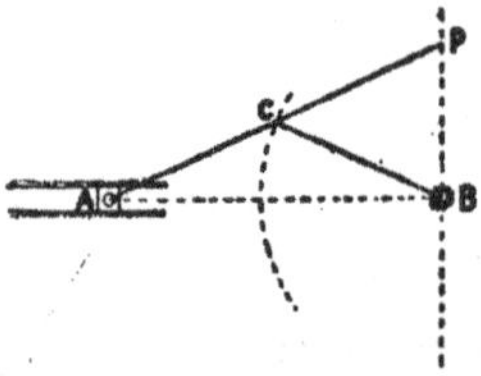

Fig. 240.—Scott-Russell parallel motion.

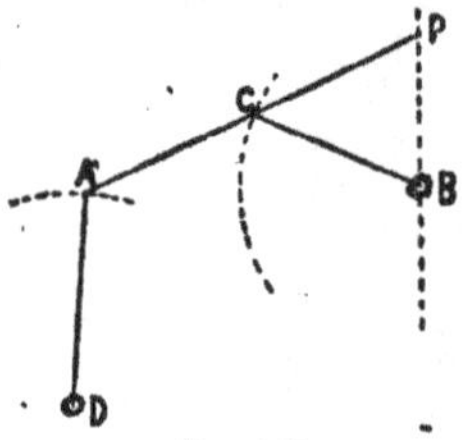

Fig. 241.

the angle in a semicircle, is a right angle. Now this is true whatever be the position of the mechanism, so that *P* always moves in the line *BP* perpendicular to *AB*, i.e. *P has a straight line motion.*

Practically, it is not always convenient to make *A* slide in guides to give it a straight line motion, and with little error to the motion of *P*, *A* may be connected by a link to a fixed centre *D* (Fig. 241), so that *A*'s motion is in the arc of a circle of centre *D*. *A*'s travel is usually small, so that it does not appreciably depart from the straight line *AB*. It will be found now that *P*'s motion is practically straight over a considerable distance above and below *B*.

This mechanism is often used in steam engine indicators, the piston rod being attached to a point between A and C by means of a short link EF, and the tracing pencil being attached at P (Fig. 242). It is usual for this purpose to alter the ratio of the arms AC, CP, CB. The position of P may be found, after settling the lengths of AD, AC, and CB, if this portion be drawn to scale and the instantaneous centre of AC found. This will be I, the point of intersection of DA and BC produced. The link AC is moving for an instant about I, so that a point in it which will be moving in a vertical line will be at the end P of the horizontal radius IP. Thus P is easily found. Or, CP may be found by calculation, thus, A and B are usually at the same height, so that AB is horizontal; the triangles ABC, ICP are therefore similar, so that

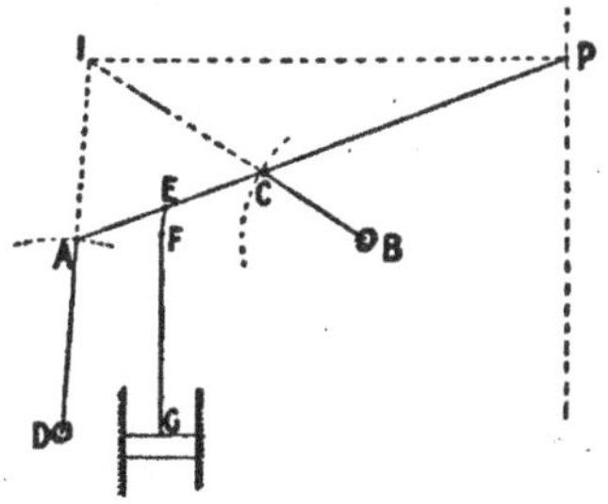

FIG. 242.—Modified Scott-Russell parallel motion applied to an indicator.

$$AC : CB = CP : IC.$$

Now if the angle PAB is small (as it usually is), IC and AC are very nearly equal to one another, so that

$$\overline{CB} \times \overline{CP} = AC^2.$$

From this equation, if CB and AC are settled first, CP may be found, thus fixing the position of P.

The **Thomson Indicator** of Messrs. Schaffer and Budenberg is fitted with this parallel motion, and the **Crosby Indicator** with one slightly modified from this. The latter instrument is shown on page 212, where its mechanism can be easily understood. The Scott-Russell motion has also been used for guiding the top of the piston rod in small engines known as **grasshopper engines**, from their peculiar resemblance to grasshoppers when working.

Watt's parallel motion was used by Watt for guiding the top of the piston rod in his beam engines. Its simplest form consists of two equal links AB, CD (Fig. 243), swinging about centres A and D, their ends B and C being connected by a link. If we first put AB and CD horizontal so that BC is vertical, and then move the links up or down, we see that B deviates to the one side of the vertical and C deviates an equal amount to the other side of the vertical. so that P, the centre of BC, will remain in the vertical. Therefore, for some distance on either side of the mean position of the mechanism, P will remain in a straight line, and this point may be attached to the top of the piston rod so as to guide it in a straight line.

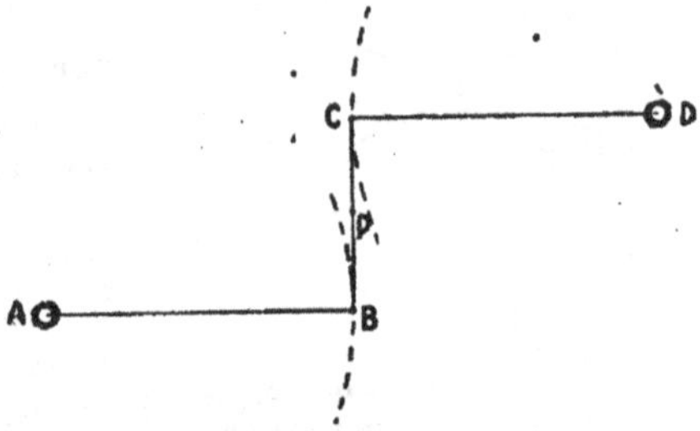

FIG. 243.—Watt's parallel motion: simplest form.

If AB and CD are unequal (Fig. 244), the deviations of B and C will be nearly inversely proportional to AB and CD; of course it is evident that the deviation of B is made greater by

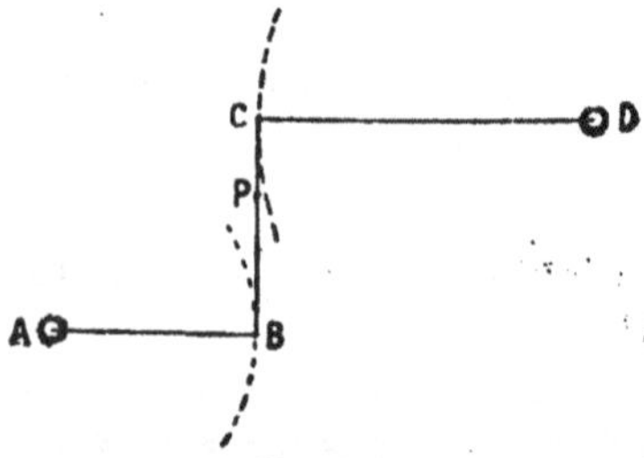

FIG. 244.

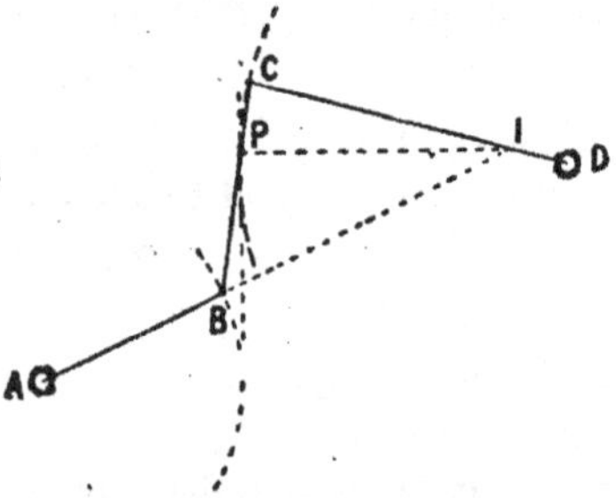

FIG. 245.—Method of finding P, using the instantaneous centre I.

decreasing the radius AB. If, therefore, P is to move in a straight line, BP and PC must be inversely proportional to AB and CD, that is,

$$BP : PC = CD : AB.$$

P may also be found by drawing the mechanism to scale and finding the instantaneous centre of the link BC when some little distance from its mean position. In Fig. 245, I is the point where AB intersects DC, and P will be where a horizontal line from I cuts BC.

This parallel motion is often extended so as to give a second point moving in a straight line. Thus, taking the simplest form of the mechanism (Fig. 246), if *AB* is equal to *CD*, and *BP* equal to *PC*, *P* will move in a straight vertical line. If *DC* be extended to *E*, making *CE* equal to *CD*, and other two bars, *EF* equal to *BC*, and *FB* equal to *EC*, be added, then *F* will also move in a straight vertical line.

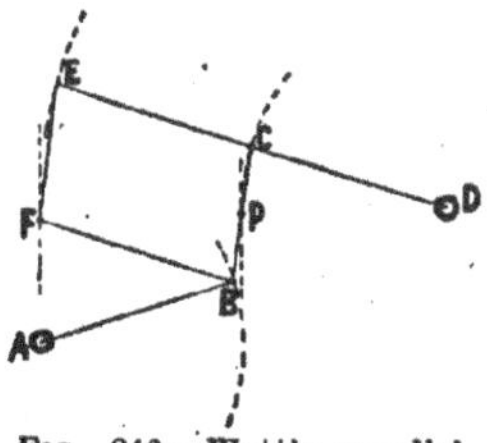

FIG. 246.—Watt's parallel motion; ordinary form.

For *EFBC* will always be a parallelogram in any position of the mechanism, so that *EF* is always parallel to *PC*; and since

$$ED = 2CD$$

and $$EF = 2CP,$$

F, *P* and *D* will always be in one straight line, and *FD* will always be double of *PD*, so that if *P* is moving in a vertical line, *F* will be also moving in a vertical line at twice *P*'s distance from *D*, *i.e.* the vertical through *A*.

FIG. 247.—Model of Watt's parallel motion applied to a beam engine.

This is the arrangement which is commonly adopted for **beam engines**, the piston rod being connected to *F* and the air pump rod to *P*, or in compound beam engines, the high-pressure piston rod to *P* and the low-pressure rod to *F*. A model of this is shown in Fig. 247.

Peaucellier straight line motion.—There are many other straight line motions, mostly approximate only, like the above instances. The **Peaucellier links** give an absolutely straight line, but owing to number of joints it cannot be used for practical purposes. It consists of a jointed parallelogram *CDEF* (Fig. 248), controlled by links *EA*, *DB* and *FB*, to fixed centres *A* and *B*, *EA* being equal to *AB*. *C* moves in a straight line perpendicular to *BA*. In Fig. 249 a model of this mechanism is shown.

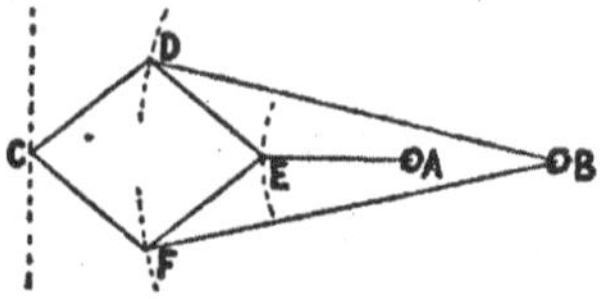

FIG. 248.—The Peaucellier links.

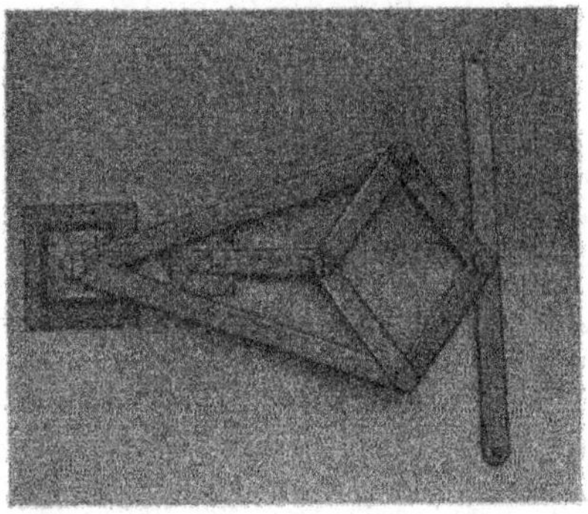
FIG. 249.—Model of Peaucellier straight line motion.

The Eccentric.—**Valves** are generally driven by means of an eccentric (Fig. 250), which consists of a circular disc of centre *B*, having a hole bored through it at centre *C* to receive the shaft. A strap *A*, is passed round the disc, a working fit, so that the disc may rotate inside of the strap. A rod connected to the strap drives the valve.

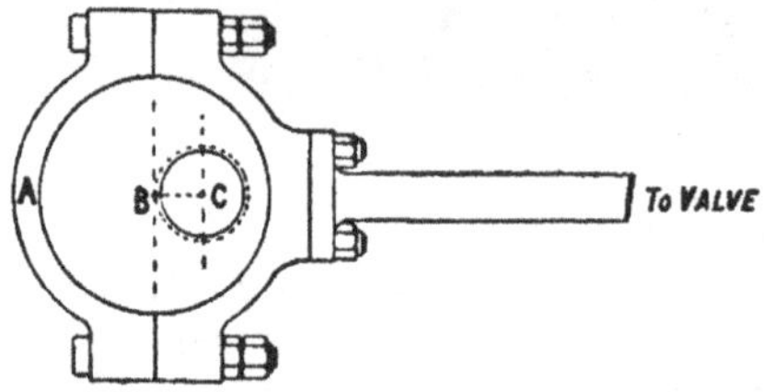

FIG. 250.—Eccentric, used for driving a slide valve.

It will be seen, by inspection of the mechanism, that it is simply equivalent to a crank of radius *BC*, the disc forming the crank pin being so large in diameter that crank cheeks are dispensed with, and the shaft hole can be bored in the crank pin. The motion will be identical to that of the crank and connecting rod; the travel given to the valve will be $2BC$.

Cams are also used for the purpose of obtaining a reciprocating motion from a rotary one. In Fig. 251, a shaped cam BC is mounted on a rotating shaft A. The rim of the cam has projections of any desired form, and, in the example shown, a lever

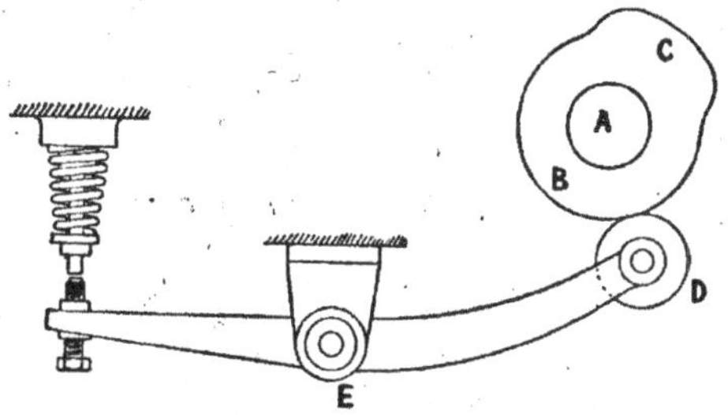

FIG. 251.—Arrangement for operating a valve by means of a cam BC.

pivoted at E carries a small roller, D, at its end bearing on the edge of the cam. The lever will be motionless so long as the roller bears on a circular portion of the edge of the cam, but when the projecting parts of the cam reach the roller, the lever will be operated. Evidently any desired motion may be given to the lever by suitably forming the edge of the cam. Cams are much used for operating the valves of gas and oil engines and can be seen in such cases in their best forms.

EXERCISES ON CHAP. XIII.

1. An engine driving a line of shafting by belts, has a belt pulley 24" diameter, that on the line shafting being 26" diameter. The engine runs at 230 revolutions per minute. A counter-shaft is driven from the line-shaft, the belt running on a pulley 3 feet diameter on the line-shaft and on one 2 feet diameter on the counter-shaft. What are the speeds of revolution of the line-shaft and of the counter-shaft, (*a*) supposing no slip of belt, (*b*) supposing 2 per cent. slip at each belt?

2. A belt, running at a speed of 2000 feet per minute, has a difference of 240 lbs. between the pulls on the tight and slack sides. What horse-power is being transmitted?

3. A line-shaft transmits 4 H.P. to a counter-shaft through a belt running on a pulley 12″ diameter on the line-shaft and one of the same diameter on the counter-shaft. The speed of the line-shaft is 150 revolutions per minute, and it is found that there is a great deal of slip. On the pulleys being replaced by others 24″ diameter each, the slip is much reduced. Explain this, giving actual figures.

4. Give sketches and description of the shafts, pulleys, etc., used in distributing to the machines the power developed by the engine in any engineer's workshop you are acquainted with.

5. Two toothed wheels, mounted on parallel shafts, are to be in gear with one another. Their speeds of rotation are to be in the ratio of 2 : 1. If the distance between the axes of the shafts is 12″, and the pitch of the teeth is to be as nearly 1″ as possible, find the numbers of teeth on each wheel.

6. Give sketches and description of the train of wheels connecting the hour axle with the minute axle in a clock. Give suitable numbers to the teeth.

7. The counter-shaft driving a turning lathe runs at a speed of 180 revolutions per minute. The largest step on the speed cone is 10″ diameter and the smallest 4″ diameter. Each pair of wheels in the back gear have numbers of teeth 15 and 45 respectively. If the belt is running on the smallest step of the countershaft cone, and the back gear is "in," what will be the surface speed, in feet per minute, of a piece of work 7″ diameter?

8. An epicyclic train consists of four wheels. *A* is fixed and has 40 teeth; *B* gears with *A* and has 20 teeth; *C* gears with *B* and has 30 teeth; *D* gears with *C* and has 15 teeth. *B*, *C* and *D* are carried on an arm revolving on the axis of *A*. Find the revolutions of each wheel if the arm is rotated once clockwise.

9. In Question 7, supposing the pitch of the leading screw to be $\frac{1}{2}$ inch, and that it takes its motion from a wheel of 20 teeth on the lathe spindle, give suitable numbers for the wheels driving the leading screw for a feed of $\frac{1}{16}$″.

10. The crank of an engine is 1 foot long, and the connecting rod $3\frac{1}{2}$ feet. Revolutions per minute, 100. Find (*a*) velocity of crank pin centre (supposed uniform), (*b*) average velocity of piston, (*c*) actual velocity of piston at 0°, 30°, 60°, 90°, 120°, 150°, and 180° from inner dead point. Plot these velocities on a time base line.

11. Draw an approximate acceleration diagram for the piston in Question 10.

12. In the parallel motion shown in Fig. 242, $AD = 1\frac{1}{4}''$, $AC = 1\frac{1}{2}''$, $BC = 1''$. *A* and *B* are on the same horizontal line. Find, both by calculation and construction, the length of *CP*, and prove your work by drawing the motion in several positions so as to show the path of *P*.

13. How would you determine the "pitch circles" and the proper "pitch of the teeth" for a pair of spur wheels? What would be

the diameter of the pitch circle of a spur wheel having 80 teeth of $\frac{3}{4}$ inch pitch? Three spur wheels A, B, C are on parallel axes, and are in gear. A has 10 teeth, B has 35 teeth, and C has 55 teeth. How many revolutions upon its axis will be made by A for every 4 revolutions of C? Why is B called an *idle* wheel, and what is its use? (1896.)

14. Describe the construction of the ratchet-brace employed in drilling holes by hand. Explain, with the aid of a sketch, how the drill is revolved always in the same direction, and also how the feed is put on, whilst the handle is worked alternately backwards and forwards. (1896.)

15. Sketch and describe an arrangement of belts and pulleys whereby a reversing motion may be obtained from a single wide pulley running on the main shaft at a constant speed in one direction only. (1898.)

16. A rope transmits 20 horse-power to a rope pulley of 8 feet diameter; draw a section of the rope in its groove. If the pulley makes 100 revolutions per minute, what is the speed of the rope in feet per minute? What is the difference of the tensile forces in the rope on the two sides of the pulley? As it is the difference between the tensile forces in a belt or rope that is important for power, why is it necessary to have any pull on the slack side? (1898.)

17. What are cone or speed pulleys? Describe the use of such pulleys in any machine with which you are acquainted.

The spindle of a wood-turning lathe can, by moving the belt on its cone pulleys, be driven at the rate of 400 revolutions per min. when at its greatest and at 100 revolutions per min. when running at its lowest speed. If the revolutions of the driving shaft are kept constant throughout, and the largest diameter of the speed cones is 20″, what must be the diameters of the smallest steps on the pulleys, the speed pulleys on the two shafts being of the same size? Sketch the pulleys in position. (1899.)

18. Describe and sketch the arrangement of the mechanism by which the saddle of a lathe is traversed by hand along the bed. If the slide rest of a screw-cutting lathe when in gear with the leading screw moves along the bed for a distance of 14″, while the leading screw makes 56 revolutions, what must be the pitch of the thread on the leading screw? (1899.)

19. Sketch and describe the arrangement of mechanism by which the tool of a planing machine is traversed across the slide of the machine at each stroke of the table. (1900.)

CHAPTER XIV.

ACTUAL MECHANICAL ADVANTAGE AND EFFICIENCY. EXPERIMENTS ON SIMPLE MACHINES.

Simple pulley block.—The simplest means we have for raising loads consists of a pulley suspended from an overhead beam, with a rope passing over it (Fig. 252). A load secured to one end of the rope may be raised by pulling on the other end. If there were no losses by friction, stiffness of rope, etc., the greatest load a man could support would be equal to his own weight, and to do this he would have to lift himself off the floor by pulling on the rope. Frictional losses always prevent so great a load from being raised. By attaching two scale pans to the rope ends, *A* and *B*, and placing loads in them, the force *P* required to steadily raise a load *W* may be found. If equal loads are placed in the pans, no movement will result, but if one load is increased until, by slightly pulling the rope so as to start motion it is found to be sufficient to maintain steadily the movement so produced, the weights in the pans will give the value of *P* required for this load *W*. In this arrangement the velocity ratio is clearly 1; the mechanical advantage for any load *W* is equal to $\frac{W}{P}$, which will always be less than 1, as *W* is always less than *P*. The imaginary load *F*, which, if placed in

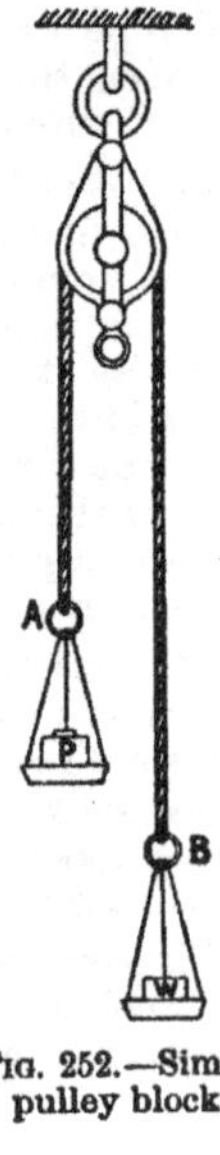

FIG. 252.—Simple pulley block.

the same pan as W, would be equivalent to the frictional resistances of the machine (p. 150), will be equal to $P - W$, for if there were no losses, a force P would raise a load P. The efficiency of the machine (p. 125) will be found by considering W to be raised one foot; P will then descend one foot.

$$\text{Useful work done on } W = W \times 1.$$

$$\text{Energy supplied to effect this result} = P \times 1.$$

$$\therefore \text{ Efficiency} = \frac{W}{P} \times 100 \text{ per cent.}$$

Plan of procedure.—In experimenting with any machine, about 10 experiments should be made with loads increasing by equal steps up to the maximum the machine can safely carry. If scale pans or hooks are used for attaching P and W, the weights of these should be included in the recorded values of P and W. In recording the results, a sketch showing the mechanism clearly should first be inserted and a description of the machine. The calculation for the velocity ratio of the machine should then be given and its result stated. If suitable, the velocity ratio should also be determined by direct measurement at the places where W and P are applied. Weigh the scale pans or hooks and state their weights separately. Oil the parts of the machine requiring lubrication, make sure that everything is running nicely and then make the experiments. Record the results in the form of a table. Curves should then be plotted on squared paper showing the relations of P to W, of F to W, and of the efficiency of the machine to W. From the first two curves equations showing the connection of P and W and of F and W may be found. The third curve will show the value of the efficiency of the machine for any load; as a rule it will be observed that the efficiency rises rapidly when the loads are small, and tends to become constant as the maximum load is approached.

As an example of the method, the following experiment is worked out in full.

Experiment on Pulley Blocks.

Pulleys used: A single pulley at *A* (Fig. 253) and another at *C*, both hung from an overhead beam; a single movable pulley at *B*, from which *W* is suspended.

These pulleys were of galvanised iron, small size, such as are used for ordinary household purposes. A thick cord attached to the pulley *A* at *D* passes down through pulley *B*, up over pulley *A*, then over pulley *C* and a scale pan is attached to its end for applying *P*. Another scale pan attached to pulley *B* serves for applying *W*. The pulley *C* would not be usually employed in practice, its present use is to keep the scale pan for applying *P* clear of the other scale pan and cords.

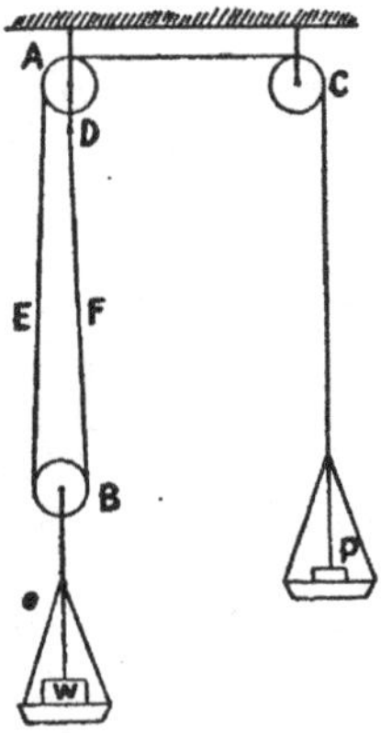

Fig. 253.—Arrangement of pulleys used in the experiment.

Velocity Ratio. If both cords *E* and *F* were raised one foot each, *W* would also be raised one foot. In the actual machine, the cord *F* does not move until after it passes *B*, so that if *E* alone is raised one foot, *W* will ascend half a foot. The pulleys at *A* and *C* merely change the direction of the rope *E*, so that raising *E* one foot causes *P* to descend one foot. The velocity ratio is therefore equal to 2. This was confirmed by actual measurement of the distance moved by *P* when *W* was raised one foot.

We may also obtain the velocity ratio by considering the machine as being free from frictional resistances. In this case, as *W* is suspended by two cords *E* and *F*, the pull in each will be $\frac{1}{2}W$; consequently *P* will be equal to $\frac{1}{2}W$ and the mechanical advantage, neglecting frictional losses, would be equal to 2. In this case (p. 128) the values of the mechanical advantage and velocity ratio are equal, consequently the velocity ratio is 2.

Two sets of experiments were made; I., with the machine unlubricated, after a period of some months' disuse; II., after oiling.

Weight of scale pan for $W = 0·68$ lb.
" " " $P = 0·68$ lb.

I. *Before oiling.*

Actual Load Raised, including Weight of Pan, W lbs.	Force Applied, including Weight of Pan, P lbs.	Calculated Load, if no Friction, W_1 lbs. $=2P$.	Effect of Friction, $W_1 - W = F$ lbs.	Mechanical Advantage, $\frac{W}{P}$.	Efficiency, $\frac{W}{2P} \times 100$ per cent.
0·68	1·14	2·28	1·6	0·6	29·8
1·68	1·88	3·76	2·08	0·89	44·7
2·68	2·68	5·36	2·68	1·0	50·0
4·68	4·42	8·84	4·16	1·06	53·0
6·68	6·02	12·04	5·36	1·11	55·5
8·68	7·42	14·84	6·76	1·17	58·5
10·68	9·12	18·24	7·56	1·17	58·5
12·68	10·72	21·44	8·76	1·18	59·0
14·68	12·22	24·44	9·76	1·2	60·0
16·68	13·72	27·44	10·76	1·21	60·7

II. *After oiling.*

0·68	0·92	1·84	1·16	0·74	37·0
1·68	1·52	3·04	1·36	1·13	56·3
2·68	2·22	4·44	1·76	1·21	60·4
4·68	3·42	6·84	2·16	1·37	68·5
6·68	4·72	9·44	2·76	1·42	70·9
8·68	5·92	11·84	3·16	1·47	73·3
10·68	7·22	14·44	3·76	1·48	73·9
12·68	8·52	17·04	4·36	1·49	74·4
14·68	9·72	19·44	4·76	1·51	75·5
16·68	11·2	22·4	5·72	1·49	74·5

These results show an increased efficiency, due to oiling, of about 15 per cent. with the highest loads.

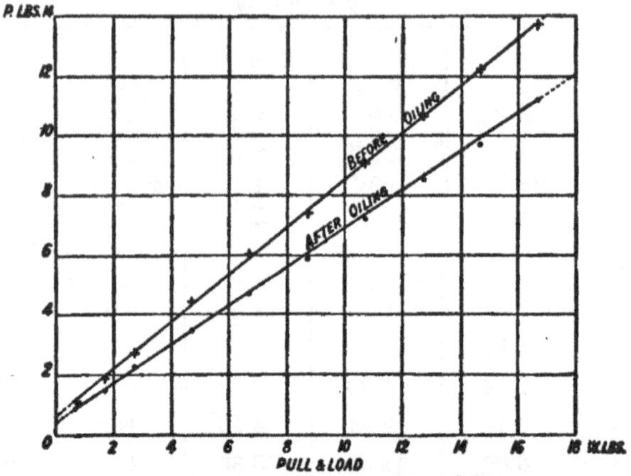

FIG. 254.—Plotted values of P and W.

On plotting the values of P and W (Fig. 254), also of F and W (Fig. 255), the points are found to lie nearly on a straight line. This line is found by stretching a thread on the paper and shifting it about until the plotted points are well divided on either side of it. A mark is then made under each end of the thread and the line drawn through it.

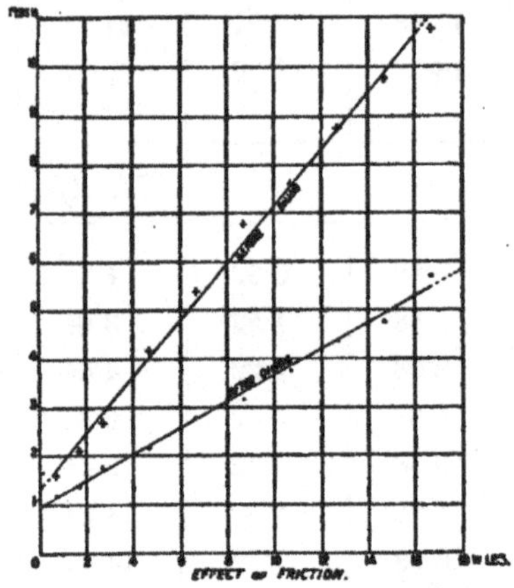

FIG. 255.—Plotted values of F and W.

Equations for the machine.—In all cases where a straight line is given when the results are plotted, the relation between the results can be expressed by an equation like

$$P=aW+b, \qquad (1)$$

where P and W are the observed quantities and a and b are constants.

If the constants are found and inserted in the equation, then this equation may be used for determining the value of P for any value of W. These constants are found from the straight line on the diagram thus : Taking the diagram showing P and W before oiling, we see from it that

$P = 2{\cdot}2$ lbs. when $W = 2$ lbs., and
$P = 13{\cdot}3$ lbs. when $W = 16$ lbs.

Fill in these values in equation (1), giving two simultaneous equations from which a and b will be found. Thus :

$$2{\cdot}2 = (a \times 2) + b. \quad \text{.......................(2)}$$
$$13{\cdot}3 = (a \times 16) + b. \quad \text{.......................(3)}$$

Solving these, we find $a = 0{\cdot}79$,
$b = 0{\cdot}62$;

so that $P = 0{\cdot}79\,W + 0{\cdot}62$ ....(4)

gives the required equation showing the connection between P and W, with the machine unlubricated.

Taking now the diagram for P and W after oiling, we see that $P = 1{\cdot}7$ lbs. when $W = 2$ lbs, and
$P = 10{\cdot}7$ lbs. when $W = 16$ lbs.

These values inserted in (1) give $1{\cdot}7 = (a \times 2) + b$..............(5)
$10{\cdot}7 = (a \times 16) + b$. ....(6)

Solving these gives $a = 0{\cdot}64$,
$b = 0{\cdot}41$;

so that $P = 0{\cdot}64\,W + 0{\cdot}41$ ....(7)

gives the relation of P and W after oiling the machine.

Taking now the diagrams showing the relation of F and W, we see that

Before Oiling.	After Oiling.
$F = 1{\cdot}3$ lbs. when $W = 0$.	$F = 0{\cdot}95$ lb. when $W = 0$.
$F = 10{\cdot}62$ lbs. when $W = 16$ lbs.	$F = 5{\cdot}24$ lbs. when $W = 16$ lbs.
$F = a'W + b'$, ...(1)	$F = a'W + b'$, ...(1)
giving $1{\cdot}3 = 0 + b'$, ...(2)	giving $0{\cdot}95 = 0 + b'$, ...(2)
$10{\cdot}62 = (a' \times 16) + b'$, (3)	$5{\cdot}24 = (a' \times 16) + b'$, (3)
and from (2) and (3)	and from (2) and (3)
$a' = 0{\cdot}58$,	$a' = 0{\cdot}27$,
$b' = 1{\cdot}3$;	$b' = 0{\cdot}95$;
$\therefore\ F = 0{\cdot}58\,W + 1{\cdot}3$. (4)	$\therefore\ F = 0{\cdot}27\,W + 0{\cdot}95$.(4)

Tabulating these equations :

Before Oiling.	After Oiling.
$P=0{\cdot}79\,W+0{\cdot}62.$ $F=0{\cdot}58\,W+1{\cdot}3.$	$P=0{\cdot}64\,W+0{\cdot}41.$ $F=0{\cdot}27\,W+0{\cdot}95.$

The unit used for P, F and W is the pound weight. These equations give all the required information about the machine under the two given conditions.

It should be noted that as the efficiency curve (Fig. 256) is not a straight line, an equation like those given above cannot be

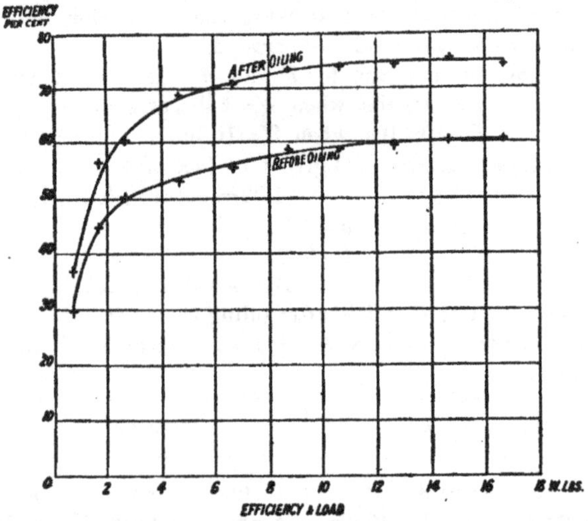

FIG. 256.—Curves of efficiency and load.

applied. Its equation is not so simple, and may be neglected meanwhile by the student.

The above should be taken as showing the method which must be employed in experimenting upon and working out the results for all the simple machines described here. It may be observed, in the example taken, that the pan and weights placed in it are not the only loads raised. The bottom pulley B is also raised, and part of the rope changes its height also. These loads need not be taken into account, as what we require to know practically is not what parts of the machine itself can be raised by a given force, but what useful load may be raised. Special cases, however, may sometimes arise in which the weights of parts belonging to the machine must be considered.

Another arrangement of pulley blocks.—Keeping the arrangement of ropes the same as in Fig. 253, a higher velocity ratio may be obtained by increasing the number of sheaves, or wheels, in each block. In Fig. 257 there are three sheaves in the upper block and two in the lower. W is now supported by 5 ropes between the upper and lower block, consequently the velocity ratio is 5. Usually the sheaves belonging to each block are mounted side by side on the same spindle.

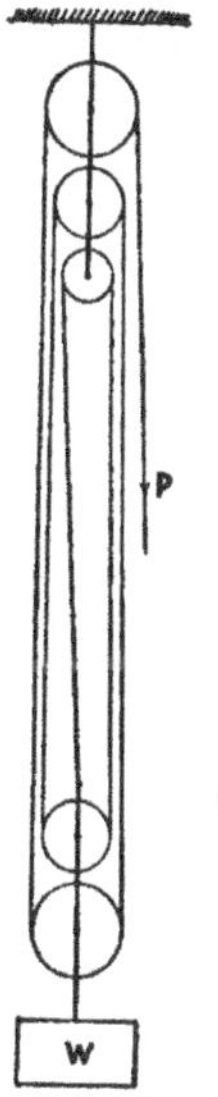

FIG. 257.—Pulley blocks.

Let W lbs. = load actually raised.

P lbs. = pulling force.

Then, if there were no frictional losses, and if the lower block and the rope had no weight, the useful load raised would be

$$W_1 = 5 \times P.$$

Therefore losses in machine $= 5P - W = F$, as before. The mechanical advantage will be $\frac{W}{P}$, and the efficiency $\frac{W}{5P} \times 100$ per cent. It will be found, on experiments being made on a set of actual blocks like this, that the efficiency is not so great as with those first given, even when well oiled. The stiffness of the rope passing so many times around the sheaves accounts for a large loss.

Weston's blocks are much used for hoisting loads. The top block (Fig. 258) contains two sheaves of different diameters, cast together so as to form one piece, the rims of each being formed so as to receive the links of the hoisting chain and to prevent it slipping. The bottom block contains one sheave only. The chain is endless and passes first round the larger top sheave, then round the bottom sheave, then round the top smaller sheave, the two ends being connected and allowed to hang loose. In raising a load, the loose chain passing from the larger top sheave is pulled; in lowering, the other is pulled.

The arrangement may be studied either by use of the principle of moments, or by calculating the velocity ratio direct.

1st, Let R = rad. of large sheave = CF

r = „ small „ = CE.

Since W is sustained by the two chains A and B (Fig. 259) the pull in each (friction being neglected) will be $\frac{1}{2}W$. If now we consider DF as a lever with fulcrum at C, we see there are three forces acting on it, at D, E and F, tending to turn it about C. For balance,

clockwise moments = anticlockwise moments,

$$(P \times CF) + (\tfrac{1}{2}W \times CE) = \tfrac{1}{2}W \times DC,$$

or,

$$(P \times R) + (\tfrac{1}{2}W \times r) = \tfrac{1}{2}W \times R$$

$$P \times R = (\tfrac{1}{2}W \times R) - (\tfrac{1}{2}W \times r)$$

$$= \tfrac{1}{2}W(R - r)$$

$$P = \tfrac{1}{2}W\left(\frac{R-r}{R}\right)$$

or,

$$\frac{W}{P} = \frac{2R}{R-r},$$

which gives the mechanical advantage with no friction. We see from this equation that the smaller the difference between R and r the greater will be this mechanical advantage, and consequently the velocity ratio which is equal to this numerically.

2nd. The same result may be obtained by considering the velocity ratio. Thus, let the upper sheaves turn round once clockwise; to do this P must

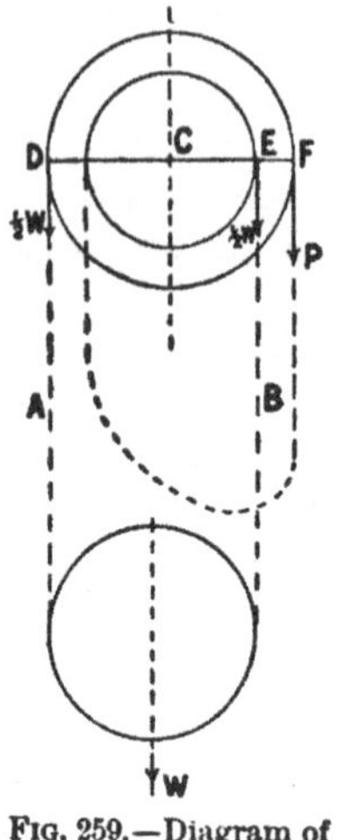

FIG. 259.—Diagram of Weston's blocks.

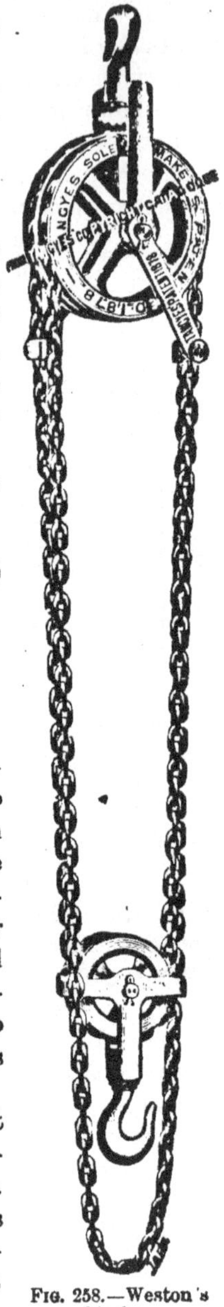

FIG. 258.—Weston's blocks

travel through a distance equal to the circumference of the larger sheave, *i.e.* $2\pi R$. A will be raised a distance equal to the circumference of the larger sheave $2\pi R$, and B will be lowered a distance $2\pi r$. So that if B were kept fixed, A would be raised a distance $(2\pi R - 2\pi r)$, and W would be raised one half of this $(\pi R - \pi r)$.

So that, distance travelled by P : distance travelled by $W = 2\pi R : (\pi R - \pi r)$;

$$\therefore \text{ velocity ratio} = \frac{2\pi R}{\pi(R-r)}$$

$$= \frac{2R}{R-r},$$

which is the same result as that found by the previous method.

Running down of the load.—Notice the question of the balance of the lever DF above, when both chains are hanging loose.

$$\text{Clockwise moments} = \tfrac{1}{2} W \times r.$$
$$\text{Anticlockwise moments} = \tfrac{1}{2} W \times R.$$

There is therefore an anticlockwise moment equal to $\frac{1}{2} W(R-r)$ tending to turn the sheaves about C and so cause the load W to run down. To prevent this, we have to depend, in this tackle, on the friction of the spindles of the sheaves in the upper and lower blocks, which must supply a clockwise moment equal to $\frac{1}{2} W(R-r)$. If, therefore, we want to raise a load easily by this tackle, we must grease the bearings of the spindles, but not too freely, as we may thereby reduce the friction so much as to cause the load to run down, that is, the blocks cease to be self-sustaining.

It can be shown that in any machine in which the removal of the pulling force does nothing to alter the magnitude of the frictional resistances, the suspended load will not run down if the efficiency of the machine is less than 50 per cent. If, however, the removal of the force causes alterations in the frictional resistances, no general statement true for all machines can be made, each case must be investigated separately.

Running down of the load when P is removed, may be prevented by fitting **a pawl and ratchet wheel** to the machine (Fig. 260). When the pawl engages the teeth, motion can only take place in the direction necessary to raise the load. When lowering the load, the pawl is raised by hand, thereby permitting the required motion to be effected.

The wheel and differential axle (Fig. 261) consists of a drum *BC*, divided into two portions having different diameters, and

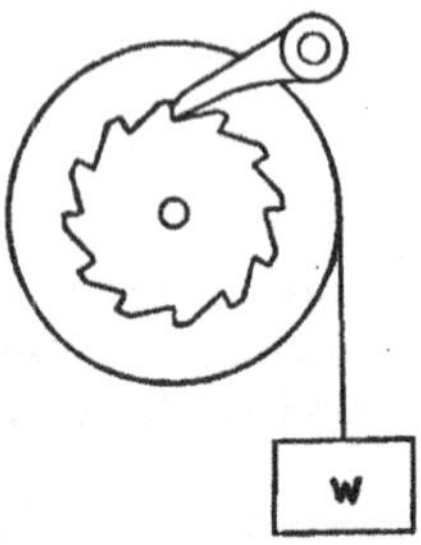

FIG. 260.—Pawl and ratchet wheel.

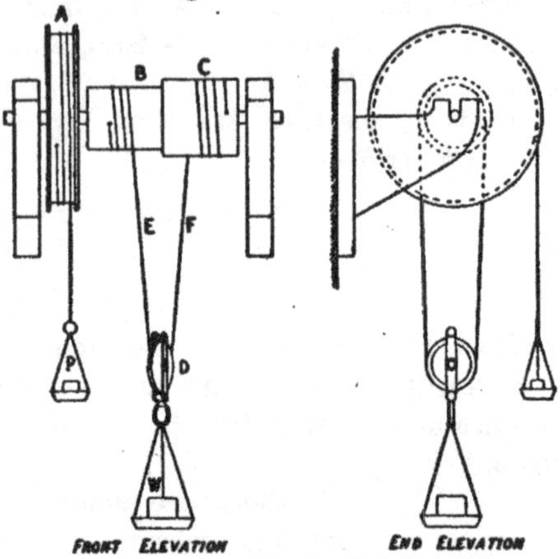

FIG. 261.—Wheel and differential axle.

mounted on the same spindle as a wheel *A*. A rope is attached to *B*, and, after a few turns round *B*, is led down through a pulley *D*, then up and turned round *C* in the opposite direction to the winding on *B*, its end being made fast to *C*. A cord made fast to *A*, passes two or three times round *A* and carries a load *P* for working the machine. The load to be raised, *W*, is carried by the pulley *D*.

Suppose the drum to rotate once. *P* will be lowered a distance equal to the circumference of the circle in which the cord sustaining *P* is wrapped. A point on the cord *E* will be lowered, and one on the cord *F* raised, the amounts in each case being equal to the circum ference at the drum of the cord circle. It will be noticed that for all three cords, the circumference must be measured, not of the drum, but of the circle at the cord centre; for the inside of the cord, in contact with the drum (Fig. 262), will have a shorter

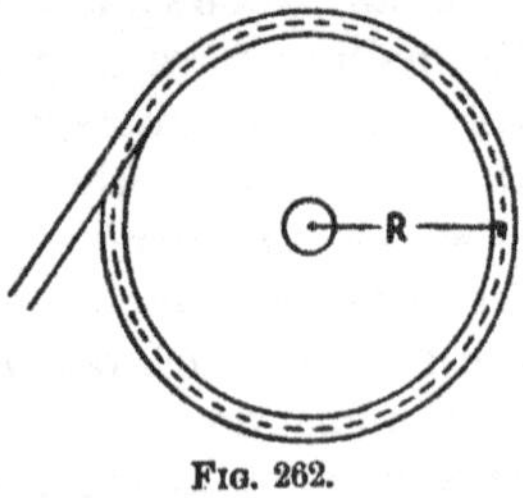

FIG. 262.

circumference than that of the outside part of the cord, and the actual distance moved by the cord will be the mean of these two, that is, the circumference of the cord centre.

To find the velocity ratio, let R_A, R_B, R_C be the radii of A, B and C, in each case measured to the centre of the cord. Let the drum rotate once, then

$$\text{Distance moved down by } P = 2\pi R_A$$
$$\text{〃 〃 down by } E = 2\pi R_B$$
$$\text{〃 〃 up by } F = 2\pi R_C.$$

Consequently, as in the Weston's Blocks, W will be raised a distance equal to $\frac{1}{2}(2\pi R_C - 2\pi R_B)$;

$$\therefore \text{ Velocity ratio} = \frac{2\pi R_A}{\frac{1}{2}(2\pi R_C - 2\pi R_B)}$$
$$= \frac{2R_A}{R_C - R_B}$$

The actual mechanical advantage will be found by experiment. From the results the effect of friction and the efficiency will be calculated as already explained.

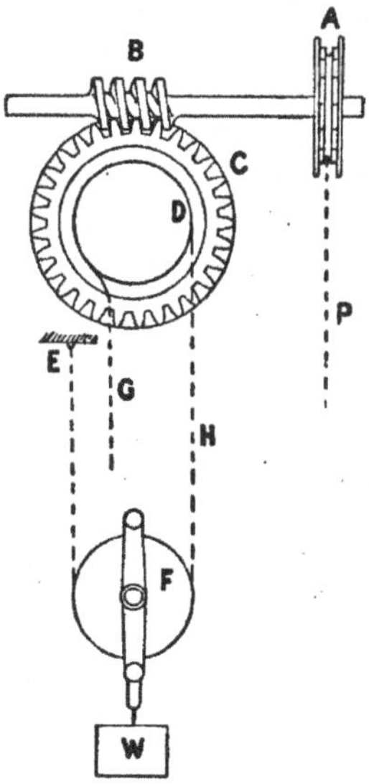

FIG. 263.—Diagram of helical blocks.

Helical blocks are often used on account of their self-sustaining qualities. The upper block consists of a chain wheel A (Fig. 263) carrying an endless chain for working the blocks; a worm B is fixed to the same spindle as A, and is driven by A; this worm drives a worm wheel C and also a chain wheel D, which is keyed on the same spindle as C. A chain is secured to the upper block at E, passes down through F, then up over D, its end hanging loose at G. The rim of D is shaped to take the links of the chain without any chance of slipping occurring, and the chain is guided on and off D. On turning A by pulling the endless chain, the wheel D either raises or lowers the chain H and so raises or lowers the load W.

To calculate the velocity ratio:—Let W ascend a distance equal to the *length* of the number of links D can take round its complete circumference. Call this length h. Then, as the chain is fixed at E, the wheel D must rotate twice. Now, as one revolution of the worm passes on one tooth of the worm wheel, therefore for two revolutions of D to occur, the revolutions of the worm must be equal to twice the number of teeth on the worm wheel. Call the number of teeth N. Then the worm, and consequently A, rotate $2N$ times in raising W through a height h. Let the *length* of the number of chain links A can take round its complete circumference be called L; then, if A rotates once, P will be lowered a distance L, and for $2N$ revolutions of A, a distance $2NL$. The velocity ratio is therefore equal to $\frac{2NL}{h}$.

The crab is often used along with arrangements of pulley blocks for raising weights. The rope from the pulleys is wound round a barrel, having a toothed wheel secured to its spindle. A pinion, secured to another spindle parallel to the barrel, gears with the toothed wheel, so that if this second spindle is turned by means of handles, a considerable velocity ratio is obtained. This arrangement is said to be **single geared.** In **double geared** crabs, an additional toothed wheel and pinion are introduced on another spindle, so that a much greater velocity ratio is obtained. The double geared crab is usually so arranged that it can be rapidly converted to single geared; light loads can then be raised more quickly than by use of the double gear. A band brake is fitted to the drum spindle for use while lowering the load, and a pawl and ratchet

FIG. 264.—Crab, arranged for performing experiments.

wheel on the same spindle prevent the load running down if the handles are released.

Fig. 264 shows such a double geared crab arranged so that experiments may be performed. The handles have been taken off and a wheel, having a rim grooved to receive a cord, substituted. The pull required to work the machine is supplied by weights placed in a scale pan at the end of this cord, the cord being led over a pulley secured overhead, so as to obtain a considerable vertical travel for the scale pan. Fig. 265 shows a diagram of the mechanism from which the velocity ratio may be calculated.

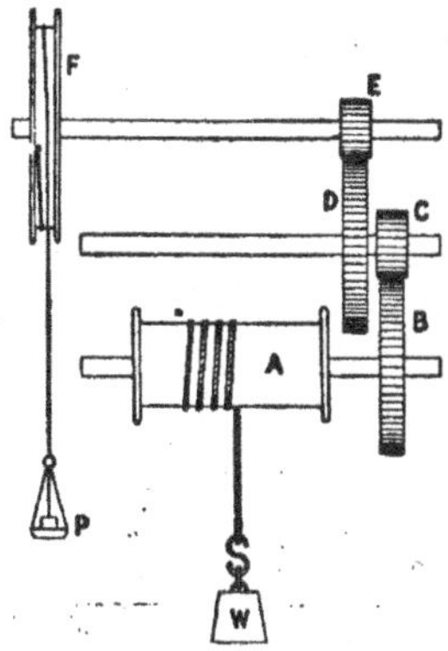

FIG. 265.—Diagram of the mechanism of the crab.

Let d_1 = the diameter to the centre of the rope on the barrel A; d_2 = the diameter to the centre of the cord on the wheel F.

Let B, C, D, E represent the numbers of teeth on the wheels as shown.

Then, if A revolves once, F will revolve $\frac{B \times D}{C \times E}$ times, and W will be raised a distance equal to πd_1. In one revolution of F, P will be lowered a distance equal to πd_2; therefore, for one revolution of the barrel A, P will be lowered a distance $\frac{B \times D}{C \times E} \times \pi d_2$. Consequently the velocity ratio will be

$$V = \frac{\frac{B \times D}{C \times E} \times \pi d_2}{\pi d_1}$$

$$= \frac{B \times D \times d_2}{C \times E \times d_1}.$$

Screw jacks are used for heavy loads requiring a small lift only. A hollow case A (Fig. 266) has a hole at its top screwed to receive a strong square threaded screw B. The load is applied at the top of this screw, on C, which is a piece free to rotate on the top of B. B is turned by a *tommy-bar* inserted into holes in the screw head as shown, and as C is free to rotate on B, the load is not turned by the rotation of the screw.

To obtain the velocity ratio :

Let R = radius at which P is applied, inches.

p = pitch of screw, inches.

Then, in one revolution of the screw, P moves a distance tangentially equal to $2\pi R$, and W moves a distance equal to p. Therefore the velocity ratio is

$$V = \frac{2\pi R}{p}.$$

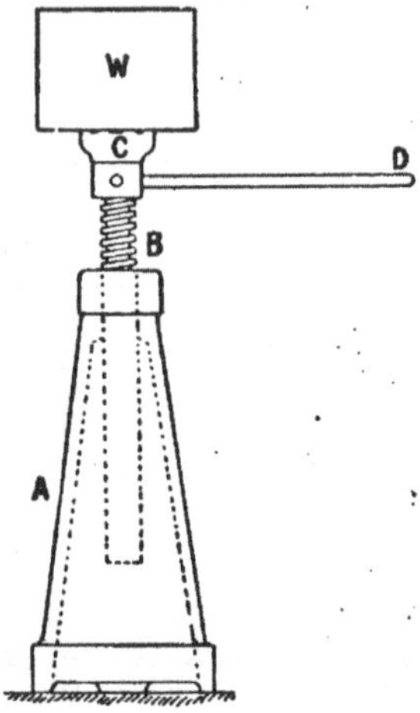

FIG. 266.—Screw jack.

By substituting a wheel with a grooved rim for the tommy bar, and attaching a cord to it, led over a pulley, a load may be hung on equivalent to P, and experiments may be made to find the mechanical advantage, effect of friction and efficiency, in the same manner as for the other machines.

EXERCISES ON CHAP. XIV.

1. In a Weston's Block the large pulley is 6″ diameter and the small one 5″ diameter, measured in each case to the chain centre. Neglect friction and find what pull will be required to raise ¼ ton.

2. A single geared crab has handles 15″ long and barrel 6″ diameter to the rope centre. The pinion has 20 teeth and the spur wheel 60 teeth. Give an outline sketch and calculate the velocity ratio.

3. A screw, 1″ pitch, is used to raise a load of 5 tons. The screw is turned by means of a toothed wheel 15″ effective radius. Calculate the pressure required on the teeth tangential to the pitch circle, supposing 50 per cent. to be lost in friction.

4. In a wheel and differential axle, the wheel is 24″ diameter and the drum has diameters of 7″ and 6″ respectively. Calculate the velocity ratio.

5. A worm and wormwheel are used for applying the twist to a piece of material under torsion test. The worm is turned by a handwheel, and the test piece is connected to the shaft on which the wormwheel is mounted. If the wormwheel has 90 teeth, how many

degrees of twist will be given to the test piece by turning the hand-wheel through 235 revolutions?

6. A lifting tackle is formed of two blocks, each weighing 15 lbs.; the lower block is a single movable pulley, and the upper or fixed block has two sheaves. The cords are vertical and the fast end is attached to the movable block. Sketch the arrangement and determine what pull on the cord will support 200 lbs. hung from the movable block, and also what will then be the pressure on the point of support of the upper block. (1896.)

7. Describe either a screw jack (pitch of screw ½″, handle 19″ long) or a simple winch for lifting weights up to 1 ton by one man. What is the mechanical advantage neglecting friction? Describe what sort of trial you would make to find its real mechanical advantage under various loads, and what sort of result would you expect to find? (1897.)

8. Describe any machine, workable by hand, for lifting weights. Give the rule for its velocity ratio. When is its velocity ratio the same as its mechanical advantage? Describe carefully how you would make tests to determine its real mechanical advantage under various loads. (1898.)

9. A machine is concealed from sight except that there are two vertical ropes; when one of these is pulled downwards the other rises. How would you find the efficiency of this lifting machine? What do we mean by the velocity ratio, and by the mechanical advantage? (1900.)

CHAPTER XV.

INDICATED AND BRAKE HORSE-POWER. ABSORPTION AND TRANSMISSION DYNAMOMETERS. FLY-WHEELS. STEADINESS OF MACHINES. MOMENTUM. IMPACT. FORCE OF BLOW. CENTIFRUGAL PUMP.

The indicator.—The student has now, from the preceding chapters, a fair idea of how energy supplied and energy delivered are measured in the case of certain machines, and what sort of results may be expected. The methods by which the same thing is done in cases of machines running continuously, and where considerable quantities of energy are being dealt with, should now be considered.

In ordinary engines with pistons reciprocating in cylinders, the energy delivered to the engine is measured from the diagram of work done on the piston. These diagrams are drawn by means of an instrument called an **indicator.** The essential parts of an indicator consist of a small cylinder which can be connected with the cylinder of the engine by means of a *cock*. A piston moving in this small cylinder is controlled by a *spring*. The elastic properties of this spring cause the piston to take up a definite position in its cylinder depending on the pressure exerted on it. The movement of the piston is communicated by *linkwork* to a *pencil*, the linkwork being so arranged that a straight line motion is given to the pencil. The pencil takes up a position corresponding with the pressure in the cylinder. A paper, stretched usually over a *drum*, is caused to move transversely under the pencil, being driven to and fro by being connected to some part of the engine so as to give a faithful copy of the motion of the piston of the engine to a reduced scale. When the pencil is pressed on the paper, the indicator being connected to one end of the engine cylinder, it will now trace a curve showing the pressure on the piston at any part of

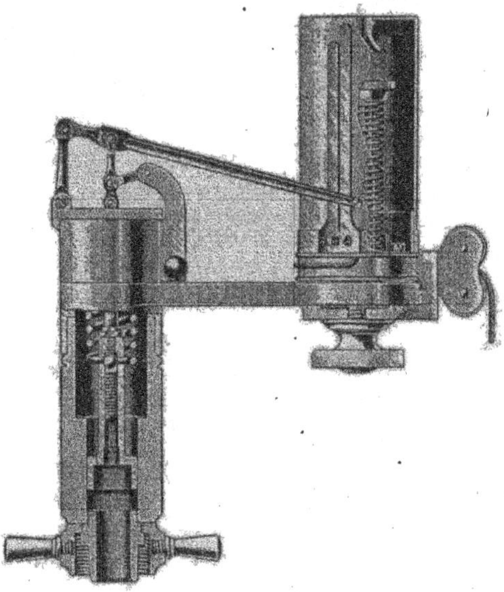

FIG. 267.—Crosby Gas Engine Indicator.

the double stroke of the engine. From this curve the average pressure on the piston may be found. A *datum line*, showing atmospheric pressure, is traced on the paper by putting the indicator cylinder in communication with the atmosphere. A small side hole in the communication cock enables this to be done. Fig. 267 shows the Crosby Gas Engine Indicator, and explains the construction of the instrument clearly.

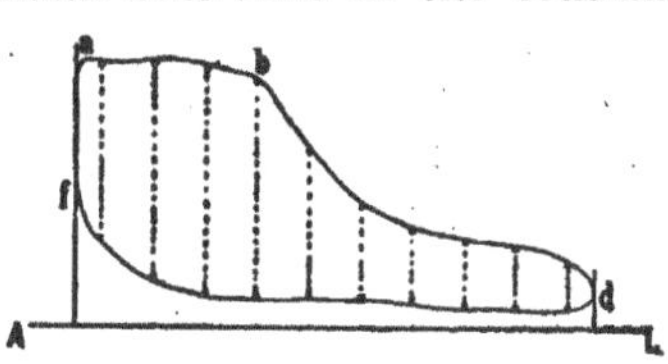

FIG. 268.—Indicator diagram for end *A* of the cylinder in Fig. 270.

Calculation of I.H.P. from the indicator diagram.—This diagram (Fig. 268) represents what the indicator might draw if connected to a steam engine cylinder. *AL* is the datum atmospheric line. *ab* shows the admission of steam at high pressure to the engine cylinder. At *b* the steam is cut off and expands during the remainder of the stroke, falling in pressure as it does so. *df* shows the back pressure on the piston during the return stroke. The average breadth of the diagram is usually found by dividing its length into ten equal parts and measuring its breadths at the centres of these parts. The sum of these breadths divided by 10 will give the average breadth. Suppose this gives 1·25″ as the average breadth. The strength of the indicator spring being known, the height on the diagram corresponding to a given pressure is known. Thus, supposing in the given case a spring had been used of such strength that 30 lbs. per square inch pressure is represented by a height of 1″ on the paper, then the average pressure on the piston will be

$$30 \times 1{\cdot}25 = 37{\cdot}5 \text{ lbs. per square inch.}$$

The average pressure on the other side of the engine piston will be found, in the same way, from the diagram drawn by the indicator when connected to the other end of the cylinder (Fig. 269).

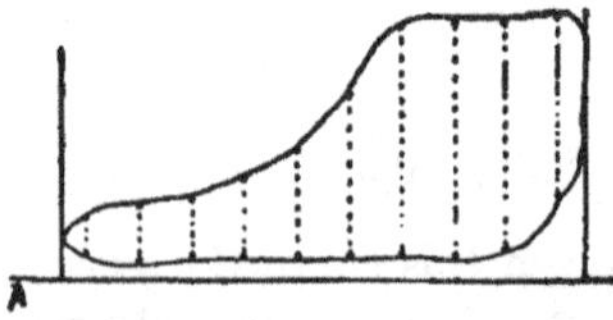

FIG. 269.—Indicator diagram for end *B* of the cylinder in Fig. 270.

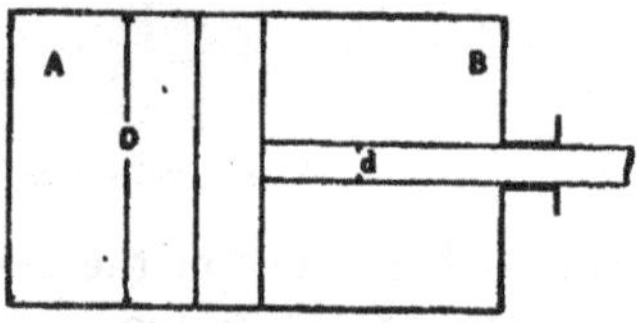

FIG. 270.—Diagram of steam engine cylinder.

Having these average pressures, the work done per stroke may be easily found. In Fig. 270,

Let p_A = average pressure for end *A* of cylinder, lbs. per sq. inch.

p_B = " " " *B* " " "

Total pressure on piston on side $A = p_A \times$ area of piston

$$= p_A \times \frac{\pi D^2}{4} = P_A \text{ lbs.}$$

Total pressure on piston on side $B = p_B \times$ (area of piston − area of piston rod)

$$= p_B \times \left(\frac{\pi D^2}{4} - \frac{\pi d^2}{4}\right)$$

$$= p_B \times \frac{\pi}{4}(D^2 - d^2) = P_B \text{ lbs.}$$

Let N = revolutions of engine per minute.

L = length of stroke in feet.

Work per stroke for side $A = P_A \times L$ ft.-lbs.

" " minute " " $= P_A \times L \times N$ ft.-lbs.

and Horse-power $= HP_A = \dfrac{P_A \times L \times N}{33{,}000}$ for side A.

and In same way $HP_B = \dfrac{P_B \times L \times N}{33{,}000}$ for side B.

The total H.P., called the **Indicated Horse-Power**, will be the sum of these,

$$\text{I.H.P.} = \frac{P_A \times L \times N}{33{,}000} + \frac{P_B \times L \times N}{33{,}000}$$

$$= \frac{L \times N}{33{,}000}(P_A + P_B).$$

Brake horse-power.—The quantity $L \times N(P_A + P_B)$ measures the whole energy in foot-pounds given to the piston in one minute. It is necessary to know not only this, but also **the energy which the engine can deliver per minute**, as from these two quantities, the energy lost in overcoming frictional resistances in the engine, and the mechanical efficiency of the engine can be calculated. If the engine is not too large, the energy delivered can be conveniently obtained by putting a **brake** on the flywheel. The simplest method is to pass a double rope round the wheel, held in position by means of loosely fitting wood blocks secured to the rope. One end of the rope is attached to a spring balance, the other end carries a load W. The direction of rotation of the wheel being as shown in Fig. 271, it will be noticed that P is helping to turn the wheel and W is opposing its rotation. The friction of the ropes on the rim communicates these forces to the wheel.

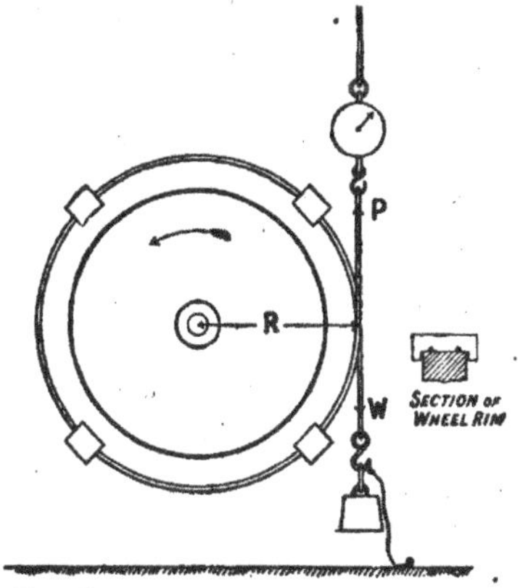

FIG. 271.—Arrangement of rope brake.

Let R = radius, in feet, to the centre of the rope, then

$$WR - PR = R(W - P)$$

will be the net opposing moment in pound-feet, if W and P are in pounds.

If now the wheel rotates once, the work done against this moment will be

$$2\pi . R(W - P) \text{ ft.-lbs.},$$

and for N revolutions per minute,

$$\text{Work per minute} = 2\pi . R . N(W - P) \text{ ft. lbs.},$$

and

$$\text{H.P.} = \frac{2\pi RN(W - P)}{33,000}.$$

This is called the **Brake Horse-Power** of the engine, written B.H.P.

Mechanical efficiency.—If the B.H.P. and I.H.P. are known, the **mechanical efficiency of the engine** may be calculated. For the I.H.P. may be taken as a measure of the energy delivered to the piston per minute and the B.H.P. as a measure of the energy produced by the engine in the same time. So that

$$\text{Mechanical efficiency} = \frac{\text{B.H.P.}}{\text{I.H.P.}} \times 100, \text{ per cent.}$$

Prony brake.—Brakes of the kind described above are called **Absorption Dynamometers.** It will be noticed, in the one described, that we depend on the pulls given to the ends of the rope

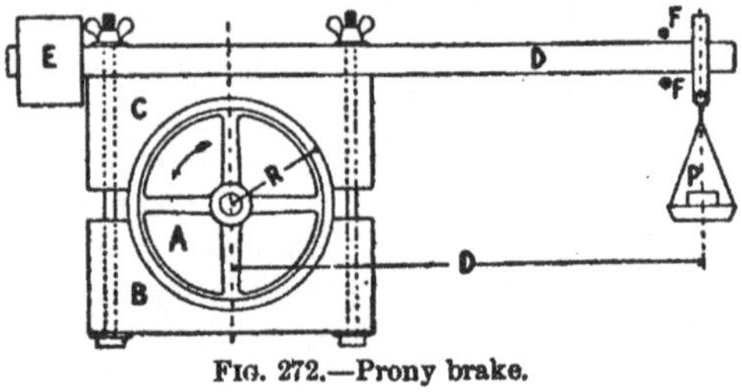

FIG. 272.—Prony brake.

for the production of frictional resistance to motion, and that these pulls bear a definite ratio to one another in any given case. If a brake like this is used for absorbing large powers, these pulls may have to be very great and make the use of a large weight necessary on the lower end of the rope. To obviate this a **Prony brake** may be used. The principle of this brake is shown in Fig. 272. A is the wheel driven by the source of power, and has two brake blocks *B* and *C* fitted to it. These blocks can be made to grip as tightly as may be required on the wheel *A* by tightening two thumbscrews on the bolts holding the blocks together. A lever *D*, balanced by a counterpoise at *E*, carries a load *P* lbs. in a scale pan at its end, and this load gives a measure of the power. The lever *D* has a limited freedom of movement between two fixed stops *F*, *F*. Let *R* be the radius of the wheel *A*, and *D* the horizontal distance between the centre of *A* and the place where *P* is applied, both expressed in feet. Let F=the total frictional forces generated all round the rim of the wheel. Then, taking moments about *A*, and remembering that the brake blocks and lever are balanced,

$$F \times R = P \times D,$$

or

$$F = \frac{P \times D}{R} \text{ lbs.}$$

Work done against F in one revolution of $A = F \times 2\pi R$ ft.-lbs., so that if N=revolutions of *A* per minute,

$$\text{B.H.P.} = \frac{F \times 2\pi R \times N}{33{,}000}$$

$$= \frac{P \times 2\pi D \times N}{33{,}000}.$$

Heating of the brake.—In absorption dynamometers the energy produced by the engine is absorbed by the frictional resistances of the brake, and is transformed into heat. It is therefore necessary to keep the flywheel cool by lubrication with soapy water, this being assisted by the air draught produced by the rotation of the wheel. Sometimes flywheels have their rims made of a channel section so as to receive a stream of water, which being whirled round by the wheel, retains its position in the rim in the same way as a whirled stone at the end of a string keeps its circular path. The water is kept continually flowing into the rim and is drained away by a sharp-

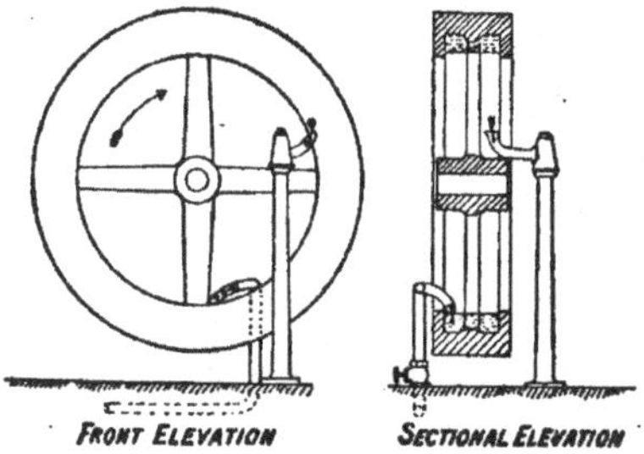

FIG. 273.—Arrangement of water-cooled flywheel.

edged scoop on the other side, and therefore keeps the rim cool. The arrangement is shown in Fig. 273.

Transmission Dynamometers are used for measuring the power delivered to a machine. They receive energy from a moving belt or otherwise, measure it, and deliver it to the machine with as little loss as possible. They usually work by measuring the torque transmitted. Let two shafts, both in the same straight line, have parallel discs mounted as shown in Fig. 274. Let the discs be connected by springs, one end secured to a pin on one disc, the other end to a pin on the other. If A is driven in the direction shown, the torque will be communicated to B by means of the springs, which will therefore extend and cause A and B to take a new relative position. This relative movement of A and B will depend on the magnitude of the torque, which may therefore be measured by observing the movement. Any convenient arrangement of linkwork to cause, by utilising the relative motion of A and B, a pointer to move over a scale will answer.

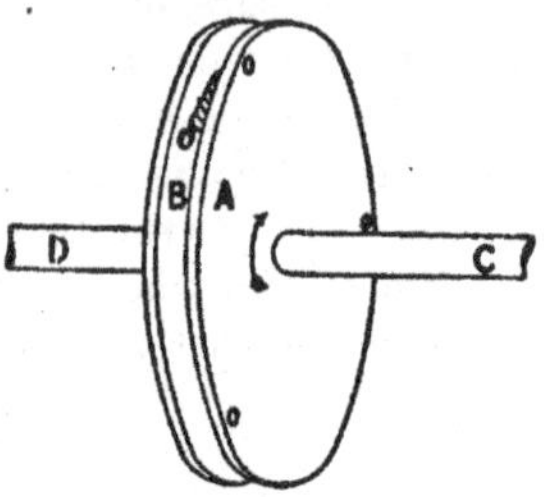

FIG. 274.—Spring transmission dynamometer.

Let T = observed torque, lb.-feet.

Work of one revolution $= 2\pi T$ ft.-lbs.,

and
$$\text{H.P. transmitted} = \frac{2\pi T.N}{33,000}.$$

Dynamometers measuring the difference in belt pulls.—In other forms of transmission dynamometers the difference in

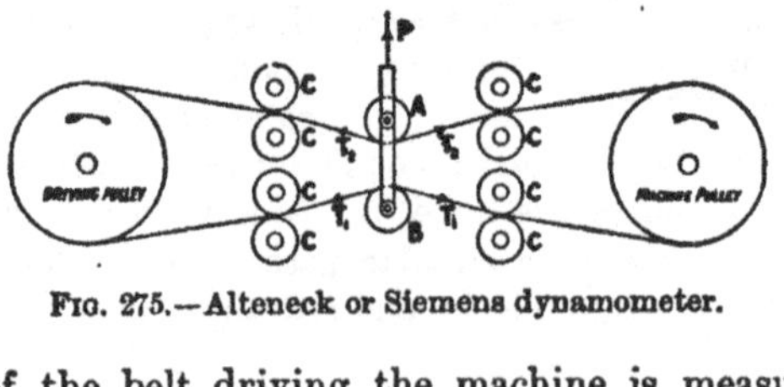

FIG. 275.—Alteneck or Siemens dynamometer.

the pulls of the belt driving the machine is measured. The belt is led over guide pulleys at A and B (Fig. 275) mounted in a frame, the angles of deflection of the belt being made equal by the pulleys *C*. If T_1, T_1 are the greater pulls and T_2, T_2 the lesser, T_1 will give a resultant downward force, R_1, greater than the resultant upward force, R_2, due to T_2, T_2. A force $P = R_1 - R_2$ will therefore be required to hold the frame in position, and if this force is measured, T_1 and T_2 may be found from it and the angles of deflection of the belt.

Let R = radius of machine pulley in feet, then

$$\text{Work of one revolution} = (T_1 - T_2)2\pi R \text{ ft.-lbs}$$

and
$$\text{H.P.} = \frac{(T_1 - T_2)2\pi RN}{33{,}000}.$$

Fig. 276 shows the **Froude, or Thorneycroft, transmission dynamometer**, in which also the difference between the tensions of the belt driving the machine is measured for the required information. *A* is the driving pulley and *B* the driven one. Two guide pulleys run loose on spindles mounted at *C* and *D* on a ⊥-shaped frame, this frame being pivoted at *F*. The belt is led from *A*, round *D*, *B*, *C*, in the manner shown. Neglecting frictional losses due to the use of the guide pulleys, the tensions of the belt will not be altered by passing round them, so we have two forces T_1 and T_1 pulling on the left-hand portion of the lever and other two, T_2, T_2, acting on the right-hand portion. A force *P* will be required to balance the moments of these pulls. Taking the resultant of the pulls on the tight side, equal to $2T_1$ acting through *C*, and also the resultant of the pulls on the slack side, equal to $2T_2$ acting through *D*, then

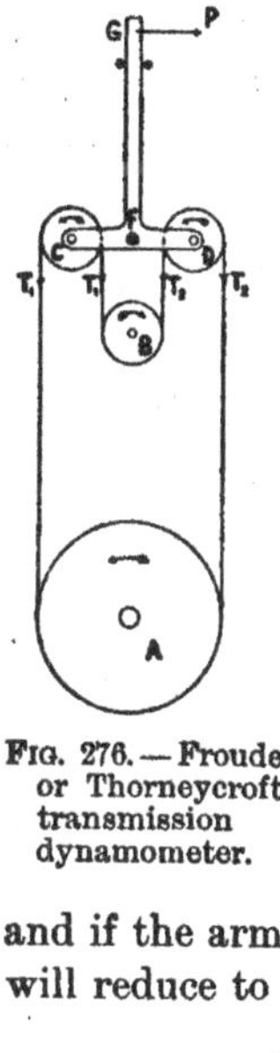

FIG. 276.—Froude or Thorneycroft transmission dynamometer.

$$P \times GF = (2T_1 \times FC) - (2T_2 \times FD),$$

and if the arms *FC* and *FD* are equal, as they usually are, this will reduce to

$$P \times GF = 2 \,.\, FC(T_1 - T_2)$$

$$\therefore\ T_1 - T_2 = \tfrac{1}{2} \,.\, P \times \frac{GF}{FC}.$$

Knowing the dimensions *GF* and *FC* in a given brake of this class, $(T_1 - T_2)$ may be calculated from the observed value of P, and if the speed of the belt, or the revolutions and the diameter of the driven pulley be known, the H.P. transmitted may be calculated in the same manner as in the preceding case.

Measurement of large powers.—The operation of measuring the energy delivered by very large engines is a very difficult one. It is not easy to imagine an absorption dynamometer suitable for attachment to a marine engine of, say, 10,000 I.H.P. In such cases, the Indicator Diagram is the sole means of estimating the power of the engines. The case of large engines driving electrical machinery is simpler. Here the electrical power delivered by the machines can be measured with considerable accuracy, and if the I.H.P. of the engine is obtained at the same time, the mechanical efficiency of the combined engine and dynamos can be calculated. Thus

Electrical power delivered by the dynamos

$$= \frac{\text{ampères} \times \text{volts}}{746}$$

horse-power = E.H.P.;

$$\therefore \text{ Efficiency} = \frac{\text{E.H.P.}}{\text{I.H.P.}} \times 100 \text{ per cent. for the combined plant.}$$

If the electrical and mechanical losses in the machine can be obtained at the same time, we have the power actually delivered by the engine to the machine from

Power delivered to machine = E.H.P. + H.P. absorbed by electrical and mechanical losses in the machine.

Fluctuations in the speed of machines.—In engines used for the purpose of converting energy supplied from a natural source into a form more suitable for practical purposes, such as the driving of machinery, the energy taken in may be received by the engine either at a uniform, or nearly uniform, rate, or it may vary considerably.

Machines such as steam, gas, and oil engines, periodically take in the energy to be converted and do not convert it at a uniform rate. Again, in many cases, such as slotting, punching and shearing machines, energy is supplied at a fairly uniform rate, but work is done by the machine periodically only. In all cases, a want of uniformity in the rate of supply of energy to a machine, or fluctuations in the rate at which energy is produced from a machine, will produce unsteadiness in the motion. Excess energy given to a machine above what is required to satisfy the demand must remain in the machine as additional kinetic energy in the parts, and these can only store additional kinetic energy by moving faster. If the demand exceeds the supply at any time, then this must be met from the stored kinetic energy in the machine, which, accordingly, must diminish its speed.

Methods of securing steady motion.—It is customary, in cases where the demand or supply fluctuates, and steadiness is aimed at, to attach to the machine some body having a great mass and capable of storing a large quantity of kinetic energy. Comparatively small fluctuations in the demand or supply are met from the store of energy in this body, with the effect that the fluctuations in speed are greatly reduced in magnitude and the machine moves more steadily.

In ordinary engines having a revolving crank shaft, the body for storing energy consists usually of a **flywheel** attached to the shaft and revolving with it. The great mass of a railway train when in motion enables a large quantity of kinetic energy to be stored and prevents small fluctuations of the energy produced by the locomotive from being perceived, although one can detect them for a few seconds as the train is starting when the speed is very slow. Paddle steamers, especially those with one cylinder only, depend on the mass of the vessel when in motion for steadiness of speed. The "kick" of the vessel as she takes in more kinetic energy can be plainly perceived. In the case of an ordinary bicycle, the mass of the rider and the machine prevent fluctuations in the energy supplied by the rider to the cranks during a revolution being observed to any extent.

Energy stored in flywheels.—We may find approximately the kinetic energy of a flywheel by considering its mass to be concentrated at the mean radius of the rim. Let v ft. per sec. be the velocity of a point at this mean radius, and m the mass of the wheel; then

$$\text{Kinetic energy} = \frac{mv^2}{2g}.$$

EXAMPLE. Suppose the mass of the rim of a flywheel is 10 tons and its mean radius is 6 ft. Find its kinetic energy when rotating 90 times per minute.

$$\text{Revolutions per sec.} = \frac{90}{60}.$$

Velocity of a point on the wheel at 6 ft. radius

$$= 2\pi \times 6 \times \frac{90}{60}$$

$$= \frac{396}{7} \text{ ft. per sec.}$$

$$\text{Kinetic energy} = \frac{mv^2}{2g}$$

$$= \frac{10 \times 396 \times 396}{64{\cdot}4 \times 7 \times 7}$$

$$= 496{\cdot}9 \text{ ft.-tons.}$$

M of a flywheel.

Let N = revolutions per minute of a flywheel.

r = mean radius of rim, feet.

m = mass of wheel.

v = velocity of a point at r radius, ft. per sec.

Then

$$v = \frac{N}{60} \cdot 2\pi r$$

$$= \frac{N \cdot \pi r}{30} \text{ ft. per sec.}$$

$$\text{Kinetic energy of wheel} = \frac{mv^2}{2g}$$

$$= \frac{m}{2g} \cdot \left(\frac{N \cdot \pi r}{30}\right)^2$$

$$= \left(\frac{\pi^2 \cdot m \cdot r^2}{1800 \cdot g}\right) \cdot N^2.$$

The coefficient of N^2 in this result is constant for a given wheel, so that we may state that the kinetic energy of a given wheel depends simply on the square of the revolutions per minute. If, therefore, we know the kinetic energy of the wheel at any given speed, say one revolution per minute, we can calculate, by simple proportion, its energy at any other speed.

Let $\mathbf{M}$ = kinetic energy of wheel at 1 revolution per min.

K = " " " at N revolutions per min.

Then,

$$\mathbf{M} : K = 1 : N^2,$$

or,

$$K = \mathbf{M}N^2.$$

Fluctuation of speed in flywheels.—We may easily calculate the fluctuation in the speed of a flywheel if we know the fluctuation in the demand for energy. Thus, suppose a wheel of mass m lbs., to have a mean radius r feet; then, if the velocity of a point at the mean radius is v_1 ft. per sec. at a given instant,

$$\text{Kinetic energy} = K_1 = \frac{mv_1^2}{2g} \text{ ft.-lbs.}$$

Suppose now that there is a demand for W ft.-lbs. of energy which has to be met by the store of energy in the wheel. The wheel will slow down while supplying this. Call v_2 the velocity of a point at the mean radius when the change is complete; then

$$\text{Kinetic energy} = K_2 = \frac{mv_2^2}{2g} \text{ ft.-lbs.}$$

The wheel loses kinetic energy $= K_1 - K_2$, and this has been transformed into W ft.-lbs. of mechanical work,

$$\begin{aligned} \therefore \quad W &= K_1 - K_2 \\ &= \frac{mv_1^2}{2g} - \frac{mv_2^2}{2g} \\ &= \frac{m}{2g}(v_1^2 - v_2^2), \end{aligned}$$

or,

$$v_1^2 - v_2^2 = \frac{2g}{m} \cdot W.$$

EXAMPLE. A wheel of mass 2000 lbs. at a mean radius of 3 feet has a speed of 180 revolutions per minute. Suppose 4000 ft.-lbs. to be abstracted from it and calculate its new speed.

$$\begin{aligned} v_1 &= \frac{180}{60} \times 2\pi r \\ &= 3 \times 2 \times \frac{22}{7} \times 3 \\ &= \frac{396}{7} = 56{\cdot}57 \text{ ft. per sec.,} \end{aligned}$$

and

$$v_1^2 = 3200,$$

and

$$v_1^2 - v_2^2 = \frac{2g}{m} \cdot W.$$

$$\begin{aligned} 3200 - v_2^2 &= \frac{64{\cdot}4}{2000} \times 4000 \\ &= 128{\cdot}8. \\ v_2^2 &= 3200 - 128{\cdot}8, \\ v_2 &= \sqrt{3071} \\ &= 55{\cdot}4 \text{ ft. per sec.} \end{aligned}$$

Let new speed $=N_2$ revolutions per minute.

$$N_2=\frac{55\cdot4\times60}{2\pi r}$$
$$=\frac{55\cdot4\times60\times7}{2\times22\times3}$$
$$=\underline{176\cdot3}\text{ revolutions per min.}$$

The wheel therefore loses 3·7 revolutions per minute while giving up 4000 ft.-lbs. of energy.

It forms a useful exercise for the student to perform some experiments on the energy of a flywheel. For this purpose, a small flywheel is mounted on a horizontal shaft which also has a drum on which a cord is coiled (Fig. 277). A scale pan is attached to the end of the cord. Loads placed in this pan and allowed to descend will rotate the flywheel and give energy to it. The cord must be attached to the drum in such a manner that it will easily free itself from the drum when the pan reaches the floor.

Fig. 277.—Experimental flywheel.

Expt. (*a*).—Place a sufficient load in the pan, and let its weight, together with that of the pan, be W lbs. Raise the pan a measured height H feet by rotating the wheel by hand so as to coil up the cord. Now allow the pan to descend, being careful not to aid it to start in any way, and observe the time of its descent, using a stop-watch for this purpose. Repeat this several times in order to obtain the average time of descent; let this be t seconds.

Potential energy lost by the descending load $=WH$ ft.-lbs.

This energy has been transformed partly into kinetic energy in the rotating wheel, partly into kinetic energy in the descending load, and some has been utilised in overcoming the frictional resistances of the bearings of the shaft.

To obtain the kinetic energy taken up by the load during its descent :

Let v = velocity acquired, feet per second.

Average velocity during descent $=\frac{1}{2}v$.

$$H = \text{average velocity} \times \text{time of descent}$$
$$= \tfrac{1}{2}v \times t;$$
$$\therefore\ v = \frac{2H}{t} \text{ feet per second, and}$$

$$\text{kinetic energy acquired by the load} = \frac{Wv^2}{2g} \text{ ft.-lbs.}$$

Neglecting frictional losses at the bearings, the kinetic energy acquired by the wheel would be

$$E_1 = \left(WH - \frac{Wv^2}{2g}\right) \text{ ft.-lbs.} \quad \ldots\ldots\ldots\ldots(1)$$

EXPT. (*b*).—Observe the number of revolutions made by the wheel while the load is being wound through the measured height H. The wheel will rotate the same number of times while the load is descending. Call this number n. Since the wheel makes n revolutions in a time t seconds, starting from rest, its average speed of rotation will be equal to $\frac{n}{t}$ revolutions per second, and its maximum speed of rotation will be $\frac{2n}{t}$ revolutions per second. Let N be the maximum speed of rotation, stated in revolutions per minute, then

$$N = \frac{2n}{t} \times 60 = \frac{120n}{t}.$$

The kinetic energy acquired by the wheel may now be stated in terms of $\mathbf{M}$, thus,

$$E_1 = \mathbf{M}N^2,$$

or

$$\mathbf{M}N^2 = WH - \frac{Wv^2}{2g},$$

giving

$$\mathbf{M} = \frac{WH - \dfrac{Wv^2}{2g}}{N^2} \text{ ft.-lbs.,} \quad \ldots\ldots\ldots\ldots(2)$$

this being the kinetic energy of the wheel when rotating with a speed of one revolution per minute, frictional losses being still neglected.

EXPT. (*c*).—To determine the energy utilised in overcoming the friction of the bearings, wind up the load W again through the same height H, and allow it to descend as in Expt. (*a*). This time, in addition to the other observations, note the total number of revolutions made by the wheel from starting to stopping. Call this number n_2.

Whole energy transformed $= WH$ ft.-lbs.

Kinetic energy acquired by descending load $= \frac{Wv^2}{2g}$ ft.-lbs.

The difference between these, viz.,

$$\left(WH - \frac{Wv^2}{2g}\right) \text{ ft.-lbs., } \qquad (3)$$

reaches the drum, and is disposed of, while the load is descending, partly in overcoming the frictional resistances and partly in giving energy to the wheel. After the load has reached the floor, the whole of the energy given to the wheel is absorbed in overcoming the frictional resistances while the wheel is coming to rest, consequently the whole of the energy reaching the drum is ultimately expended in overcoming the frictional resistances during n_2 revolutions. Assuming that the frictional resistances are constant during the experiment, the energy required to overcome them during one revolution will be

$$\frac{WH - \frac{Wv^2}{2g}}{n_2} \text{ ft.-lbs.};$$

and during the time when the load is descending, in which period the wheel rotates n times, the energy required will be

$$\left(WH - \frac{Wv^2}{2g}\right)\frac{n}{n_2} \text{ ft.-lbs. } \qquad (4)$$

The equation showing the disposal of energy will now be

$$WH = \mathbf{M}N^2 + \frac{Wv^2}{2g} + \left(WH - \frac{Wv^2}{2g}\right)\frac{n}{n_2},$$

from which the final value of **M** may be found, giving

$$\mathbf{M} = \frac{WH - \frac{Wv^2}{2g} - \left(WH - \frac{Wv^2}{2g}\right)\frac{n}{n_2}}{N^2}. \qquad (5)$$

EXPT. (*d*).—Now perform a series of experiments, using different loads and heights, and observe in each case all the particulars required as mentioned above. Tabulate your observations thus:

EXPERIMENT ON ENERGY OF FLYWHEEL.

No. of Experiment.	Load, W lbs.	Height W descends, H feet.	Time of descent, t seconds.	Revolutions, W descending n.	Revs., total, n_2.

The results must now be worked out, and are best stated in tabular form in the following manner:

RESULTS OF EXPERIMENT.

No. of Experiment.	Velocity acquired by load, $v=\frac{2H}{t}$ feet per sec.	Speed acquired by wheel, $N=\frac{120\,n}{t}$ revs. per min.	N^2	Energy transformed, WH foot pounds.	Kinetic Energy acquired by W, $\frac{Wv^2}{2g}$, foot pounds.	Energy spent in Friction, $\left(WH-\frac{Wv^2}{2g}\right)\frac{n}{n_2}$ foot pounds.	**M.**

Find from the last column the average value of **M**, which will now be very near to the true value.

Momentum.—Momentum is possessed by a body when in motion; it is proportional to the mass of the body and to its velocity jointly, and is measured by the product of these.

$$\textbf{Momentum} = mv.$$

Units of momentum will be stated by giving the units of mass and velocity employed; thus, if the pound is used for the unit of mass, and one foot per second is the unit of velocity, then

$$\text{Momentum} = mv, \text{ pound-foot-seconds.}$$

EXAMPLE. Find the momentum of a body of mass 100 lbs. when it has a velocity of 5 feet per second.

$$\text{Momentum} = mv$$
$$= 100 \times 5 = 500 \text{ lb.-ft.-secs.}$$

Forces generating momentum.—Suppose a body, of mass m pounds, to be acted on by a force P pounds during a time t seconds, and that the body is at first at rest. An acceleration a feet per second per second will be produced, such that

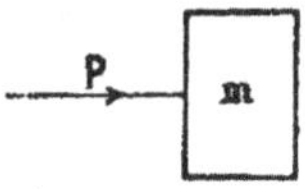

$$P = \frac{ma}{g} \text{ lbs.} \qquad \text{(1)}$$

FIG. 278.

Since P acts for a time t seconds, the velocity of the body at the end of this time will be

$$v = at \text{ feet per second};$$
$$\therefore \quad a = \frac{v}{t} \text{ feet per second per second.}$$

And from (1), by substitution,

$$P = \frac{mv}{gt} \qquad \text{(2)}$$

Now mv is the momentum possessed by the body at the end of the time t seconds, consequently $\frac{mv}{t}$ will be the momentum it acquires each second, and **the force generating momentum will be numerically equal to the momentum generated per second divided by** g. This relation is very useful when considering **impulsive forces**, *i.e.* forces which act during a very small interval of time.

A given momentum can be possessed either by a body of large mass having a small velocity, or by a body of smaller mass having a larger velocity. Suppose A is a body of mass M and velocity v, and that B is another body of mass m and velocity V, the velocities having been produced from rest by equal forces, P, P, acting during the same time t.

$$P = \frac{Mv}{gt}, \text{ for the body } A; \text{ [from (2) above]}$$
$$P = \frac{mV}{gt}, \text{ for the body } B;$$
$$\therefore \quad \frac{Mv}{gt} = \frac{mV}{gt},$$

or,

$$Mv = mV.$$

It may therefore be stated that **equal forces, acting during the same time on bodies of different masses, generate equal momenta.**

For example, consider the forces acting on a shell and gun when the gun is discharged. Neglecting effects produced by the inertia of the powder gases, it is evident that forces equal to one another act forward on the shell and backward on the gun during the time that the shell remains in the gun barrel, these being due to the pressure of the powder gases. The forces being equal, and the times of action being also equal, equal momenta will be generated in the gun and in the shot, assuming of course that the gun is so mounted, or suspended, that it is free to move backwards.

FIG. 279.

Let M = mass of gun, lbs. (Fig. 279).

v = backward velocity of gun, ft. per sec.

m = mass of shell, lbs.

V = forward velocity of shell, ft. per sec.

Then
$$Mv = mV,$$
which enables the velocity of either gun or shot to be calculated if the other quantities are known.

Impulsive forces.—Imagine a body in motion to possess a momentum equal to M, which is abstracted by the body encountering a uniform resistance P. If this is accomplished in t seconds, then

$$P = \frac{M}{gt}.$$

It will be noticed that if t becomes very small, P will become very large, and in fact, the force will be **impulsive.** Notice also that, if the resistance encountered is not uniform, we may still find its average value from this equation. This force we may call, in the case of impulsive action, the **average force of the blow.** It is quite impossible, in most cases of impulse, to state exactly what the actual reactions are at any instant, and it is very convenient to be able to calculate, at anyrate, their average value.

EXAMPLE. A hammer head, 2 lbs. mass, moving with a velocity of 40 feet per second, is arrested in $\frac{1}{200}$th second by meeting an obstacle. Calculate the average force of the blow.

$$\text{Momentum of hammer} = 2 \times 40$$
$$= 80 \text{ lb.-ft.-secs.}$$
$$\text{Momentum changed per second} = 80 \div \frac{1}{200}$$
$$= 16{,}000 \text{ lb.-ft.-secs.,}$$

and
$$\text{average force of blow} = P = \frac{16{,}000}{32{\cdot}2}$$
$$= \underline{500} \text{ lbs. nearly.}$$

Change of momentum.—Momentum depends on the velocity of a body; and, as velocity has direction, momentum will also be a directed quantity. Change of momentum in any given case must therefore be measured by taking the change in the body's velocity. The method of ascertaining this has already been described in Chap. XII. Having found the change in momentum and its direction, the force required will act in the same line.

EXAMPLE. Suppose a stream of 180 bullets per minute to impinge at 90° to a plate with a velocity of 1000 feet per second, and then to drop vertically downwards. If each bullet has a mass of 1 ounce, what is the reaction of the plate?

The change of velocity in this case will be at 90° to the plate and will be equal and opposite in sense to the velocity of the bullets.

$$\text{Change in velocity} = 1000 \text{ feet per second.}$$
$$\text{Mass reaching plate per second} = \frac{180}{60} \times \frac{1}{16} = \frac{3}{16} \text{ lb.}$$
$$\text{Reaction of plate} = \frac{mv}{g} = \frac{\frac{3}{16} \times 1000}{32{\cdot}2}$$
$$= \underline{5{\cdot}8} \text{ lbs.}$$

If a jet of water be substituted for the bullets, the problem will be similar.

EXAMPLE. Suppose a jet to have a sectional area of $\frac{1}{100}$th square foot and a velocity of 200 feet per second.

$$\text{Mass reaching plate per second} = \frac{1}{100} \times 200 \times 62{\cdot}5$$
$$= 125 \text{ lbs.}$$
$$\text{Reaction of plate} = \frac{mv}{g}$$
$$= \frac{125 \times 200}{32{\cdot}2} = \underline{776} \text{ lbs.}$$

Centrifugal pump.—Suppose we have a hollow wheel with openings at *A*, *A* (Fig. 280), for water to flow into it, and that the wheel has blades arranged radially, so that if the wheel is at

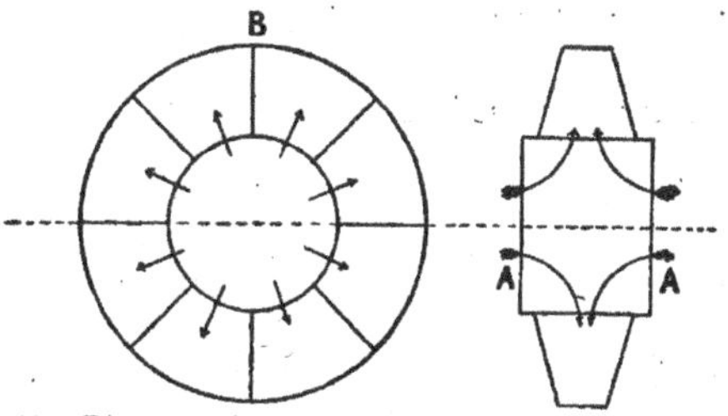

FIG. 280.—Diagram of a centrifugal pump with radial blades.

rest, the water will flow outwards and be discharged all round the rim of the wheel in radial jets. Assuming that the radial velocity of the water is kept constant while passing along the blades, there will be no change of momentum whatever, and consequently no reaction required from the blades. If, however, we set the wheel in rotation and assume that the water still enters the blade portion of the wheel in radial lines, it will be discharged at the outer circumference with not only radial velocity, but also a tangential velocity of an amount equal to that of the rim of the wheel. This must be the case, for the water can only pass through the wheel by sliding along the blades, and consequently at any part of the blade will have a tangential velocity equal to that of the blade there.

Let V = the velocity of the rim of the wheel, feet per second, then Change in tangential velocity of water = V, being zero at the inner part of the blade and V at the outer.

Let M = mass of water, lbs., flowing per second, then

Change of tangential momentum = MV, and

$$\text{Force which the wheel has to exert to produce this} = \frac{MV}{g} \text{ lbs.}$$

This force may be imagined concentrated at the rim of the wheel, so that

$$\text{Work done per second} = \frac{MV}{g} \times V$$

$$= \frac{MV^2}{g} \text{ foot-lbs.}$$

Horse-power which must be applied to the wheel

$$=\frac{MV^2}{g}\times\frac{60}{33,000}.$$

The arrangement is that of the **ordinary centrifugal pump.** In practice the blades are generally curved to avoid shocks and to allow the water to flow as quietly as possible through the wheel. The only difference this will make to the above calculation is in the work done. For if the water still enters with zero tangential velocity and is discharged with V feet per second tangential velocity, then the tangential force required at the rim of the wheel to do this will be $\frac{MV}{g}$ lbs.

Let V_p = tangential velocity of the wheel rim, then

$$\text{Work per second}=\frac{MV}{g}.\ V_p \text{ foot-lbs.}$$

$$\text{Horse-power required}=\frac{MV}{g}.\ V_p.\frac{60}{33,000}.$$

Example. 80 cubic feet of water per minute pass through a centrifugal pump, the velocity of the rim in a tangential direction being 25 feet per second. Suppose also that the water enters with no tangential velocity and is discharged with one of 20 feet per second.

Then, tangential force required at wheel rim

$$=\frac{80\times 62\cdot 5}{60}\times 20\div 32\cdot 2=51\cdot 7 \text{ lbs}$$

$$\text{Work done per second}=51\cdot 7\times 25$$

$$=1292 \text{ ft.-lbs.}$$

$$\text{H.P. required}=\frac{1292\times 60}{33,000}=\underline{2\cdot 35}.$$

EXERCISES ON CHAP. XV.

1. A friction dynamometer has a single load of 60 lbs. weight, the arm of which is 30″. Revolutions 100 per minute. Find the Brake Horse-power.

2. In testing an engine for B.H.P. a brake similar to that shown in Fig. 276 was used. The speed of the engine was found to be 200 revolutions per minute; the pull of the spring balance 11 lbs., and the dead load 70 lbs. The brake wheel was 5 feet diameter. Calculate the B.H.P.

3. In Question 2, calculate the heat developed by the brake. State the result in British Thermal Units per minute.

4. In Question 2 the Indicated Horse-power was found to be 6·5. Calculate the mechanical efficiency of the engine, and also the energy lost per minute in overcoming frictional resistances in the engine.

5. The mean diameter of a flywheel is 12 feet. Its mass is 8 tons, and it is running at 90 revolutions per minute. Find its Kinetic Energy.

6. Suppose the flywheel in Question 5 gives up 100,000 ft.-lbs. of energy, what speed will it have?

7. A flywheel has a mean radius of 3 feet and a normal speed of 120 revolutions per minute. It is required to supply 2000 ft.-lbs. from its store of energy while slowing down to 118 revolutions per minute. What mass of rim is required?

8. A truck, mass 10 tons, moving with a velocity of 4 feet per second, comes into collision with fixed buffers and is stopped in $\frac{1}{4}$ second. What is the average force of the blow?

9. The mass of the moving parts of a steam hammer is 1000 lbs., and the hammer head reaches the work with a velocity of 20 feet per second and is brought to rest in $\frac{1}{100}$ second. Calculate the average force of the blow.

10. A hammer head of $2\frac{1}{2}$ lbs. moving with a velocity of 50 ft. per second, is stopped in 0·001 second. What is the average force of the blow? (1897.)

11. A ship of 2000 tons, moving at 3 knots, is stopped in one minute; what is the average retarding force? Neglect the motion of the water. One knot is 6080 feet per hour. (1898.)

12. A car weighing $2\frac{1}{2}$ tons and carrying 40 passengers, the average weight of each of them being 145 pounds, is travelling on a level rail at the rate of 6 miles an hour. What is the momentum in engineer's units? If the propelling force be withdrawn, what average force in pounds must be exerted to bring the car to rest in two seconds? And supposing the force to be constant, what distance would the car travel before it came to rest? (1899.)

13. Explain the use of the flywheel in any machine with which you are acquainted. To what class of machines is such a wheel usually applied? What is the kinetic energy in a wheel revolving at 150 revolutions per minute, if the wheel loses 5000 ft.-lbs. of energy when its speed is reduced to 147 revolutions per minute? (1899.)

14. If a gun delivers 400 bullets per minute, each weighing 0·5 oz. with 2000 feet per second horizontal velocity; neglecting the momentum of the gases, what is the average force exerted upon the gun? (1900.)

15. A flywheel is required to store 12,000 ft.-lbs. of energy as its speed increases from 98 to 102 revolutions per minute; what is its kinetic energy at 100 revolutions per minute? (1900.)

16. A man and his bicycle weigh 170 lbs.; he has a speed indicator (not a mere counter). When going at 10 miles an hour on a level road he suddenly ceases to pedal, and in 15 seconds finds that his speed is 8 miles an hour. What is the force resisting motion? (1901.)

17. A machine is found to have 300,000 ft.-lbs. stored in it as kinetic energy when its main shaft makes 100 revolutions per minute; if the speed changes to 98 revolutions per minute, how much kinetic energy has it lost? (1901.)

18. The flywheel of a gas engine has a mass of 1000 lbs. at a mean radius of $2\frac{1}{2}$ feet and runs at 200 revs. per min. The supply of gas to the engine is stopped at a given instant and it is found that the flywheel comes to rest in $\frac{3}{4}$ minute. Calculate (*a*) the energy stored in the wheel at first; (*b*) the number of revs. while coming to rest; (*c*) the energy abstracted from the wheel per revolution, assuming this to be the same for every revolution.

CHAPTER XVI.

CENTRIFUGAL FORCE. EXAMPLES OF WANT OF BALANCE IN ROTATING BODIES. BURSTING EFFECT IN FLYWHEELS. GOVERNORS.

Motion in a circle.—If a body be attached to one end of a string, the other end held in the hand, and the body whirled round in a circle, it will be noticed that a pull along the string has to be resisted. Suppose the body m to be at A (Fig. 281), the natural tendency is for it to move in the line AB, tangential to the circle, and the pull P, applied to the string AC by the hand at C, is required to overcome this tendency and to cause the body to move in the circular path. The inertia of the body causes it to resist this pull with an opposite force F equal to P. P is called the **Central Force** (sometimes **Centripetal Force**), and F is called the **Centrifugal Force.** Evidently, if the string is let go, these forces would cease to exist, and the body would move off in a path tangential to the circle.

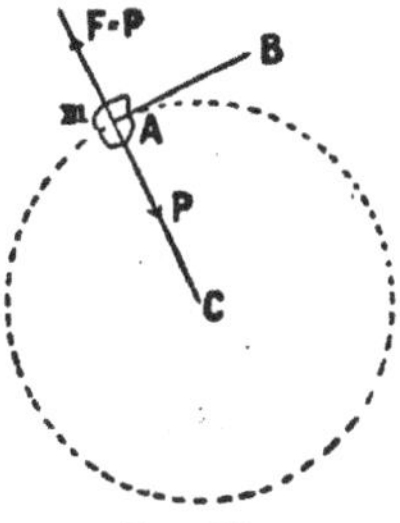

FIG. 281.

Since a central force is required to overcome the inertia of the body, it follows that there must be an acceleration caused by it, towards the centre of the circle.

Let v = the velocity of the body in its circular path, feet per second,

r = the radius of the circle, feet,

then it can be shown that the acceleration towards the centre of the circle is given by

$$a=\frac{v^2}{r} \text{ feet per second per second. } \quad \text{.................(1)}$$

Again, $P=\dfrac{ma}{g}$;

$$\therefore \quad P=\frac{mv^2}{gr}, \quad \text{..(2)}$$

the units being pounds or tons weight depending on the unit of mass used for measuring m.

This equation (2) may be used to calculate the central force, or the centrifugal force, acting on a rotating body.

EXAMPLE. Calculate the central force required to cause a body of mass 100 pounds to whirl in a circle 4 feet diameter 120 times per minute.

Here
$$v=\frac{120}{60}\times\pi d$$
$$=2\times\pi\times 4=8\pi \text{ ft. per sec.}$$
$$P=\frac{mv^2}{gr}$$
$$=\frac{100\times(8\pi)^2}{32{\cdot}2\times 2}$$
$$=\underline{982} \text{ lbs.}$$

Want of balance in machines.—The effect of the centrifugal force caused by a revolving mass in a machine running at a high speed may be very serious, producing disturbances which may cause the bearings to give out in a very short time and possibly also to hammer other parts of the machine to pieces.

EXAMPLE. In a machine running at 1800 revolutions per minute, there is an *unbalanced* mass of 1 pound at a radius of 1 foot. Find the pull on the bearings due to the centrifugal force.

$$v=\frac{1800}{60}\times 2\pi r$$
$$=30\times 2\times\pi\times 1=60\pi \text{ ft. per sec.}$$
$$\text{Centrifugal force}=\frac{mv^2}{gr}$$
$$=\frac{1\times 60\times 60\times\pi\times\pi}{32{\cdot}2\times 1}$$
$$=\underline{1120} \text{ lbs.}$$

This force, constantly directed from the centre, will revolve with the shaft, causing pressures on every part of the bearings during one revolution. The pressures on the bearings may be got rid of by fastening a mass of one pound in the same plane of revolution as the given one, and exactly opposite it at a radius of one foot. The centrifugal forces of the two masses will balance one another (Fig. 282) without any aid from the bearings, and the machine is now said to be **balanced**. Usually the data in practice are more complicated than in the simple case described here.

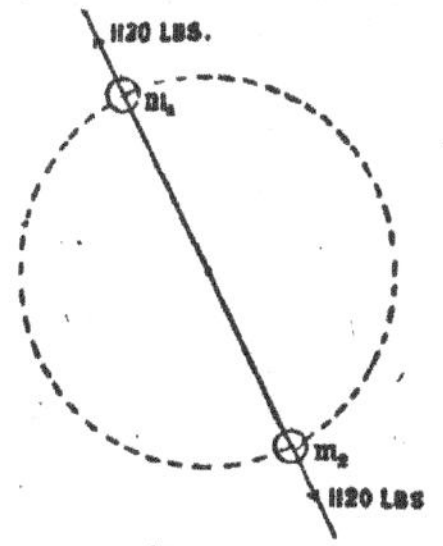

FIG. 282.

Fig. 282A shows an apparatus which may be used for detecting any want of balance in rotating bodies. Four discs are mounted on the same shaft, which is driven by a small electro motor, the whole being mounted on a frame suspended by means of three chains. Masses may be attached to the discs, and, if unbalanced, will produce vibrations in the frame when rotation occurs. If the masses are in balance, then the frame will remain steady at all speeds of rotation. This apparatus is due to Professor Ewing.

FIG. 282A.—Apparatus for experiments on the balancing of rotating bodies.

Bursting effect in flywheels.—The effects of centrifugal force have also to be considered in designing flywheels, as their rims are put under tensile stress thereby, when the wheel rotates. Fig. 283 shows the rim of a rotating flywheel; every piece of material in the rim will produce a centrifugal force directed from the centre as shown. The case is analogous to a cylindrical shell under internal pressure (p. 80), and we may treat it in the same manner.

Let m = mass of rim, in lbs. per foot circumference,
v = velocity of rim, feet per second,
r = mean radius of rim, feet.

Then, centrifugal force per foot circumference $=\dfrac{mv^2}{gr}$ lbs.

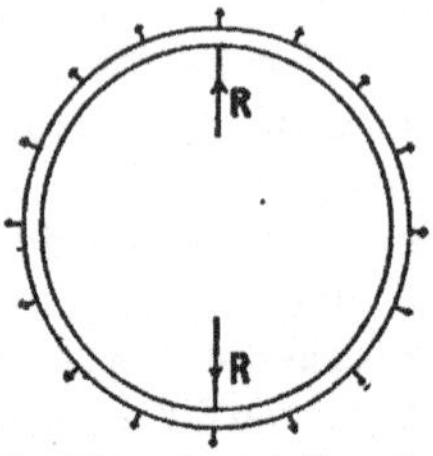

FIG. 283.—Rim of a rotating flywheel.

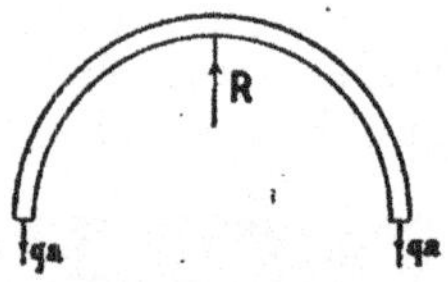

FIG. 284.

The resultant centrifugal force for half the wheel, corresponding to $(p \times d)$ in the boiler question, will be

$$R=\frac{mv^2}{gr}\times 2r$$

$$=\frac{2mv^2}{g} \text{ lbs.}$$

Let a = area of section of rim in square inches,
q = tensile stress on a, lbs. per square inch.

Then $R=2qa$ (Fig. 284),

or

$$\frac{2mv^2}{g}=2qa$$

$$q=\frac{mv^2}{ga} \text{ lbs. per square inch.}$$

Notice that m, the mass of rim per foot length is given by

$$m=\text{mass of material per cubic foot}\times 1\times\frac{a}{144}.$$

Therefore $q=\dfrac{\text{mass per cubic foot}\times v^2}{144 . g}$ lbs. per square inch.

This equation shows that **the tensile stress due to centrifugal**

force is independent of the sectional area of the rim and of the radius of the wheel.

EXAMPLE. A cast-iron wheel is run at a rim speed of 80 feet per second. If the material has a mass of 450 pounds per cubic foot, find the tensile stress due to centrifugal force.

$$q=\frac{450\times v^2}{144\times g}$$

$$=\frac{v^2}{10}\text{ nearly,}$$

$$=\frac{80\times 80}{10}=\underline{640}\text{ lbs. per sq. inch.}$$

Centrifugal governors.—Engines supplying motive power usually have to run at a fairly constant speed. This is managed

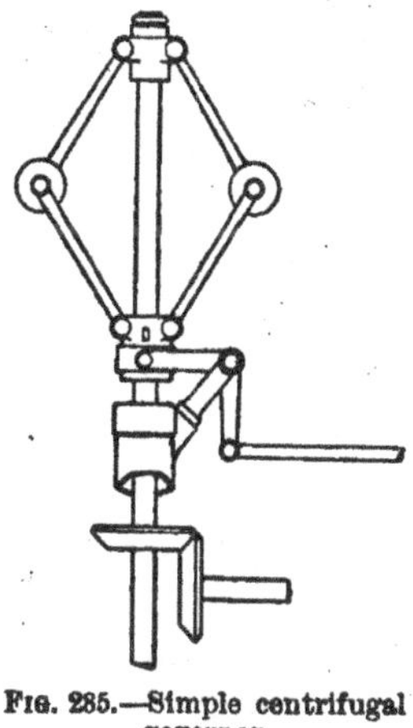

FIG. 285.—Simple centrifugal governor.

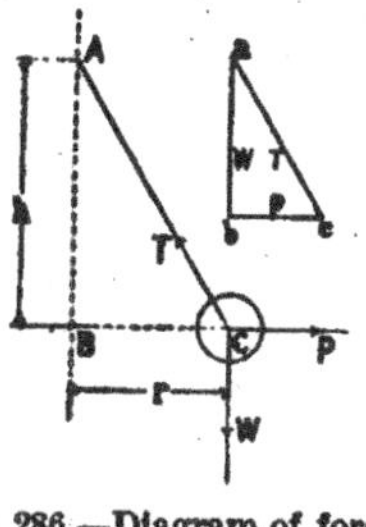

FIG. 286.—Diagram of forces acting on one ball.

by means of a **governor**, which controls the supply of steam, gas, or oil, by opening or closing a valve. The governor usually consists of two revolving balls mounted on two arms attached to a shaft driven by the engine (Fig. 285). The balls move out to a larger radius of rotation if the speed increases, and move inwards if the speed falls, and this movement controls the supply valve. Let us consider the forces acting on one of the balls. There will be its weight, W, the pull of the arm, T, and centrifugal force, P (Fig. 286). If the ball is rotating at a steady speed it will keep at a constant radius r from the axis AB, and P, W and T will balance. Drawing the triangle of forces for these forces, as at abc, it may be seen that this triangle is similar to the triangle ABC. Therefore

$$P : W = r : h,$$

or

$$\frac{Wv^2}{gr} : W = r : h;$$

$$\therefore \frac{Wv^2}{gr} . h = Wr,$$

or

$$hv^2 = gr^2;$$

$$\therefore h = g\frac{r^2}{v^2}. \qquad (1)$$

Let N = revolutions per minute of the ball, then

$$v = \frac{2\pi r \times N}{60},$$

and

$$v^2 = \frac{4\pi^2 r^2 N^2}{60 \times 60}. \qquad (2)$$

Substituting in (1) we get

$$h = g\frac{r^2 \times 60 \times 60}{4\pi^2 r^2 N^2}$$

$$= \frac{3600g}{4\pi^2} \times \frac{1}{N^2}$$

$$= \text{a constant} \times \frac{1}{N^2}. \qquad (3)$$

This shows that h **is independent of the mass of the ball and of the length of the arm, and depends only on the reciprocal of the square of the revolutions per minute in this type of governor.**

EXAMPLE 1. Find h for a governor of the above described type when making 60 revolutions per minute. Take $g = 32$.

$$h = \frac{3600g}{4\pi^2} \times \frac{1}{N^2}$$

$$= \frac{900 \times 32 \times 7 \times 7}{22 \times 22 \times 60 \times 60}$$

$$= \underline{0{\cdot}809} \text{ feet, or } \underline{9{\cdot}708} \text{ inches.}$$

EXAMPLE 2. In this governor, if the speed be increased to 61 revolutions per minute, find the vertical movement of the balls.

$$h = \frac{3600g}{4\pi^2} \times \frac{1}{61 \times 61}$$

$$= 0{\cdot}784 \text{ feet, or } 9{\cdot}408 \text{ inches;}$$

$\therefore$ Vertical movement of the balls $= 9{\cdot}708 - 9{\cdot}408 = \underline{0{\cdot}3}$ inch.

This movement of 0·3″ at the governor is transmitted by levers and connecting links to the valve controlling the supply of working stuff and partially closes this valve; the speed of the engine will in consequence be reduced to the required value again.

FIG. 287.—Loaded governor used for controlling the speed of an oil engine.

Governors are usually loaded by means of a spring or weight as shown between the balls in Fig. 287. This is done in order that they may be run at a higher speed than is possible in an unloaded governor, and the sensitiveness and power of controlling the valve are increased by so increasing the speed.

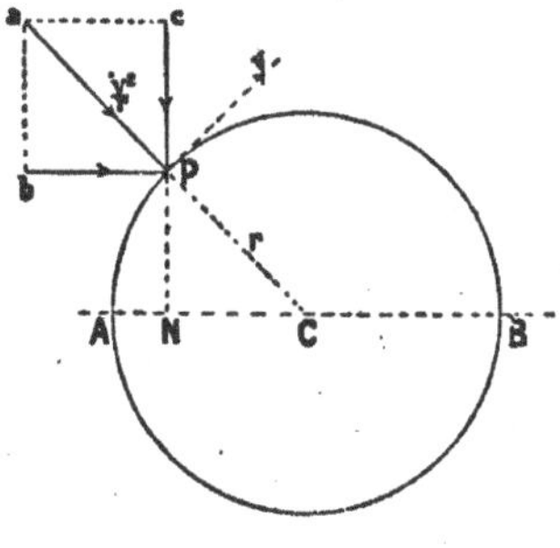

FIG. 288.—N, the projection of P on AB, has simple harmonic motion.

Simple harmonic motion.—Considering a point P (Fig. 288) rotating in a circle of radius r, with a uniform velocity v, we have seen that its acceleration towards the centre of the circle, C, is $\frac{v^2}{r}$. Taking components of this acceleration parallel and at right angles to AB when P is in any position, it may be seen that its horizontal acceleration, represented by bP, will be equal to $\frac{bP}{aP}$ multiplied by $\frac{v^2}{r}$; or, since the triangles aPb and PNC are similar,

$$\text{horizontal acceleration} = \frac{NC}{PC} \cdot \frac{v^2}{r}; \text{ and } PC = r,$$

$$\therefore \text{ horizontal acceleration} = NC \cdot \frac{v^2}{r^2} \quad \ldots\ldots\ldots\ldots (1)$$

$$= \text{a constant} \times NC.$$

That is, the horizontal acceleration is proportional to the distance, in feet, of N from C. When N is to the left of C, this acceleration is directed towards the right; and towards the left when N is to the right of C; that is, the acceleration is always directed towards C. As the point P rotates in the circumference of the circle, N will vibrate to and fro in AB, and its motion is called **simple harmonic**. If a body of mass m lbs. vibrates with N, a force will have to act on it to give it the required acceleration, this force being found from

$$R = ma \text{ poundals},$$

a being the acceleration $NC\frac{v^2}{r^2}$ found above; therefore

$$R = NC \cdot m \cdot \frac{v^2}{r^2} \text{ poundals}, \quad \ldots\ldots\ldots\ldots (2)$$

and, as can be seen by inspection, will be proportional to NC (the rest of the expression being constant), and will be always directed towards C. Consequently, if a body vibrates with simple harmonic motion, a force must act on it, directed always towards the middle of the swing, and proportional to the distance of the body from the middle (Fig. 289).

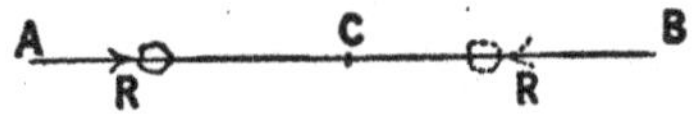

FIG. 289.

If we know the time that the body takes to travel from A to B, we can easily find R in any given case; for in Fig. 288 the point P travels a distance equal to half the circumference of the circle, *i.e.*, πr, while N travels from A to B. Let the time taken for this be t seconds. Then

$$t = \frac{\text{distance}}{\text{velocity}},$$

$$= \frac{\pi r}{v},$$

or, $$\frac{1}{t}=\frac{1}{\pi}\cdot\frac{v}{r},$$

and, $$\frac{1}{t^2}=\frac{1}{\pi^2}\cdot\frac{v^2}{r^2},$$

or, $$\frac{v^2}{r^2}=\frac{\pi^2}{t^2};$$

and, from (2), $$R=NC\,.\,m\,.\,\frac{\pi^2}{t^2}.\ \ldots\ldots\ldots\ldots\ldots\ldots\ldots\ (3)$$

We may alter this equation so as to give the time of a swing, thus,
$$\frac{R}{NC}\times\frac{1}{m}=\frac{\pi^2}{t^2},$$

or, $$t^2=\frac{\pi^2\times m\times NC}{R},$$

and, $$t=\pi\sqrt{\frac{m\times NC}{R}},\ \ldots\ldots\ldots\ldots\ldots\ldots\ldots\ldots\ (4)$$

which gives the time of a swing in seconds, if we know the force R acting on m when it is at a distance NC from the middle of its swing.

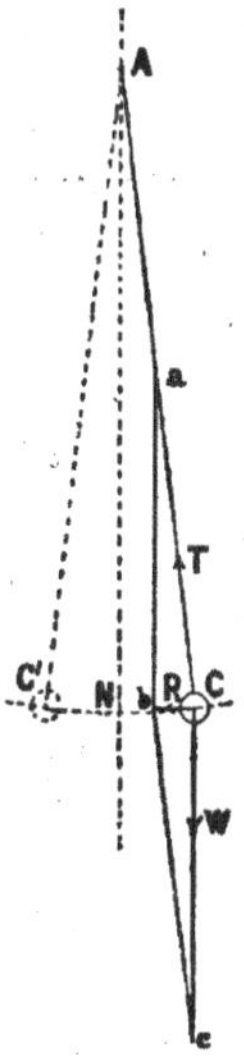

FIG. 290.—Simple pendulum.

Simple pendulum.—Referring to Example 3 on p. 27 we saw that the force, required to pull the weight out of the vertical, was almost exactly proportional to the distance from the vertical, provided this is not too great. Consequently, if the weight be released, it will vibrate to and fro with simple harmonic motion. The arrangement constitutes a **simple pendulum**. Call l the length of the suspending wire in feet; then from the diagram of forces (Fig. 290) acting on m,
$$\frac{R}{W}=\frac{R}{mg}=\frac{bC}{ab}=\frac{NC}{AN}.$$

Now if NC is small compared with l, AN and AC will be practically equal to one another, therefore
$$\frac{R}{mg}=\frac{NC}{l}\text{ very nearly, or }\frac{l}{mg}=\frac{NC}{R}.$$

Substituting this in the time equation (4) for one swing, from C to C', we get

$$t=\pi\sqrt{\frac{m}{g}\cdot\frac{l}{m}},$$

or,

$$t=\pi\sqrt{\frac{l}{g}}, \qquad \text{(5)}$$

as the time of swing of a simple pendulum.

It will be noticed that this result is independent of the mass of the body and of the distance from C to C'. For this to be true, the body should be of small dimensions and the distance CC' small also. The suspending wire or cord should be very light, so that its weight may be, as above, neglected.

The **compound pendulum** consists of any body vibrating about a horizontal axis under the influence of its own weight. An *equivalent simple pendulum* can easily be found by suspending a small bullet by means of a fine thread from the same axis, and adjusting the length of the thread until both compound and simple pendulums swing together in the same time.

Expt.—The simple pendulum may be used for roughly determining the value of g. Thus, arrange a small bullet to swing through a small angle at the end of a fine thread 3 or 4 feet long; take the time of, say, 100 swings; call this T seconds; then time of one swing

$$t=\frac{T}{100} \text{ seconds.}$$

g can now be calculated, from

$$t=\pi\sqrt{\frac{l}{g}},$$

or,

$$t^2=\pi^2\frac{l}{g},$$

and,

$$g=\frac{\pi^2 . l}{t^2}.$$

Expt.—Hang a helical spring from a fixed support, and attach a body of known mass to its lower end by means of a fine cord 5 or 6 inches long. When the arrangement is at rest, pull the body slightly downwards, and then let go. The body will vibrate vertically, and as the force acting on it will always be proportional to the distance from the middle of the vibration, the motion will be simple harmonic. Observe approximately the time during which vibrations can be perceived while the body is coming to rest. Now arrange a vessel containing water so that the hanging body is immersed. Start vibrations as before, and notice that each vibration is executed in a longer time and that the body comes sooner to rest. Repeat the experiment using a vessel containing oil. These experiments show the effect of fluid friction in altering the time of vibration and in stilling down or *damping out*, as it is called, the vibrations.

Substitute a long piece of rubber for the helical spring. It will now be found that the body will only make one or two vibrations in coming to rest. The great molecular friction in the rubber rapidly damps out the vibrations.

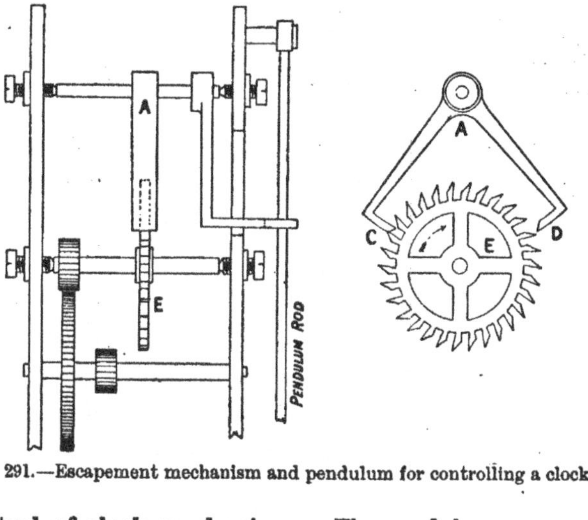

FIG. 291.—Escapement mechanism and pendulum for controlling a clock.

Control of clock mechanisms.—The pendulum, on account of the uniform time in which it executes small vibrations, is used to control the working of clocks. This it does by permitting one tooth only of a wheel driven by the clock to pass while it makes each swing. The **escapement mechanism**, as it is called, consists of an anchor-shaped pallet *CAD* (Fig. 291) mounted so as to vibrate with the pendulum. The pallet engages with the teeth of the escape wheel *E*, which is driven by the clock, and does so in such a way that one tooth escapes past *C* while the pendulum swings from right to left, and again another past *D* while the pendulum swings from left to right. The rotation of *E* is therefore controlled by the pendulum, and the clock consequently moves at a uniform rate. Energy is lost by the pendulum during each swing, due to the frictional resistances of the atmosphere and of the pivot on which it is mounted. This is made good by shaping the teeth of *E* and the edges of the pallet at *C* and *D* in such a way that an impulse is given to the pendulum each time a tooth is sliding on the pallet edge while escaping.

A **spiral spring** and **balance wheel** are substituted in some clocks and in all watches for the pendulum. The inertia of the wheel rim takes the place of the inertia of the pendulum bob; the resistance of the spring (which will be directly proportional to the angle turned through from its mean position) takes the place of the weight of the pendulum bob. When the wheel is set vibrating under the control of the spring, it will experience a torque from the spring directly proportional to its angle from the mean position. Consequently it will make simple harmonic vibrations. Impulses are given to it, and control to the clock effected by means of a small lever pivoted at *A* (Fig. 292) on the same axis as the pallet, which engages with the escape wheel as before. Watch and clock control movements take many different forms, two alone being described here.

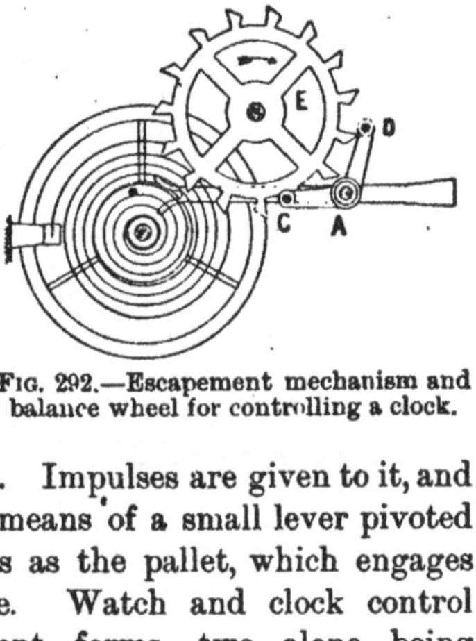

FIG. 292.—Escapement mechanism and balance wheel for controlling a clock.

EXERCISES ON CHAP. XVI.

1. A body, whose mass is 20 lbs., rotates round an axis at a radius of 9″, with a velocity of 40 feet per second. Calculate the pull on the axis.

2. A disc rotates on a shaft 120 times per minute. A wrought-iron pin, mass 5 lbs., projects from the disc, its radius being 12 inches. Find the mass required to balance the pin at a radius of 4 inches.

3. A cylindrical drum, 12″ long, has equal masses of 10 lbs. each attached to its ends at radii of 9″. Looking at the end elevation of the drum, the masses appear on the same diameter, on opposite sides of the centre. Calculate the rocking couple set up when the drum rotates 300 times per minute.

4. A cast-iron wheel, with solid rim in one piece, the material of which has an ultimate tensile strength of 8 tons per square inch, is run at a gradually increasing speed. What will be the speed of the rim in feet per second when the wheel bursts?

5. In a common unloaded governor, calculate the vertical height of the cone of revolution when the balls are rotating 60 times per minute. What will be the change in this height if the speed rises to 62 revolutions per minute?

6. A point rotates in the circumference of a circle of 6″ radius with a velocity of 10 feet per second. The plane of the circle is vertical. Find the horizontal component of the acceleration of the point when it is in positions differing by 30° round the complete circumference, and plot these on a time base.

7. Find the length of a simple pendulum to beat seconds. Take $g=32{\cdot}2$.

8. A railway coach, mass 20 tons, runs round a curve of 1,600 feet radius at a speed of 45 miles per hour. Calculate the centrifugal force.

CHAPTER XVII.

HYDRAULICS. WATER PRESSURE AND PRESSURE MACHINES.

Some properties of fluids.—Fluids are substances which are not able to offer permanent resistance to any forces, however small, which tend to change their shape. **Fluids are either liquid or gaseous**; gases possess the property of indefinite expansion, liquids do not. Thus, a small quantity of gas introduced into a perfectly empty vessel will at once expand and occupy the whole of the vessel, while a small quantity of liquid in the same circumstances will simply lie at the bottom of the vessel. Gases exist either as *vapours*, or as so-called *perfect gases*. The perfect gas was supposed to exist under all conditions as a gas, but it is now well known that all gases can be liquefied by great pressure and cold. **A vapour may be defined as a gas near its liquefying point, and a perfect gas as the same substance far removed from its liquefying point.**

Some liquids are more easily able to change their shapes than others. Liquids which change their shapes with difficulty are said to be the more viscous, the property being called **viscosity.** *Mobile* liquids change their shape very easily; thus, chloroform is used for delicate spirit levels on account of the extreme ease with which the bubble can change its position, chloroform being very mobile. Other liquids, such as cylinder oils, treacle, pitch, shoemakers' wax, are very viscous, but all change their shape if given sufficient time. As we have already seen in Chap. VII., change of shape is always produced by shearing forces. If equal compressive stresses are applied to all the faces of a cube, the body will become smaller, but will remain cubical; but if shear stresses be applied, the shape changes. It follows, therefore, that if shearing stresses be applied to a fluid, it will not remain at rest, but will change its shape, and therefore, **if the fluid is at rest, there can be none but normal stresses acting anywhere on or in it.**

Stress on horizontal immersed surfaces.—Since there can be no shearing stress in a fluid at rest, and since friction is always brought about as a shearing stress, it follows that when a liquid such as water is at rest, there can be no frictional forces contributing to preserve the equilibrium of any portion of it. Suppose, in a tank of water (Fig. 293), we think of the equilibrium of a vertical column of it standing on a horizontal base of one foot square. The forces acting will be the weight of the column, which, if the depth is D feet, will be equal to the volume of the column multiplied by $62\frac{1}{2}$, the weight of a cubic foot of water nearly, so that

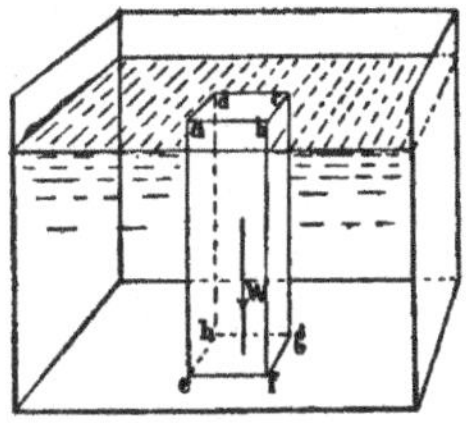

FIG. 293.—Equilibrium of a column of water.

$$W = D \times 1 \times 1 \times 62\tfrac{1}{2} = 62{\cdot}5 . D \text{ lbs.}$$

There will also be stresses on each vertical side of the column, everywhere directed perpendicular to the sides, these being due to the pressures from the surrounding water, but as there can be no friction between the surrounding water and the sides of the column, these stresses merely serve to keep the column in shape and do not help in any way to balance the vertical force W. W is balanced by the upward stresses on the base of the column, due to the pressure from the bottom of the tank. Consequently, the total force on the base, which is one square foot in area, will be equal to $62{\cdot}5 . D$ lbs. Any other horizontal square foot at the same depth will have a similar and equal pressure on it. If, therefore, we have a horizontal area, A square feet, at a vertical depth D in water, the total pressure on it will be found by multiplying the pressure per square foot by the area A, or

$$P = 62{\cdot}5 . D . A \text{ lbs.}$$

If the liquid is not water, but some other which weighs w lbs. per cubic foot, then

$$P = w \, . \, D \, . \, A.$$

We notice from this that **the pressure on a horizontal area depends directly on its depth in the liquid and is proportional to it**; at double the depth the pressure on a given horizontal area will be doubled, and so on. It will be observed that the shape of a tank does not influence the pressure on its bottom. So long as the area of the bottom and the depth of liquid are kept the same, the pressure will be unaltered. The student should keep clear of the error of supposing that the weight of water in the tank gives the pressure on the bottom. This is not the case, as may be seen in the three examples shown in Figs. 294-6.

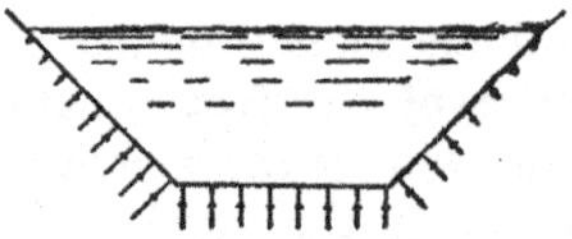

FIG. 294.

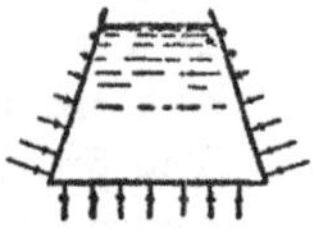

FIG. 295.

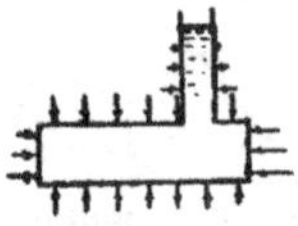

FIG. 296.

As a consequence of the pressures of the sides on the liquid being everywhere perpendicular to the sides, in the first case shown, the vertical upward components of these pressures sustain some of the weight of the contained water; in the second case, the vertical downward components of these pressures give more pressure to the bottom and so compensate for the diminished weight of water in the tank; in the third case, the pressure of the closed top has to be allowed for. In every case, the total pressure on the bottom will be found as already stated, by finding first the pressure on a square foot at the given depth and multiplying this by the actual horizontal area.

Stress on inclined immersed surfaces.—Suppose, now, we think of the column as standing not on a horizontal base, but on an *inclined* one, forming the sloping upper face of a wedge ABC, (Fig. 297) making, for simplicity, an angle of 45° with the horizontal. The pressures on this wedge will be, as before, perpendicular to its faces. Let P be the resultant pressure on the face AC, R the upward pressure on the bottom. For equilibrium of the wedge a third force Q, is required, which will act perpendicular to the face *AB* and, if we neglect the weight of the wedge itself, will be equal to *R* by the triangle of forces. That is to say, at a given depth in a liquid, the stress on a vertical area is the same as on a horizontal one. It can be shown that this is true for all areas sloping at any angle. This is usually stated as the principle that **fluids transmit stresses equally in all directions.** It should be noted that in the case of liquids in tanks subjected to pressures due to their own weight, that as the stress varies directly as the depth, a vertical square foot anywhere will not have uniform stress, but one which varies from the top to the bottom. The stress at any point on the square foot will be that due to the vertical column of water above it and is stated as the pressure which would be exerted on a square foot embracing that point if the stresses were uniform. Thus, the stress at a point 4 ft. deep in water will be $4 \times 62{\cdot}5 = 250$ lbs. per square foot on any plane—horizontal, vertical, or inclined.

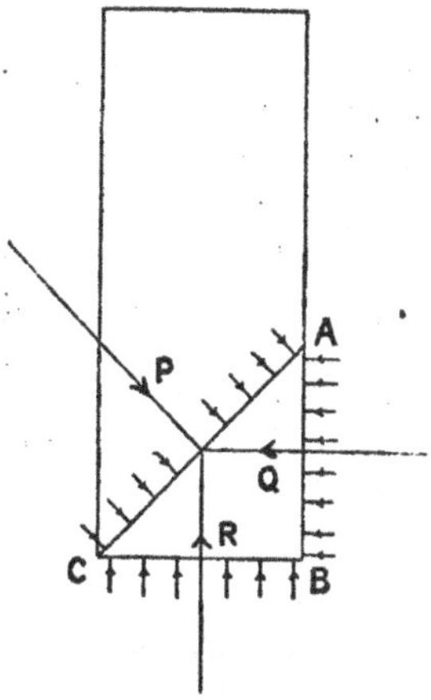

FIG. 297.—Equilibrium of the wedge *ABC*.

Pressures on the sides of a tank.—This principle enables us to calculate the pressures on the *sides of a tank*. Taking a rectangular tank as shown (Fig. 298) the stress on the bottom at a depth *AB* feet will be, for water,

$AB \times 62{\cdot}5$ lbs. per square foot.

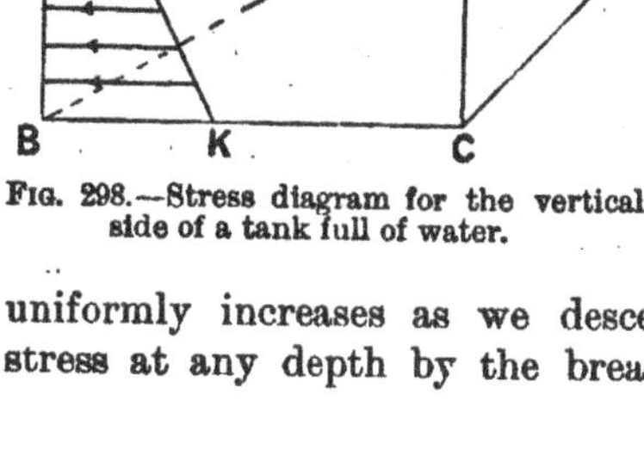

FIG. 298.—Stress diagram for the vertical side of a tank full of water.

The stress at the surface level will be zero, and as it uniformly increases as we descend, we may represent the stress at any depth by the breadth of a triangular diagram

ABK at that point, the base *BK* being made equal to $AB \times 62 \cdot 5$. The average stress on any of the vertical sides will obviously be one half of this maximum stress, and we may find the total pressure on any of the sides by multiplying the area of that side by this average stress.

EXAMPLE. A rectangular tank, 10 feet long, 6 feet broad, 4 feet deep, is full of fresh water. Calculate the resultant water pressure (*a*) on the bottom, (*b*) on one end, (*c*) on one side.

(*a*) Resultant pressure on bottom = average stress × area of bottom
$= (62 \cdot 5 \times 4) \times (10 \times 6)$
$= \underline{15,000}$ lbs.

(*b*) Resultant pressure on one end = average stress × area of end
$= (62 \cdot 5 \times 2) \times (6 \times 4)$
$= \underline{3000}$ lbs.

(*c*) Resultant pressure on one side = average stress × area of side
$= (62 \cdot 5 \times 2) \times (10 \times 4)$
$= \underline{5000}$ lbs.

Average stress on an immersed area.—The average stress on *any immersed area* can be shown to be **the stress at the depth of its centre of area.** Thus, for the rectangular sides of the tank just considered, the average stress is that occurring half way down, at the position of the centre of area of the rectangle. In other cases where the areas are not rectangular, the centre of area must be found first, or in some cases, the area may be split into others more conveniently dealt with.

EXAMPLE. The vertical end of a tank, 6 feet long, with sloping sides, is shown in Fig. 299. If full of water,

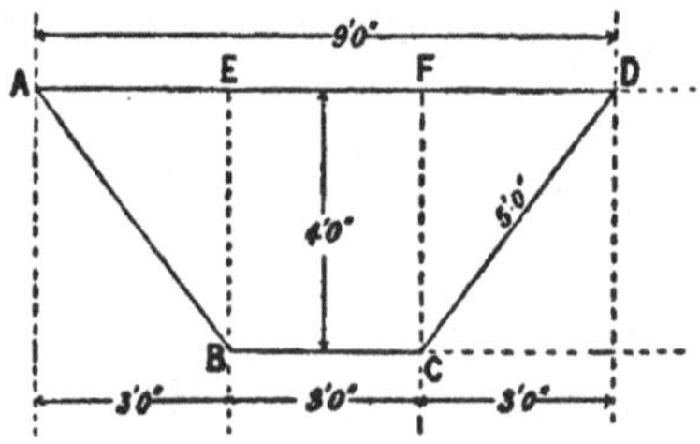

FIG. 299.—End elevation of tank.

Resultant pressure on bottom = average pressure × area of bottom
$= (62 \cdot 5 \times 4) \times (3 \times 6)$
$= \underline{4500}$ lbs.

Resultant pressure on one sloping side = average pressure × area of side

$= (62{\cdot}5 \times 2) \times (5 \times 6)$

$= \underline{3750}$ lbs.

In dealing with the ends, draw two verticals *BE* and *CF*, through *B* and *C*, and consider the end as made up of two equal triangles, *ABE* and *DCF*, and a rectangle *BCFE*.

Resultant pressure on end

= pressure on triangles + pressure on rectangle

$= 2\{(62{\cdot}5 \times \frac{4}{3}) \times (3 \times 2)\} + \{(62{\cdot}5 \times 2) \times (4 \times 3)\}$

$= 1000 + 1500$

$= \underline{2500}$ lbs.

Centre of pressure.—The resultant pressure on an immersed horizontal area acts at its centre of area, and in the case of a vertical rectangular area, having one edge in the surface, at $\frac{2}{3}$rds the depth of water from the surface. The case of other more complicated areas cannot be dealt with here. The point at which the resultant pressure on a surface acts is called the **Centre of Pressure.**

Retaining wall for water.—The overthrowing action of water pressure on a retaining wall can now be easily understood.

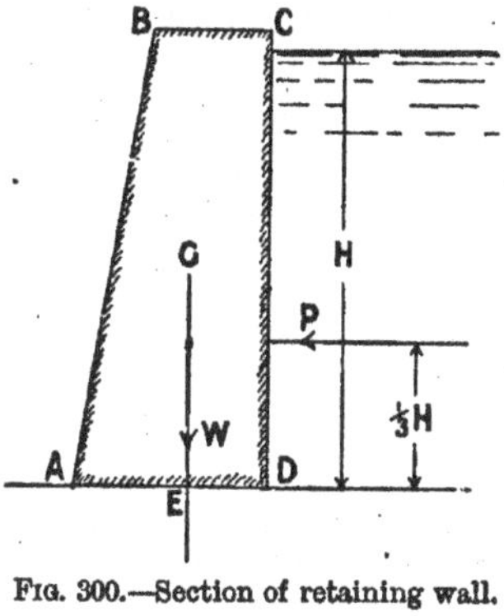

Fig. 300.—Section of retaining wall.

It is usual to consider a portion of the wall one foot long, on the assumption that whatever happens to it will equally happen to every other portion. This being so, the resultant pressure *P* (Fig. 300) will be given by

P = average pressure × wetted area

$= (62{\cdot}5 \times \frac{1}{2}H) \times (H \times 1)$

$= \frac{1}{2} \times 62{\cdot}5 \times H^2$ lbs.

P will act at a distance $\frac{H}{3}$ from the bottom, and will tend to make the wall turn about *A*, its moment about *A* being *P* multiplied by $\frac{H}{3}$ lb.-feet. This moment will be resisted by the weight, *W* lbs., of the portion of the wall under consideration, acting at its centre of gravity *G*. The moment of *W* about *A* will be opposite to that of *P*, and will be *W* multiplied by *AE* lb.-feet. If the moment of *P* is less than the moment of *W*, the wall will not be overthrown. For safety the overthrowing moment will always be considerably less in practice than the maximum resisting moment.

It is usual to test in this way. Draw a section of the wall to scale, and show *P* and *W* in their proper positions. Find the resultant *R*, of *P* and *W*, by the parallelogram of forces. Divide the base *AD* (Fig. 301) into three equal parts at *F* and *K*. If *R* passes within the middle part *FK*, the wall is safe, and if outside, it is not strong enough.

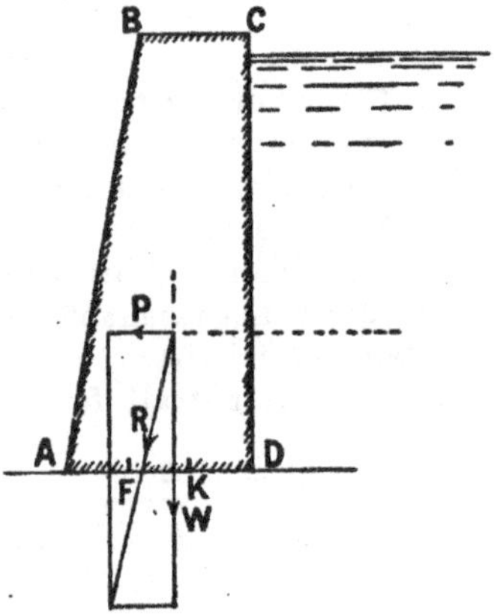

FIG. 301.—"Middle third" test for safety.

Forces acting on a floating body.—When a body, such as a ship, is floating in water, it is subjected to two resultant forces—its weight and the resultant water pressure on its sides and bottom. Consider the ship as floating at rest in still water. Its weight will be a downward vertical force *W* (Fig. 302), acting through *G*, the centre of gravity of the ship. The resultant water pressure balances *W*, and therefore must be an upward vertical force $R = W$, and in the same straight line with *W*. This force *R*, due to the buoyant effect of the water, is called the **buoyancy**.

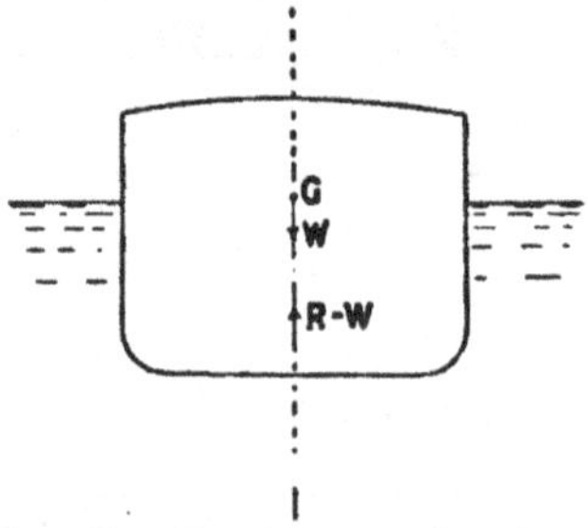

FIG. 302.—Equilibrium of a body floating at rest in still water.

Imagine for a moment that the surrounding water becomes solid, and so can preserve its shape, and let the vessel be lifted out, leaving a hole in the water which it exactly fits. Pour water into this hole until it is full to the surface level, and then let the surrounding water become liquid again (Fig. 303). The pressures on the water poured in will be exactly the same as when the vessel occupied the hole, and their effect is similar—the weight of the water poured in is supported. This being the case, we see that the weight of the water poured in and the weight of the vessel must be equal to one another, as each is equal to R, the resultant pressure from the surrounding water. Further, R must act through the centre of gravity of the water poured in, in order to support its weight. This point is called the **centre of buoyancy**—the point through which the resultant pressure of the water acts. Since R acts through G when the vessel is in the hole and through B when water is in, it follows that G and B must be in the same vertical line. We may state, therefore, that **when a vessel is floating at rest in still water, the weight of the vessel is equal to the weight of the water displaced, and that the centres of gravity of the vessel and of the displaced water are both in the same vertical line.**

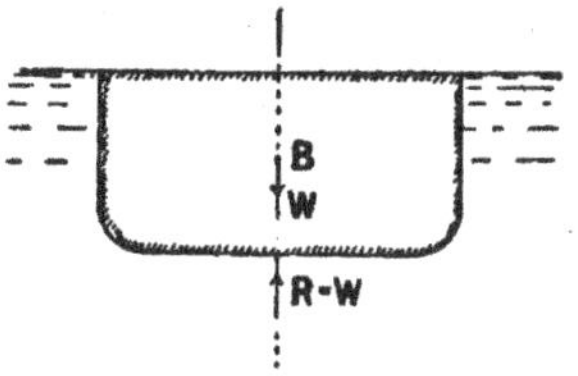

FIG. 303.

Specific gravity by experiment.—Since specific gravity is defined as the weight of a substance compared with the weight of an equal volume of water, it may be seen now how to determine experimentally the specific gravity of a body heavier than water.

EXPT.—First weigh the body in the pan of an ordinary balance. Then suspend it from the balance beam by a fine thread and weigh it again, this time the body being immersed in water at a temperature of 60° F. The weight this time will be diminished by the buoyancy of the water, which we have seen is equal to the weight of the water displaced, *i.e.*, the weight of a volume of water equal to the volume of the body. This loss of weight of the body while in water, therefore, gives the weight of an equal volume of water.

Let W_1 = weight of body in air,

W_2 = " " water,

$W_1 - W_2$ = weight of an equal volume of water,

and specific gravity $= \dfrac{W_1}{W_1 - W_2}$.

EXAMPLE. An iron rivet weighs 0·365 lbs. when weighed in the balance pan and 0·320 lbs. when weighed in water at 60° F.;

$$\text{the loss of weight} = 0{\cdot}365 - 0{\cdot}320$$
$$= 0{\cdot}045 \text{ lbs.},$$

and this is the weight of an equal volume of water. The specific gravity is therefore

$$\frac{0{\cdot}365}{0{\cdot}045} = \underline{8{\cdot}1}.$$

Pressure of the atmosphere.—Air possesses weight, and consequently the atmosphere exerts pressure on all bodies on the earth. If we were to consider a vertical column of the atmosphere, one foot square in section and reaching upwards to the limit of the atmosphere, we could calculate the pressure produced by its weight on its base, but the problem is complicated by the fact that the atmosphere is not equally dense at all parts of its height, but diminishes in density to zero. The pressure due to the weight of the atmosphere can easily be measured in the following manner.

EXPT.—Take a glass tube closed at one end and open at the other, about 36″ long, and fill it with mercury. The open end being inserted in a cup of mercury, and the tube being held vertically, the level of the mercury inside the tube will fall until it stands at a height h inches above that in the cup (Fig. 304). At A, the pressure inside the tube is that due to a column of mercury h high, and is equal to w multiplied by h lbs. per square inch, w being the weight of a cubic inch of mercury. The pressure on the surface of the mercury in the cup will be equal to this and is produced by the weight of the atmosphere. This apparatus constitutes the **common barometer**.

FIG. 304.—Barometer.

The average height of the mercury column is 30 inches, and as mercury weighs nearly 0·49 pound per cubic inch, this will give a pressure of 0·49 multiplied by 30, or 14·7 pounds per square inch; and this may be taken as the average pressure of the atmosphere on all bodies on the surface of the earth.

Boyle's Law for perfect gases.—The experiments of Boyle and others on the connection between pressure and volume of gases, show that **the pressure varies inversely as the volume, provided the temperature remains unaltered.** For example, if we have a volume v_1 cubic feet of air at a pressure p_1 pounds per square inch and change the volume to v_2 and the pressure to p_2 without change of temperature, then

$$p_1 : p_2 = v_2 : v_1,$$

or

$$p_1 v_1 = p_2 v_2.$$

Notice that the pressure must be **absolute**, that is, **measured from zero, not from atmospheric**, in applying Boyle's Law.

Suppose we have a cylinder fitted with a piston of area one square inch and stroke l inches, and allow air at a pressure p_1 pounds per square inch to enter it. The piston will be moved forward and work may be done to the amount of p_1 multiplied by l, the result being in inch-pounds. If the air supply is stopped at some part of the stroke before the end is reached, the pressure will fall as the piston moves on, but the expanding air will continue to do work. Using Boyle's Law, the pressures at various parts of the expansion may be found and a diagram plotted, from which the average pressure may be found. Thus, suppose the stroke is 12″ and air is supplied at 60 lbs. per square inch, absolute, and cut off at $\frac{1}{3}^{rd}$ stroke.

$$v_1 = 4 \text{ cubic inches.}$$

$$p_1 = 60 \text{ lbs. per sq. inch.}$$

$$p_1 v_1 = 60 \times 4 = 240.$$

The product of any other pressure and corresponding volume must be 240, so arranging a table of volumes differing by 1 cubic inch, we may calculate the corresponding pressures.

p lbs. per sq. inch.	v cubic inches.	p lbs. per sq. inch.	v cubic inches.
60	4	26·6	9
48	5	24	10
40	6	21·8	11
34·3	7	20	12
30	8		

Using these numbers, the diagram in Fig. 305 has been plotted. The average pressure and work done may now be found as described in Chaps. X. and XV.

Lift pump.—The **common lift pump** depends for its action on the pressure exerted by the atmosphere. A cylinder

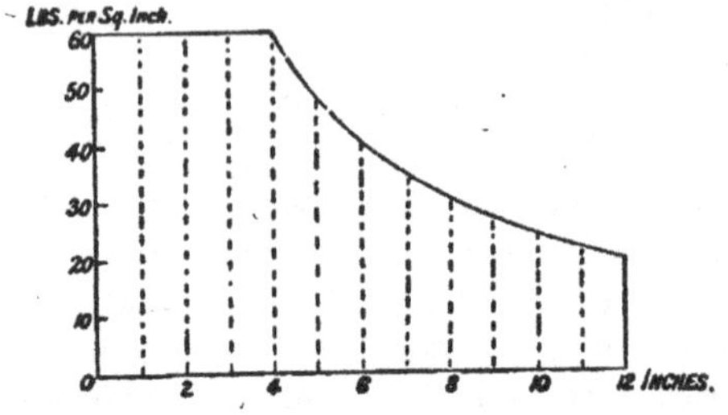

FIG. 305.—Diagram showing expansion of air in accordance with Boyle's law.

at *A* (Fig. 306) contains a piston, or **pump bucket** as it is called, fitted with a valve opening upwards. The cylinder is connected by a pipe *C*, with a foot valve at its bottom, to a cistern of water *E*. On the up-stroke of the bucket, the pressure of the air contained in *C* falls and the atmospheric pressure on the water in *E* causes some of the water to flow into the pipe *C*. On the down-stroke the valve *D* closes and the valve *B* opens. No water can pass *D* now, and air will be expelled through *B*. On the next up-stroke *B* will close again and *D* will open, and more water will flow into *C*, and this process repeated again and again will ultimately bring water into the cylinder, when it will pass *B* and be discharged through *F*. The process of starting in this manner is long and can be hastened by first charging the pump cylinder and pipe

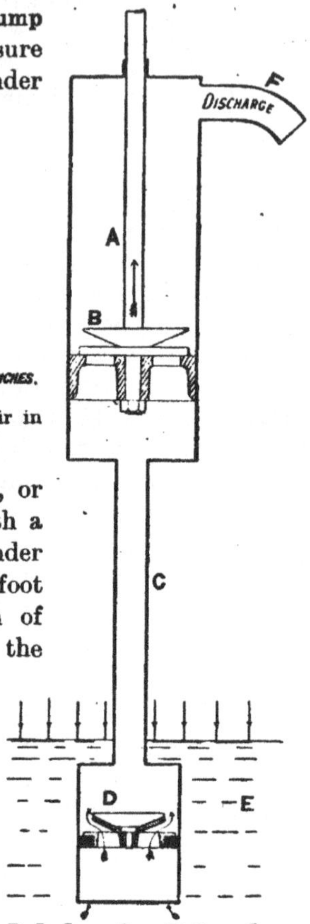

FIG. 306.—Lift pump.

C with water through a plug placed near the top of the suction pipe.

Head of water.—Water under pressure is often spoken of as being under a **head**. Head is the height from the point considered in the water to the surface level. The connection between head and stress is easily seen from the principles already discussed; thus, if H is the head in feet from the surface level of a tank at A (Fig. 307) to a pump at B, then the fluid stress on the pump piston will be $62{\cdot}5H$ lbs. per square foot. In general, if w is the weight of the liquid in pounds per cubic foot, H the head in feet, and p the fluid stress in pounds per square foot, then

$$p = wH.$$

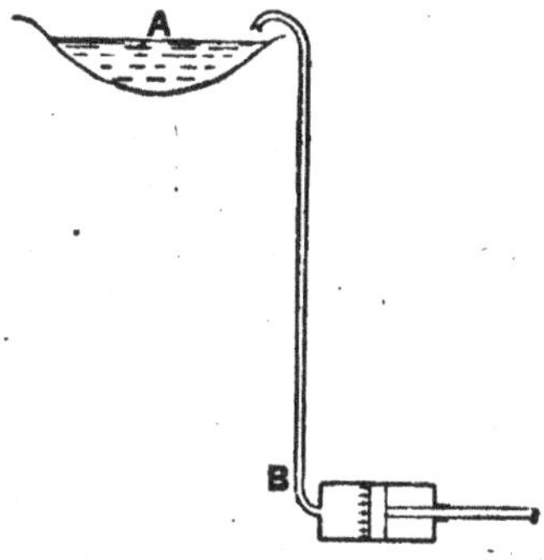

FIG. 307.

The total force on the pump piston will be found by multiplying its area in square feet by p.

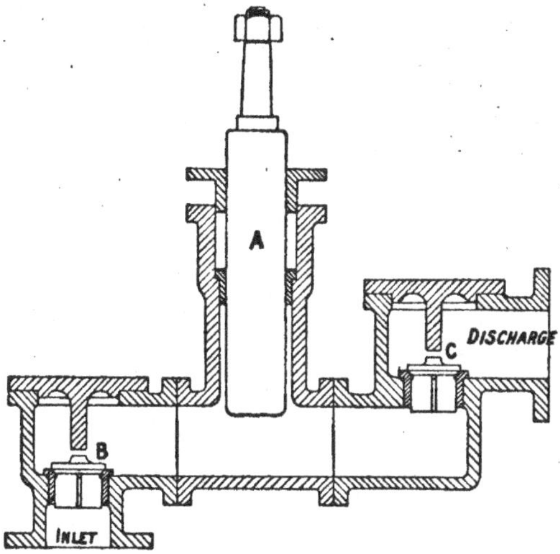

FIG. 308.—Force pump.

Force pumps.—Pumps are used for many purposes and have

different forms depending on the conditions. Fig. 308 shows a pump suitable for forcing water against a considerable pressure, such as would be the case in feeding a boiler. The plunger *A* is solid, without valves; on its up-stroke water enters through the inlet, passing the suction valve *B*; on the down-stroke *B* closes and *C* opens, allowing the water to pass into the discharge.

Hydraulic pumps are used for supplying water under high pressure for power purposes. They may be either belt-driven from a line shaft, or direct-coupled to a steam engine or other source of power. The object is to deliver a steady stream of water to the pipes at high pressure, say from ½ ton to 2 tons per square inch. To secure a fairly steady flow of water there are usually three pumps driven from a crank shaft having three cranks placed at angles of 120° to each other. The pumps are usually single acting, that is, they deliver water during one stroke and take in water during the next, the action being confined to one side of the piston only. The arrangement mentioned therefore insures that at least one pump shall always be delivering water to the mains.

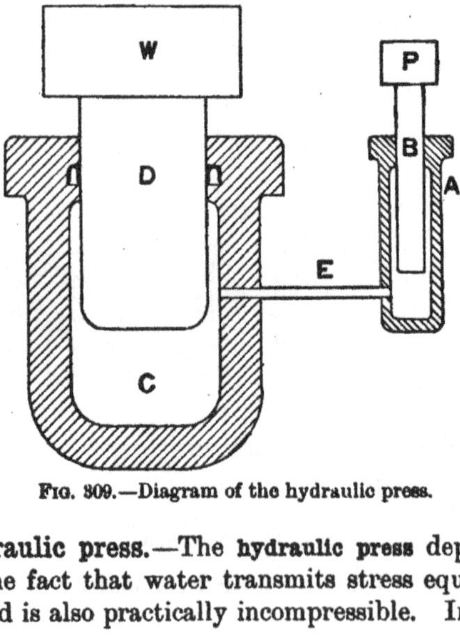

FIG. 309.—Diagram of the hydraulic press.

The hydraulic press.—The **hydraulic press** depends for its action on the fact that water transmits stress equally in every direction and is also practically incompressible. In Fig. 309, in which all details have been omitted, A is a pump having a plunger B, say d_1 inches diameter, and carrying a load P. C is a large cylinder with a ram, or cylindrical piston D, say d_2 inches diameter, carrying a load W. The pump cylinder and the large cylinder are connected by a pipe at E. Due to the load P on the plunger B, a stress will be produced on the contained water which occupies the whole of the space in the cylinders not taken up by the plunger and the ram, this stress being

$$\frac{P}{\text{area of plunger}}=\frac{P}{\frac{\pi d_1^2}{4}}=\frac{4P}{\pi d_1^2}=p \text{ say.}$$

The stress p will be transmitted to all parts of the water, and will exert a pressure on the bottom of the ram D, tending to raise it, the resultant pressure being p multiplied by $\frac{\pi d_2^2}{4}$. W will be equal to this, neglecting friction.

$$W=p\times\frac{\pi d_2^2}{4},$$

or

$$p=\frac{4W}{\pi d_2^2}.$$

We see, therefore, that

$$p=\frac{4P}{\pi d_1^2}, \text{ and also}=\frac{4W}{\pi d_2^2},$$

or

$$\frac{W}{P}=\frac{d_2^2}{d_1^2}.$$

The mechanical advantage (without friction) of the arrangement is therefore equal to the ratio of the squares of the diameters of the ram and the pump plunger. For example, if d_1 is 1 inch, and d_2 10 inches, then if P is 1 ton, W would be 100 tons.

It will be observed, also, that if P is allowed to descend, W will be raised a much smaller distance. Suppose, for example, that the area of the pump plunger section is 1 square inch, and that the ram sectional area is 100 square inches; then, if P descends 1″ it will deliver 1 cubic inch of water to the other cylinder. This cubic inch spread over the area of 100 square inches, will give a movement of $\frac{1}{100}$″ to the ram. So we see that the velocity ratio of the arrangement will be 100.

Leather packings.—Leakage past the ram is prevented by

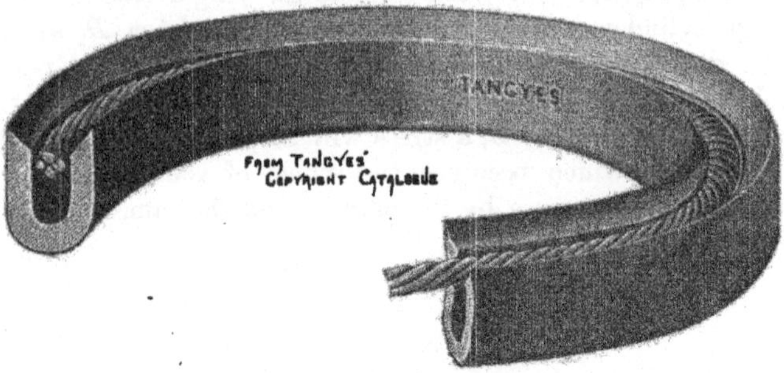

FIG. 310.—U leather packing.

means of leather packing. A leather ring, of U section, shown separately in Fig. 310, is inserted in a turned recess, as illustrated in Fig. 309. Water leaking upwards from the cylinder enters the interior of the ring and presses it firmly against the sides of the ram and the recess; leakage past this place is effectually prevented. Two other forms of leathers,

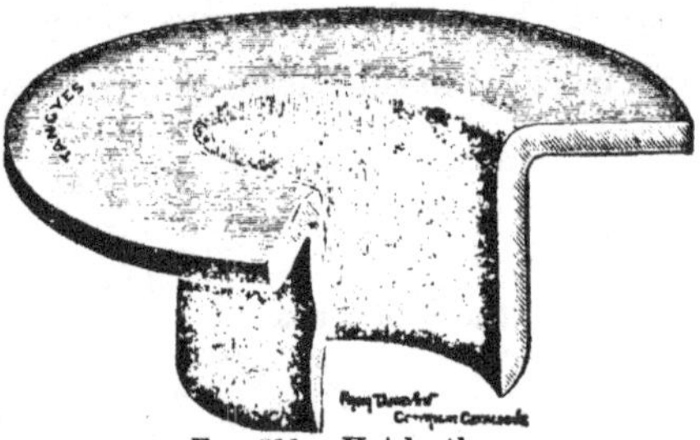

FIG. 311.—Hat leather.

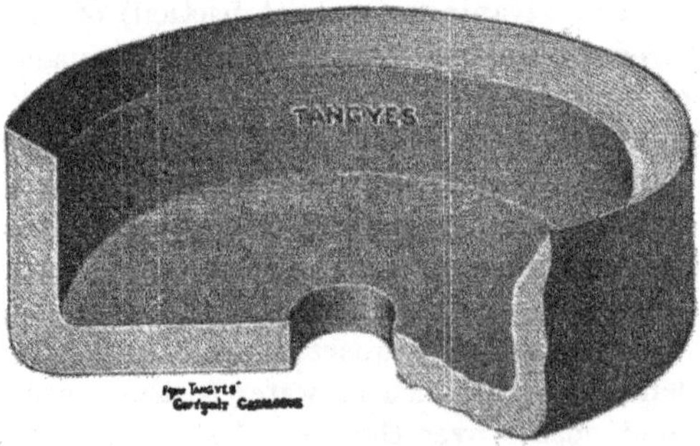

FIG. 312.—Cup leather.

hat and cup, are shown in Figs. 311 and 312. There is always a considerable loss by friction at these packings.

The hydraulic accumulator is used in connection with all hydraulic power plants. Its functions are to absorb the work done by the pumps when the presses, cranes, lifts, or other machines are at rest, and therefore taking no water, and also to prevent the water pressure exceeding a given maximum. It consists of a *hydraulic cylinder* (Fig. 313) placed upright and connected to both *pumps and machines* to be driven. The *ram* is loaded with heavy weights, and rises, when the water pressure is applied, against the resistance of these. When the ram approaches the top of its stroke it works a *tappet arrangement* connected to the throttle valve of the pump engine, or to the belt striking gear in a belt-driven pump, and so stops the pumps. The maximum working pressure of water which can be obtained is determined by the weights placed on the ram. Let W be this weight in tons and d the diameter of the accumulator ram in inches; then the water pressure

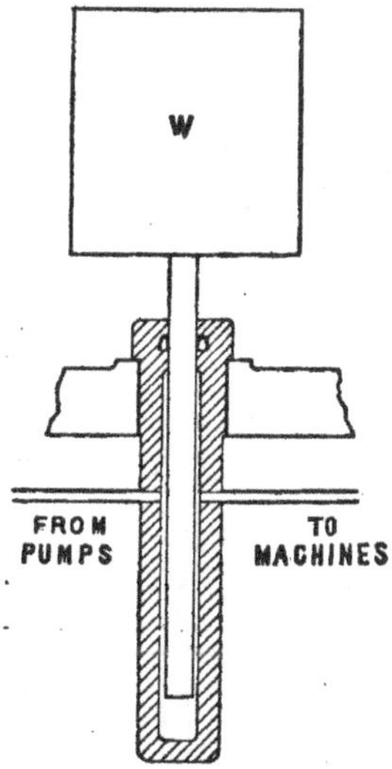

FIG. 313.—Diagram of the hydraulic accumulator.

$$p=\frac{W}{\frac{\pi d^2}{4}}=\frac{4 \cdot W}{\pi d^2} \text{ tons per square inch,}$$

neglecting the loss by friction of the ram leathers. The ram will not begin to rise until the water pressure attains this value. Suppose the accumulator ram is *up* at the top of its stroke, and that the rise has been H feet. The work done by the pumps in raising it will be, neglecting friction, WH foot-tons. If now one of the hydraulic machines, such as a crane, be started, it draws its water supply at first from the accumulator, the weights of which in consequence descend, giving up some of their stored energy to the crane. Soon after starting, the descending weights release the tappet arrangement, and the pumps start off again, delivering water to the machine direct until it is stopped, when the water from the pumps again goes into the accumulator and raises the ram. The arrangement, as will be seen from the above description, prevents any damage being done through stopping the hydraulic machines while the pumps are still working. Without the accumulator such stoppage, as water is practically incompressible, would have to be accompanied by a simultaneous stoppage of the pumps, which could not easily be accomplished.

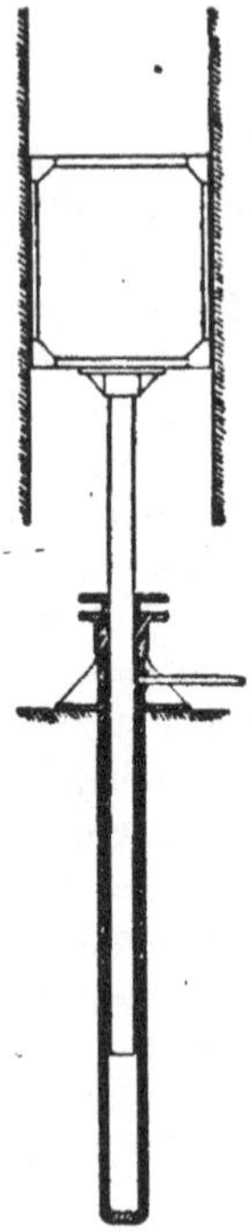

FIG. 314.—Simple hydraulic lift.

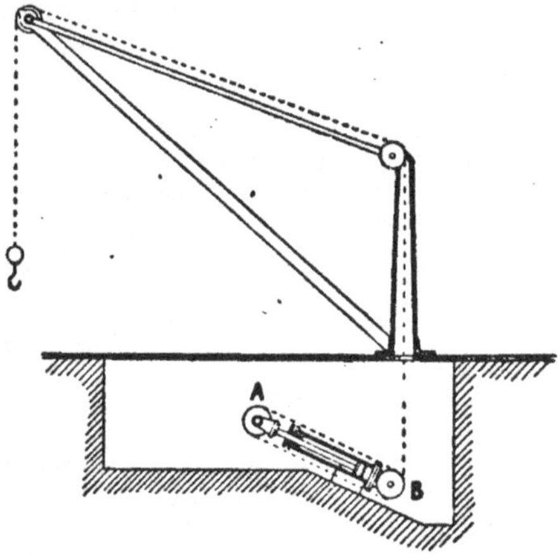

FIG. 315.—Hydraulic crane.

Hydraulic lift.—Fig. 314 shows a simple form of hydraulic lift. The cage is secured direct to the top of the ram of a vertical hydraulic cylinder. Water entering the cylinder raises the ram and so also the load. If d is the diameter of the ram in inches, and p the water pressure in pounds per square inch, then the total load which can be lifted, neglecting friction, will be $p \times \frac{\pi d^2}{4}$ lbs.

Hydraulic cranes are much used. Their action will be understood from Fig. 315. The chain sustaining the load passes along the tie, and down the interior of the hollow post to a hydraulic

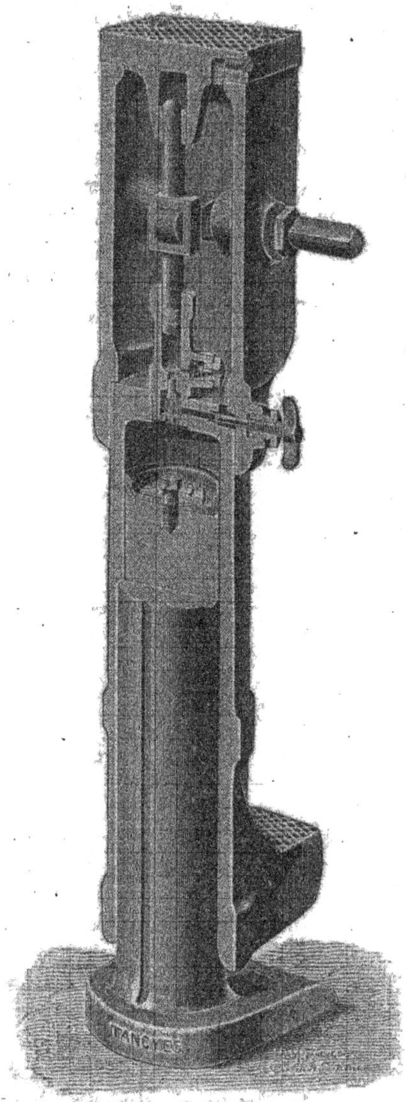

FIG. 316.—Hydraulic jack.

cylinder situated in a pit. This cylinder has a piston ram with a chain wheel *A* at its outer end, and another chain wheel *B* is mounted on the base of the cylinder. The chain is secured to the side of the cylinder, passes over *A*, then back over *B* and thence up the post. The object is to obtain a larger travel of the chain for a small movement of the ram; in the present example, the chain raising the load will move twice as fast as the ram. The cylinder of a hydraulic crane may be placed in any convenient position and the chain led to it. Usually another cylinder is fitted for revolving the post, so as to swing the jib to any convenient position for raising a load.

The hydraulic jack, Fig. 316, consists of a hydraulic cylinder inverted and working on a stationary ram. The cylinder contains a small pump operated by a lever outside. Water contained in the upper portion of the cylinder is forced, on working the outside lever, into the lower part, and so raises the cylinder and any load which may be placed on its top. These machines are very convenient for raising heavy loads through a short distance ; a large velocity ratio is possible. Fig 317 shows a small hydraulic jack arranged with levers for experimental purposes. Loads

FIG. 317.—Experimental hydraulic jack.

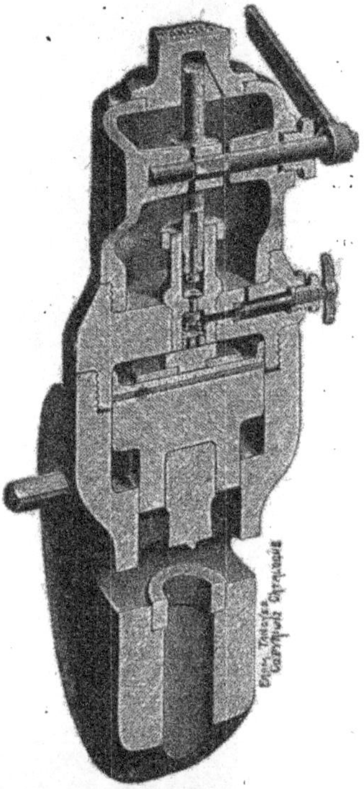

FIG. 318.—Hydraulic punching bear.

up to 300 pounds can be conveniently applied to it. The mechanical advantage, effect of friction and efficiency of this machine are determined by the methods described in Chapter XIV.

The hydraulic punching bear (Fig. 318) is a convenient tool operating on the same principle as the jack. The machine is entirely self-contained, and is consequently portable and may be set up in the most suitable position for performing any work.

EXERCISES ON CHAP. XVII.

1. A tank, 12 feet long, 8 feet broad, and 5 feet deep, is full of sea water (64 lbs. per cubic foot). Calculate the pressures on the bottom, on one side and on one end of the tank.

2. Fresh water stands to a depth of 6 feet on one vertical side of a wall 20 feet long. Calculate the resultant water pressure on the wall and its overthrowing moment.

3. If the section of the wall in Question 2 is rectangular, material 120 lbs. per cubic foot, and the height of the wall is 7 feet, what should be its thickness in order to have a righting moment of twice the overthrowing moment?

4. 2 cubic feet of air at an absolute pressure of 15 lbs. per square inch are compressed till the pressure is 95 lbs. per square inch absolute. Assuming no change of temperature, find its volume.

5. What is the maximum height that a common lift pump may be placed above the level of the supply water?

6. A pump, the diameter of the plunger being 2″, forces water against a pressure of 700 lbs. per square inch. If its stroke is 6 inches, and it makes 40 effective strokes per minute, how much work is done per minute? Suppose the efficiency to be 60 per cent., what horse-power is absorbed in driving the pump?

7. The bottom of a water tank measures 7′ in length and 3′ 4″ in width. When the tank contains 900 gallons of water, what will be the depth of the water, and what would be the pressure on the bottom, on each side and end of the tank respectively? One gallon of water weighs 10 lbs. One cubic foot weighs 62·3 lbs. (1897.)

8. An accumulator is to work at 700 lbs. per sq. inch, the ram is 10″ diam., what must be the total load? If the lift is 8 ft., what is the total store of energy when the weight is up? What is the total store of pressure water, that is, the extra amount due to the ram being up? Sketch the leather packing of an hydraulic press. (1897.)

9. Sketch in section and describe the action of the ordinary lifting pump. In such a pump the pump rod is $\frac{3}{4}$ inch in diameter, and the pump barrel is 5 inches in diameter, while the spout at which the water is delivered is 20 ft. above the surface of the pump bucket when the latter is at its lowest point; what would be the maximum tension on the pump rod in the upstroke of the pump, neglecting the weight of the pump rod and the pump bucket (the weight of a cubic foot of water is 62·5 lbs.)? (1896.)

10. Name the chief physical properties of a liquid, and show in what respect a liquid differs from a gas and from a solid. How is the pressure of water on the vertical sides of a tank calculated? (1898.)

11. A water tank is 10′ long, 10′ wide, and 10′ deep. When it is filled with water, what will be the force with which the water acts on one side of the tank? (1898.)

12. Water at 750 lbs. per square inch pressure acts on a piston 1 square foot in area, through a stroke of 1 foot; what is the work that such water does per cubic foot and per gallon? If a hydraulic company charges 18 pence for a thousand gallons of such water, how much work is given for each penny? (1898.)

13. Distinguish between the velocity ratio and the mechanical advantage of a machine.

In a hydraulic lifting-jack the ram is 6″ in diameter, the pump plunger is $\frac{7}{8}$″ diameter; the leverage for working the pump is 10 to 1. What is the velocity ratio of the machine? Experimentally we find that a force of 20 lbs. applied at the end of the lever lifts a weight of 8500 lbs. on the end of the ram. What is the mechanical advantage of the machine? What is the efficiency of the machine? (1899.)

14. Describe the construction and action of an ordinary suction pump for raising water from a well. If 200 gallons of water are raised per hour from a depth of 20 feet, and if the efficiency of the pump is 60 per cent., what horse-power is being given to the pump? (1899.)

15. A hydraulic crane is supplied with water at a pressure of 700 lbs. per sq. inch, and uses 2 cubic feet of water in order to lift 4 tons through a height of 12 feet. How much energy has been supplied to the crane, and how much has been converted into useful work? (1899.)

16. Sketch and describe the construction and working of any hydraulic accumulator with which you are acquainted. If an accumulator has a ram 20″ diam. with a lift of 15′, and the gross weight of the load lifted is 130 tons, what is the pressure of water per square inch and the maximum energy in ft.-lbs. stored in the accumulator, neglecting friction? (1900.)

17. A rectangular area has sides 18″ (horizontal) and 22″ (vertical); the uppermost edge is 5 ft. vertically below the surface of still water: what is the total pressure on the area? If the water is at rest only at the surface, is the pressure greater or less than before? (1900.)

18. A single-acting hydraulic engine has three rams, each of 3 inches diam.; common crank, 3 inches long; pressure of water above that of exhaust, 100 lbs. per sq. inch; 100 revolutions per minute; no slip of water. What is the horse-power? If this engine does 2·15 horse-power usefully by means of a rope, what is the efficiency? (1901.)

CHAPTER XVIII.

HYDRAULICS. FLOW OF WATER. WATER MOTORS.

The laws of fluid friction have already been explained, and the student should refer to the statement of them in Chap. XI. before studying the following.

Discharge from an orifice.—One of the simplest examples of the flow of water is to be found when a jet discharges through a sharp edged circular hole in a thin plate. Let *ab* (Fig. 319) be such a hole formed in the vertical side of a tank, *WL* being the surface level, giving a steady head *H* over the orifice *ab*. Outside some boundary such as *cde*, the water will be moving comparatively very slowly, and a particle will only begin to acquire important velocity when it crosses to the orifice side of the boundary. A particle situated at *d*, will move straight for the orifice and go out; one at *c* will move downwards along the tank side until it arrives at *a*, having then a considerable velocity. Its inertia prevents it from turning the corner at *a* sharply, it can only do so in a curve, and the same thing will occur to a particle situated at *e*, moving out at *b*.

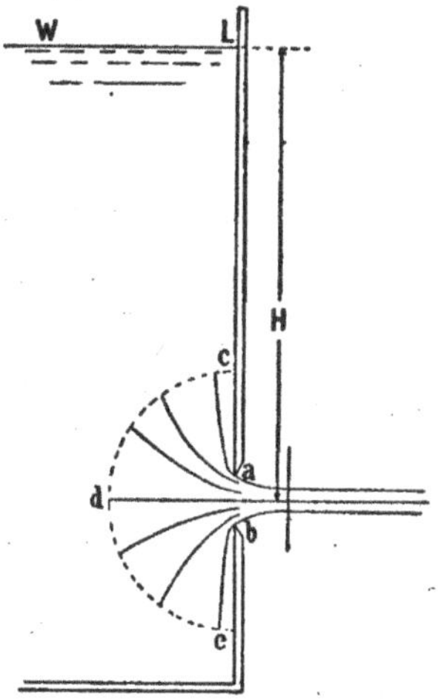

FIG. 319.—Flow of water through an orifice.

Particles crossing the boundary at other places will move in curved paths, somewhat like those shown, towards the orifice. The general effect of all these movements is to cause the jet to contract in cross sectional area after the orifice is passed and it is only at some little distance from the orifice that parallel flow is attained. The section at which contraction is complete is called the **contracted vein**; it is obvious that the water has attained its maximum velocity at this place. In obtaining the quantity discharged there are two things which must be allowed for—contraction and viscosity. It has been found experimentally that, for orifices like the one under consideration, the area of the jet at the contracted vein is 0·64 of the area of the orifice. This is called the **coefficient of contraction.** It has also been found that the actual velocity is 0·97 of that calculated from a purely theoretical basis; this number is called the **coefficient of velocity.** The product of these two is called the **coefficient of discharge**; thus

$$c = 0{\cdot}64 \times 0{\cdot}97 = 0{\cdot}62.$$

Velocity of Discharge.—Theoretically, the velocity of the jet may be calculated from a consideration of the energy of a pound of water at various places. Thus, one pound of water at the surface level *WL* will have potential energy due to its elevation *H* over the orifice; this potential energy will be equal to $1 \times H$, or *H* foot-lbs. At *d*, on the same level as the orifice, this potential energy will have been changed to energy of another kind, as the pound of water is no longer elevated over the orifice. This other kind of energy may be called *pressure energy* and the expression for it may be obtained from a consideration of the work done by the pressure on a pound of water at *d* in displacing it. Thus, let the volume of one pound of water be *v* cubic feet, and the pressure acting on it *p* pounds per square foot. Imagine the pound of water to be contained in a cylinder

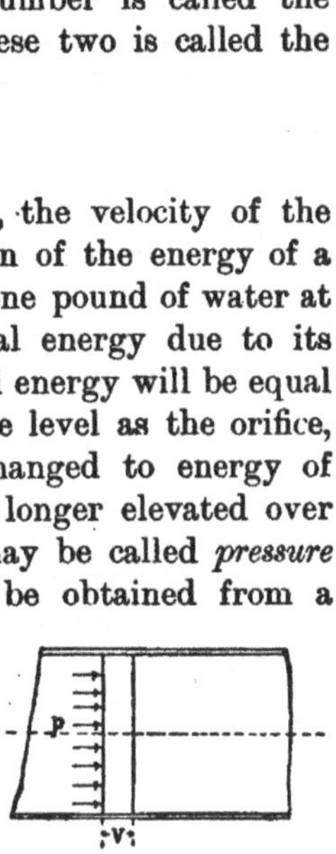

FIG. 320.

one square foot in sectional area (Fig. 320); it will then occupy a length of the cylinder equal to *v* feet. The work done by *p* in displacing the pound of water a distance *v* will be

p multiplied by v foot-pounds, and this is taken as the measure of the pressure energy. Therefore

Pressure energy of one pound of water at d in the tank

$$=p \times v \text{ ft.-lbs.},$$

and if w = weight of one cubic foot of water,

$$v=\frac{1}{w} \text{ cubic foot};$$

$$\therefore \text{ pressure energy}=pv=\frac{p}{w} \text{ foot-lbs.}$$

The potential energy H foot-lbs. at WL will have been changed into $\frac{p}{w}$ foot-lbs. of pressure energy at d, so that

$$H=\frac{p}{w}.$$

This statement will be very nearly true if there is only a very small velocity of descent from WL to d, as then the losses due to fluid friction having to be overcome will be very small. The pressure energy possessed by one pound of the water at d will be gradually converted into kinetic energy as the particle moves towards the orifice, and will be completely converted when the particle has attained its maximum velocity, that is, at the contracted vein. Let V = the velocity at the contracted vein, then

$$\text{Kinetic energy of one pound of water there}=\frac{1 \times V^2}{2g}$$

$$=\frac{V^2}{2g} \text{ foot-lbs.}$$

Assuming for a moment no losses by fluid friction between d and the contracted vein, the pressure energy lost must be equal to the kinetic energy acquired, so that

$$\frac{p}{w}=\frac{V^2}{2g};$$

or, since $H=\frac{p}{w}$, $H=\frac{V^2}{2g}$;

or $V^2=2gH$,

and $V=\sqrt{2gH}.$

The actual velocity V_a will be 0·97 of this, allowing for fluid friction, so that

$$V_a = 0{\cdot}97\sqrt{2gH}.$$

Quantity discharged from orifice.—Let Q be the quantity of water discharged per second, in cubic feet, then Q will be the volume of a stream V_a feet long, this being the length discharged in one second, and of cross sectional area equal to 0·64 . A, A being the area of the orifice in square feet.

$$\therefore\ Q = 0{\cdot}64 . A . V_a$$
$$= 0{\cdot}64 \times 0{\cdot}97 \times A \times \sqrt{2gH}$$
$$= 0{\cdot}62 . A . \sqrt{2gH} \text{ cubic feet per second,}$$

or writing c, the coefficient of discharge for 0·62,

$$Q = c . A . \sqrt{2gH} \text{ cubic feet per second.}$$

Experimental hydraulic apparatus.—Fig. 321 shows the whole arrangement of an experimental hydraulic apparatus,

FIG. 321.—Arrangement of apparatus for hydraulic experiments.

means being provided for measuring the quantity of water discharged from various shaped orifices in thin plates. An upright cylindrical vessel made of wrought iron can be supplied with water at a pressure of from 15 up to 100 pounds per square inch. The water enters the vessel at the top, centrally, being controlled by a valve outside, and the pipe is carried down inside the vessel a little more than half way. This pipe is perforated near its end with a large number of small holes and is fitted with gauze baffle sheets to still down whirling and eddies as far as possible. For small heads, a glass water gauge with a scale of feet divided into tenths is provided. This shows the head of water above the orifice level. For larger heads, a pressure gauge of the Bourdon type is attached to the vessel at the level of the orifice; this gauge is divided both in pounds per square inch and in feet head. When using heads of 2 feet or more, a cock which puts the top of the tank into communication with the atmosphere is closed. A quantity of air is thus entrapped above the water in the tank and is compressed as the pressure rises. When the desired pressure is attained, by adjusting the water regulating valve, this air cushion enables it to be maintained with great constancy. In fact the vessel plays the part, when used thus, of an ordinary air vessel. This arrangement had to be adopted in the apparatus shown, as the pressure in the ordinary water mains was very variable, sometimes zero, and it was found necessary to put down a small Worthington pump for supplying experimental water. The tank air vessel effectually stops pulsations from the pump. The jet is discharged from the left side of the vessel and is caught by the sloping baffle plate shown and directed into a rectangular tank below, whence it falls into the measuring tank shown on the extreme left.

An actual experiment.—Some results are given, obtained with this apparatus, using a circular orifice in a thin brass plate 0·25″ diameter.

EXPERIMENT ON THE DISCHARGE FROM AN ORIFICE.

Pressure lbs. per square inch.	Head H feet.	Quantity of water discharged in time t seconds Q cubic feet.	Duration of test t seconds.	Actual quantity per second $=\frac{Q}{t}$ cubic feet.	Calculated quantity per second $=0·62 . A . \sqrt{2gH}$
20	46	7·2	626	0·0115	0·0115
30	69·12	7·2	510	0·0141	0·0141
40	92·16	7·2	445	0·0162	0·0163
50	115·2	7·2	400	0·0180	0·0182
60	138·4	7·2	365	0·0197	0·0200

It will be seen that the calculated results agree fairly well with the experimental ones. The principal difficulty in carrying out such experiments lies in the measurement of the dimensions of the orifice. In the above experiments, the hole was fitted to a standard $\frac{1}{4}''$ diameter cylindrical gauge, so that its diameter was known with considerable accuracy. Other orifices which are very interesting to experiment with, chiefly on account of the beautiful forms the jet takes, are triangular and square shaped. There is great difficulty, however, in obtaining their areas to any degree of accuracy.

Flow over gauge notches.—Water flowing along a stream may be measured in two different ways depending on the magnitude of the stream. If small, the most convenient method

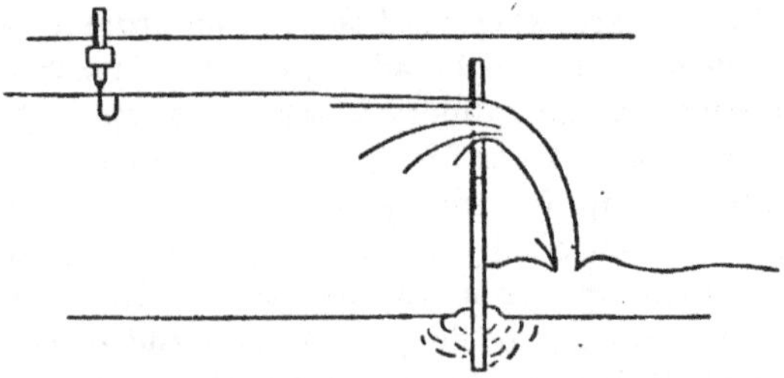

FIG. 322.—Arrangement of a weir for measuring the quantity of water flowing along a stream.

is to place a **weir** across the stream with a **gauge notch** formed in a thin plate at one part of the weir for the water to tumble over (Fig. 322). From the head of water and the dimensions of the notch, the quantity flowing may be calculated. If the stream is large, then soundings must be taken across one section of it so that the shape of the bed and the area of the cross section of the stream may be found. The velocity is then measured at various places by means of an instrument with blades like a propeller, the revolutions of which depend on the velocity of the stream and are registered by means of an attached counter. The average velocity being found, this multiplied by the cross sectional area will give the volume flowing per second. In the case of a straight stream with an ordinary river bed, the maximum velocity would be found near its centre, a little below the surface.

An experimental weir is easily arranged. In Fig. 321 the long box has a brass plate bolted to its end with a notch cut in it. Either of two plates can be fitted, one with a rectangular notch, the other with a V notch, the angle of the V being 90° (Fig. 323).

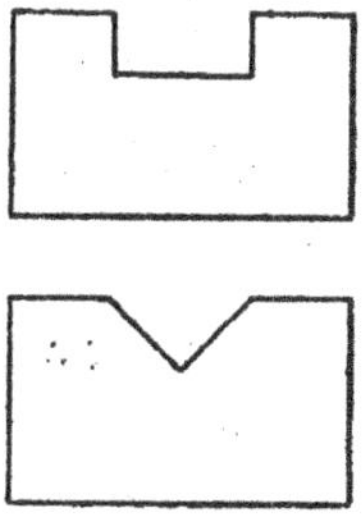

FIG. 323.—Rectangular and V gauge notches.

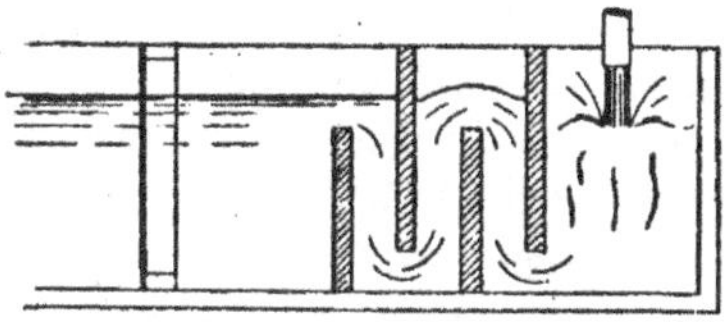

FIG. 324.—Baffle plates in the experimental apparatus for stilling down eddies.

Water is supplied at the other end of the long box and has to pass several baffle plates arranged as shown in the section (Fig. 324), the last one being a sheet of wire gauze. These still down all eddies, so that the water reaches the weir with steady motion. After passing the weir, the water falls into the measuring tank at the left-hand end of the apparatus. The measurement of the head of water above the lower edge of the notch is effected by means of a **hook gauge.** This consists of a round brass rod *AB* (Fig. 325), having a hook of brass wire fixed to it and brought to a sharp point at *C*. The rod *AB* can slide vertically in a tube at *D* which is clamped securely to a fixed support. The tube *D* is split at the top along one side and has a vernier cut on it, a scale of inches divided into tenths being cut on *AB*. In use, the rod *AB* is drawn up until the point at *C* just breaks the water surface. From the scale reading the head of water above the weir may be obtained. The hook gauge, both in the experimental tank and in actual practice must be placed a considerable distance from the weir, as the water surface always slopes slightly downwards as the weir is approached.

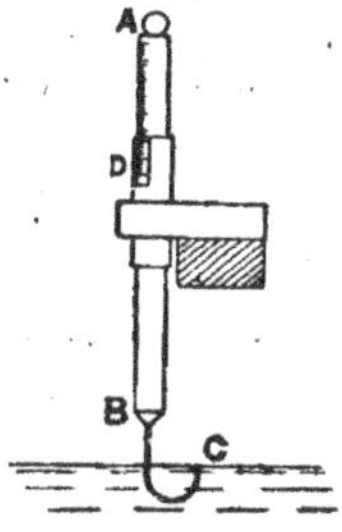

FIG. 325.—Hook gauge.

Calculation of discharge.—There is considerable difference in the formulae used to calculate the flow over V and rectangular notches. In the case of the V notch, the stream lines and the general shape of the jet remain similar to one another with all heads, while in the rectangular notch a change of head produces dissimilar jets. Fig. 326 is from a photograph of water falling over a triangular gauge notch and Fig. 327 shows a plan of the same. The curved paths taken by the various particles remain similar with all heads, and when this is the case, Prof. James Thomson has shown that the quantity flowing depends on the $\frac{5}{2}$th power of the head. A coefficient has to be introduced from experimental data, thus

$$Q=c\,(h)^{\frac{5}{2}},$$

where

Q = cubic feet flowing per second,
h = head of water over notch in feet,
c = coefficient of discharge = 2·635.

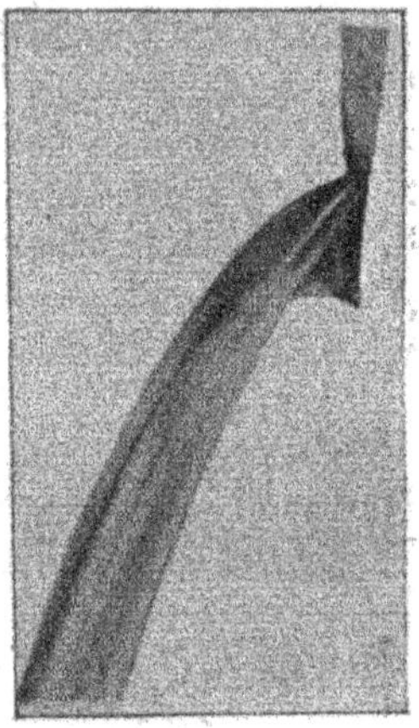

Fig. 326.—Water flowing over a V notch.

In the case of the rectangular notch, a portion near its centre will have parallel *stream lines* as seen in the plan and elevation

Fig. 327.

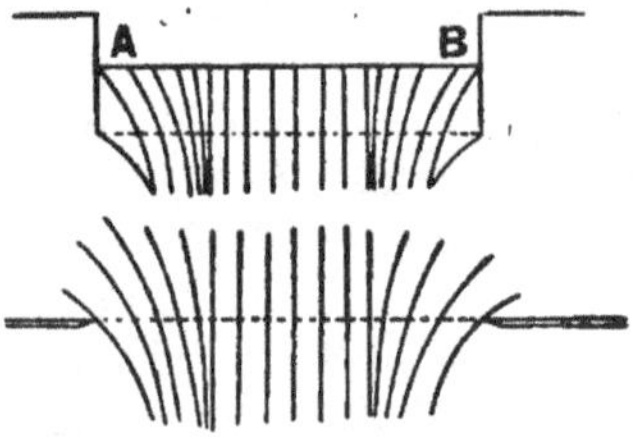

Fig. 328.—Water flowing over a rectangular notch. Two end contractions.

(Fig. 328) and near its edges A and B, curved stream lines. Alteration of head produces alterations in the shape of these stream lines, so that a formula has to be used containing a term applying to the straight portion in the middle and another term applying to the contracting portion. Thus,

$$Q = 3{\cdot}33\left(L - \tfrac{1}{10}nh\right)h^{\frac{3}{2}},$$

where Q = cubic feet flowing per second,
L = length of notch in feet,
h = head in feet,
$n=2$ for a notch as in Fig. 328,
$n=1$ " " " Fig. 329,
$n=0$ " " " Fig. 330.

n simply means the number of *end contractions* of the weir. Evidently there are two in Fig. 328, one in Fig. 329, and none in Fig. 330.

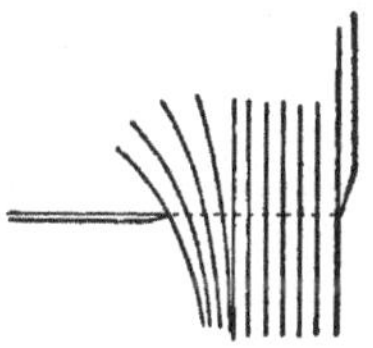

FIG. 329.—Rectangular notch; one end contraction.

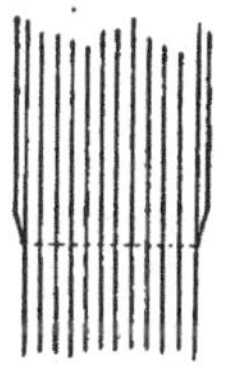

FIG. 330.—Rectangular notch; no end contractions.

Flow through pipes.—Let A and B in Fig. 331 be two tanks at different levels, connected by a pipe the length of which is L feet. Let the water in the tanks be preserved at a constant difference of level H feet. Water will flow from A into B, losing as it does so H foot-pounds of energy per pound of water. This loss is made up of three quantities:

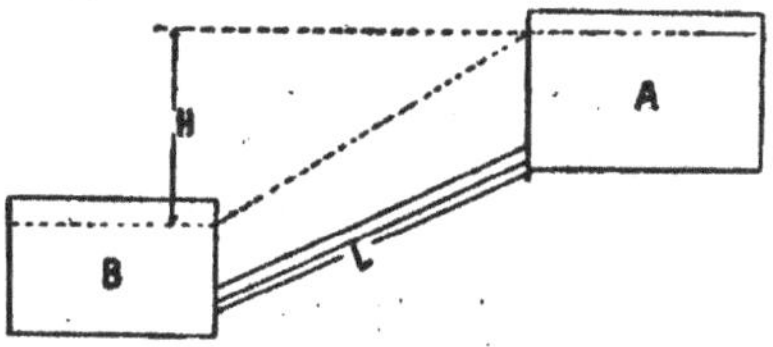

FIG. 331.—Virtual slope of a pipe.

(*a*) the kinetic energy possessed by the water entering B is lost in producing surging of the water in B;

(*b*) the water entering the pipe from A loses energy by the production of eddies in the pipe;

(*c*) energy is lost in overcoming frictional resistances to motion in the pipe.

In practice (*c*) alone is considered, as the loss under this heading in a long pipe is much greater than the others. The loss of head per foot length of the pipe will be

$$i=\frac{H}{L}.$$

This quantity is called the **virtual slope** of the pipe.

The flow of water in pipes has been shown by Prof. Osborne Reynolds to be of two different kinds, *steady flow* and *eddy flow*. Up to a certain critical velocity, which depends on the temperature, the flow is steady, and the resistance is proportional to the velocity of the water; at velocities above the critical one, the water breaks up into eddies and the resistance is proportional to some power of the velocity, this power being 1·7 for very smooth pipes, 1·722 for lead pipes, and 2 for rough pipes.

Many experimenters have observed the flow of water in pipes and there are several formulae representing the results obtained. That due to Prof. Unwin is

$$i=\frac{H}{L}=\frac{cv^{n_1}}{d^{n_2}};$$

in which d is the diameter of the pipe in feet; $c=0{\cdot}0004$, $n_1=1{\cdot}87$ and $n_2=1{\cdot}4$ for riveted wrought-iron pipes which are fairly smooth; and $c=0{\cdot}0007$, $n_1=2$ and $n_2=1{\cdot}1$ for very rough pipes. This formula may be used for pipes of from one foot up to four feet in diameter.

EXAMPLE. How much water will flow per second from a reservoir through a pipe 1 foot diameter, 5000 feet long, the fall of surface level being 25 feet? The inside of the pipe is fairly smooth.

Unwin's equation gives

$$\frac{H}{L}=\frac{0{\cdot}0004\times v^{1{\cdot}87}}{d^{1{\cdot}4}},$$

$$\frac{25}{5000}=0{\cdot}005=0{\cdot}0004\times v^{1{\cdot}87},$$

$$v=\left(\frac{0{\cdot}005}{0{\cdot}0004}\right)^{\frac{1}{1{\cdot}87}},$$

$$\log v=\frac{1}{1{\cdot}87}\log 12{\cdot}5,$$

$$v=\underline{3{\cdot}4}\text{ feet per second.}$$

Let Q = quantity flowing per second, cubic feet;
A = cross sectional area of pipe, square feet.

$$Q = v \times A$$
$$= 3{\cdot}4 \times \frac{\pi d^2}{4}$$
$$= 3{\cdot}4 \times 0{\cdot}7854$$
$$= \underline{2{\cdot}67} \text{ cubic feet per second.}$$

In addition to the losses due to fluid friction in the straight parts of a pipe, any sudden enlargement or contraction, or bends in the pipe, will cause the flowing water to break up into eddies, and will therefore produce further losses. Any change of section should be gradual, and all bends should be made very easy, merging gradually from the straight to the curved portion of the pipe.

Expt.—Arrange a horizontal glass pipe AC, Fig. 332, contracted at B and having a short branch D attached about the

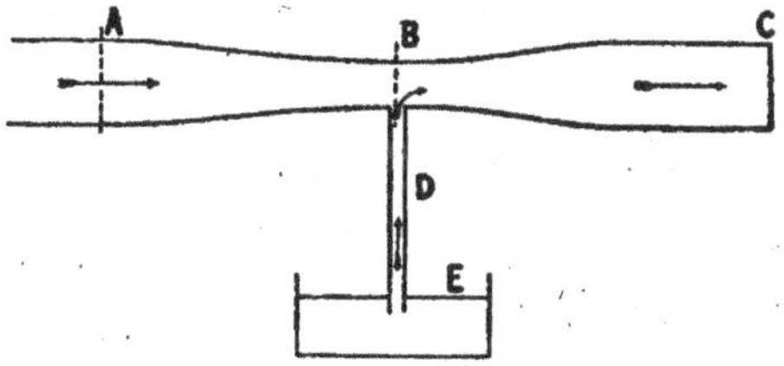

Fig. 332.—Apparatus showing change of pressure in water flowing along a pipe of varying section.

middle of the contracted part. Let D dip into a vessel E containing water coloured with red ink. Connect the end A, of the horizontal pipe, by means of a rubber tube to a water tap. The other end, C, should be over a sink, into which the water passing along the pipe AC will be discharged. Now turn on the water. In a few seconds steady flow will occur in AC and it will be found that the coloured water in E will pass up D and join the flowing water in AC. The explanation of this is as follows. Since the water is flowing steadily in AC, equal quantities must pass every section in the same time, consequently the velocity of the water must be greater at B than at A or C. By the principle of the conservation of energy, and neglecting frictional losses, **the sum of the pressure energy and the kinetic energy of a given quantity of the flowing**

water must remain constant. But the kinetic energy is greater at B than at A or C, for the velocity is maximum there, therefore the pressure energy must be less at B than it is at A or C. Now the pressure at C is that due to the atmosphere, and therefore, the pressure at B must be less than atmospheric. It follows that the pressure of the atmosphere on the surface of the water in E will cause a flow from E up D.

Thomson's jet pump.—This experiment illustrates the action of Thomson's jet pump, a section of which is shown in Fig. 333. Water enters at A, and the reduced pressure in the contracted part at B enables a flow of water to take place from a tank situated on a lower level, to which the pump is connected at D. The mixed water is discharged into the atmosphere at C.

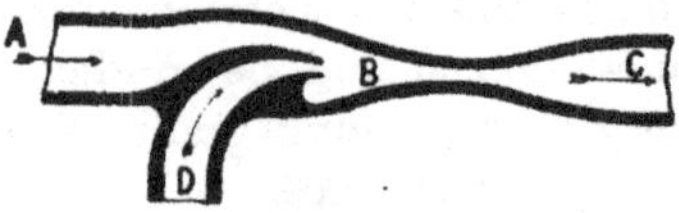

FIG. 333.—Diagram of Thomson's jet pump.

Conversion of the energy of water.—Let us now consider what happens when water falls from a height into a pond of water at rest. Suppose one pound of water to overflow from a cistern A (Fig. 334), of surface level H feet above the surface level of a pond B, into which the water falls. In A, the pound of water possesses potential energy equal to H ft.-lbs. This energy is gradually changed into kinetic energy during the fall, until at the surface level of B the whole of the potential energy has been transformed into kinetic energy, which, if the velocity of the water is v feet per second there, will be equal to $\frac{v^2}{2g}$ foot-pounds.

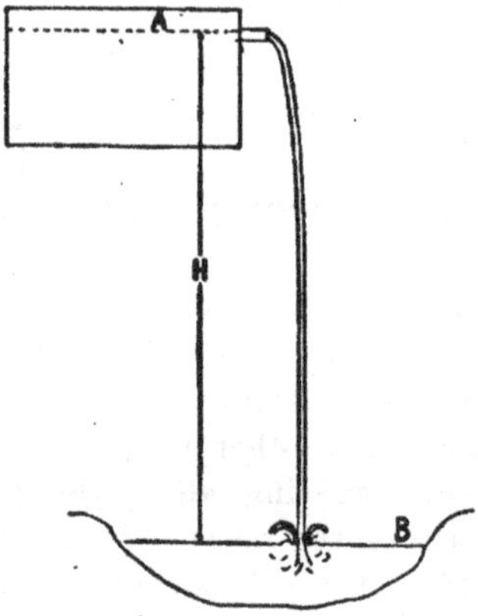

FIG. 334.

$$\text{Potential energy lost} = \text{kinetic energy gained, or } H = \frac{v^2}{2g}.$$

The water in B will be disturbed by the water entering it, but presently quietens down again, that is, the whole of the kinetic energy of the pound of water has been dissipated in creating disturbances in *B*, and none has been utilised in producing useful work. Useful work may be derived from the *H* foot-pounds of potential energy available by permitting the water to descend through a pipe, thereby producing pressure energy at the level of *B*, which may be converted into mechanical work by driving the pistons of a water engine. Or, the energy available may be utilised by means of a water-wheel of which there are three varieties—over-shot, breast-shot, and under-shot.

In the **over-shot wheel** (Fig. 335) water is brought to the top of the wheel, which has buckets fastened all round its rim; the water enters these buckets and remains in them until the wheel, turned by the extra weight of water on one side, has brought the buckets into such a position that the water is spilled out. In large wheels such as this it is usual to have teeth secured to the inner rim of the wheel and the power is taken from the wheel by a shaft carrying a pinion gearing with these teeth. This is to prevent the large stresses which would be introduced into the arms of the wheel if the power were taken from the shaft on which the wheel turns.

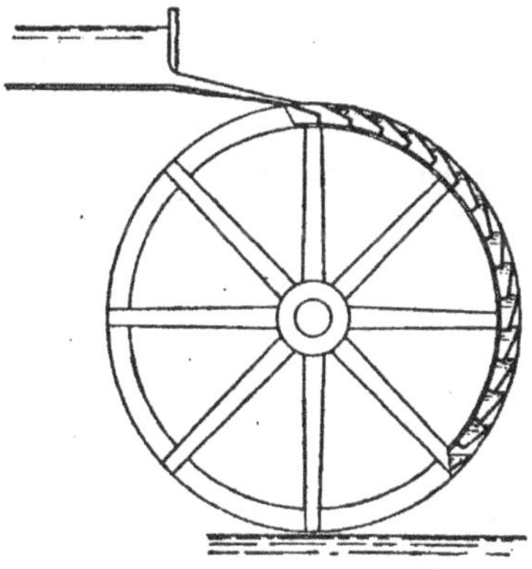

FIG. 335.—Over-shot water wheel.

In **breast-shot wheels** the water enters the buckets half-way up and remains in them until the bottom of the wheel is nearly reached, when it is spilled out.

In **under-shot wheels** the water is allowed to acquire as much velocity as possible before reaching the wheel and is then allowed to impinge on the blades. In this last case the change is from kinetic energy to mechanical work, in the others the mechanical work is done directly by the gravitational effort on the water in the buckets.

Turbines are machines used for converting the energy of water coming from a height into mechanical work. These are of two kinds—one in which the energy of the water is partly pressure and partly kinetic in passing through the machine, these being called **reaction turbines**; and another kind in which the energy of the water is wholly kinetic on reaching the machine, these being called **impulse turbines.**

The turbine consists of a wheel having blades running in a casing furnished with guide blades. The entering water is guided by these blades so as to have tangential velocity and consequently tangential momentum. This momentum is abstracted during the passage through the wheel by the action of the curved blades on the wheel. Consequently, pressure is exerted on the rotating wheel, and work is done thereby. Reaction wheels, in which the water has its energy partly in the pressure form, must run full of water; in impulse wheels, on the other hand, the pressure of the water is atmospheric or nearly so, and the water slides along the blades in comparatively thin streams.

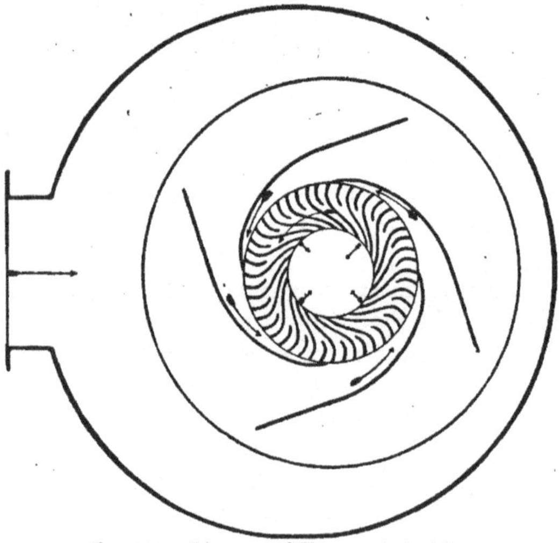

Fig. 336.—Diagram of Thomson's turbine.

In **Thomson's turbine** (Fig. 336) the water enters the wheel at its outer circumference, being guided by four blades which may be adjusted to suit varying quantities of water passing ; it then passes through the wheel, moving inwards, and is discharged at the inner circumference. The shape of the guide blades and of the wheel blades at the outer circumference is such that *the velocity of the entering water relative to the wheel is along the wheel blade* ; consequently the water enters without shock. At the inner circumference the shape of the blade is such that *the water leaves with radial velocity only.*

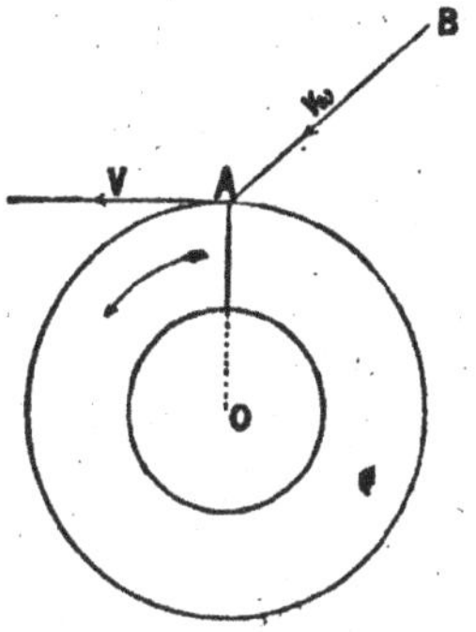

Fig. 337.—Turbine wheel having radial blades.

Wheel having radial blades.—Suppose we have a wheel with radial blades, as in Fig. 337, and that the velocity of the blade at A is V. Let AB be the direction of the guide blades, then the direction of the water moving towards the wheel will be along BA. For the water to enter the wheel without shock, its velocity relative to the wheel must be along AO. Give the wheel at A and a particle of water just leaving the guide blade each a velocity equal to V, to the right. This will stop the wheel. Let P (Fig. 338) be the particle of water. It will now have a velocity V represented by PC, to the right, and a velocity V_w along the direction of the guide blade. The resultant of these is V_r, represented by PD, and this will be the relative velocity of water and wheel blade. PD will evidently be along the blade at A if we give the proper value PE to V_w, the velocity of the water along the guide blade.

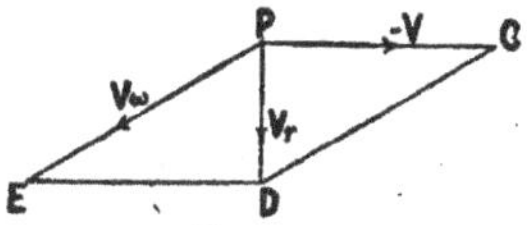

Fig. 338.

Wheel having curved blades.—A similar construction will give the proper velocities if the wheel blade is curved at A instead of being radial. This is shown in Figs. 339 and 340. PD equal to V_r is here the relative velocity of water and wheel, and is directed along the blade at A. PE equal to V_w is the velocity of the water along the guide blade. At O, the outlet, the relative velocity of the water and the blade must be along the blade, that is, along FO, and for radial discharge, the direction of this velocity must be chosen to suit the speed of the wheel blade there. Let V_0 be the tangential velocity of a point on the blade at O, and let this be represented by PG in Fig. 341. Stop the wheel as before, by giving P, and a particle of water at P, velocities equal to V_0, but in the opposite sense; this is shown by PH equal to V_0. A particle of water at P will now have a radial velocity v represented by PK at 90° to PH and also a velocity V_0 represented by PH. The resultant of these, V_r equal to PL,

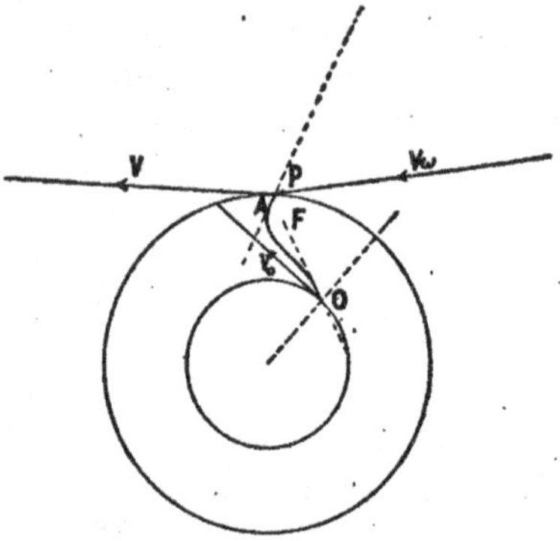

FIG. 339.—Turbine wheel having curved blades.

will be the relative velocity of water and blade at P, and consequently the direction of the blade at the inner circumference must be along LP.

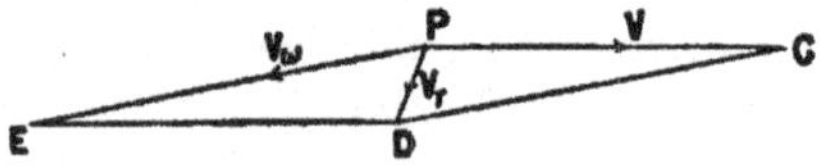

FIG. 340.

Horse-power of wheel.—The entering velocity V_w being now known, and the exit velocity being radial, we may easily find the momentum changed by passage through the wheel. Thus, let

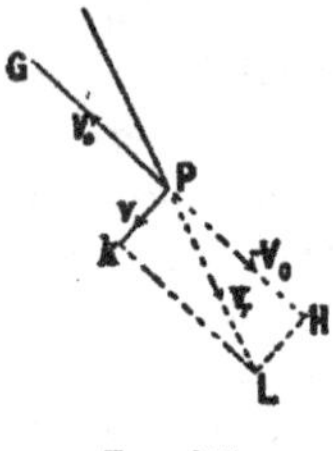

FIG. 341.

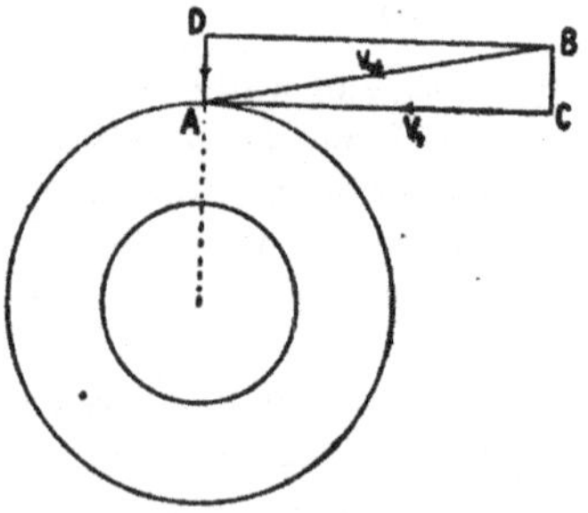

FIG. 342.

BA, equal to V_w, be the velocity of the entering water in Fig. 342; resolving this into tangential and radial components, we find the tangential velocity V_t represented by CA. This disappears on passage through the wheel, consequently the tangential momentum changed per pound of water is equal to $1 \times V_t$. Let m lbs. of water pass per second, then

$$\text{Change of momentum per second} = m\,V_t, \text{ and}$$

$$\text{Force given to wheel at } A \text{ in consequence of this} = \frac{m\,V_t}{g} \text{ lbs.}$$

Therefore, Work done per second $= \dfrac{m\,V_t}{g} \, . \, V$ ft.-lbs., where V is, as before, the velocity of the wheel at A, and

$$\text{Horse-power} = \frac{m\,V_t}{g} \, . \, V \, . \, \frac{60}{33{,}000}.$$

EXAMPLE. Suppose 500 lbs. of water per second to be delivered to a wheel with a tangential velocity of 40 feet per second. The velocity of the wheel rim is 35 feet per second. The water leaves the wheel radially. What horse-power can be developed?

$$\text{Pressure on wheel} = \frac{500 \times 40}{32{\cdot}2} \text{ lbs.}$$

$$\text{Work per second} = \frac{500 \times 40}{32{\cdot}2} \times 35 \text{ ft.-lbs.}$$

$$\text{H.P.} = \frac{500 \times 40}{32{\cdot}2} \times 35 \times \frac{60}{33{,}000} = \underline{\underline{39{\cdot}5}}.$$

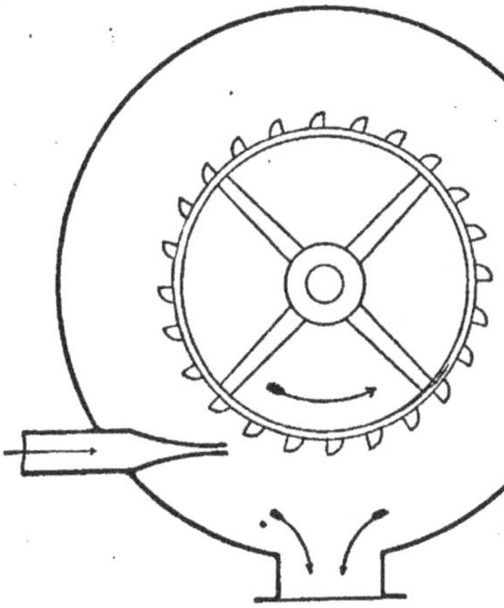

FIG. 343.—Diagram of the Pelton wheel.

The **Pelton wheel** is an example of an impulse wheel. It consists of a wheel running in an outer casing (Fig. 343), and having blades or buckets arranged round its rim. A jet of water impinges on the buckets, gives up its momentum to the wheel, and escapes by a pipe from the lower part of the casing. The shape of the buckets is shown in plan in Fig. 344; the jet impinges centrally, divides and circles round the curved bucket, and is then discharged. The shape is semi-circular in plan, and is such that the maximum amount of momentum is abstracted from the water. If the bucket were at rest, the water would be directed backwards with a velocity equal to its original one, v_1. The whole change of momentum per pound of water would be $2 \times v_1$, and the pressure on the bucket due to this would be $\frac{2v_1}{g}$ lbs. If the wheel had such a speed of rotation that the velocity of the bucket was equal to that of the jet, no momentum would be changed, and the pressure would be zero. In either case, no work would be done. At a speed of rotation such that the buckets have a velocity half that of the jet, these conditions giving the theoretical maximum efficiency, the water would leave the bucket with little or no velocity relative to the earth, and consequently would have a maximum quantity of energy abstracted from it. The whole momentum $1 \times v_1$ of a pound of water in the jet would be changed. The efficiency of such a wheel would be 100 per cent., only the imperfect action of the water reaching the buckets, due to their different inclinations caused by the rotation of the wheel, and the interference of one bucket just entering the jet with the supply going to another, prevent this. An efficiency of about 80 per cent. can be attained.

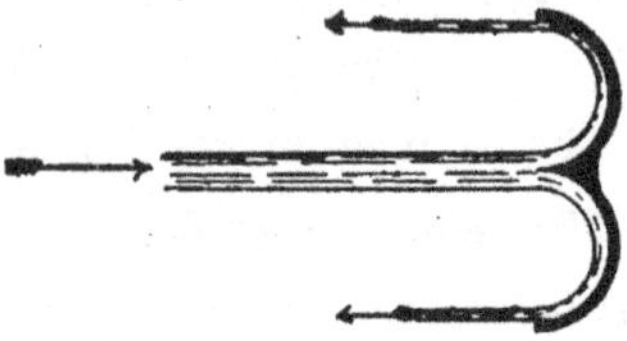

FIG. 344.—Plan of bucket.

Experimental Pelton wheel.—In Fig. 321 a small Pelton wheel may be observed at the extreme right of the apparatus. This is connected to the upright tank so that water can be supplied to it at any pressure up to 100 lbs. per square inch. A speed counter driven by a small worm and worm wheel counts the revolutions of the wheel. A pulley on the wheel shaft has a brake fitted to it by which the horse-power of the wheel can be measured. The water is discharged from the wheel casing, when done with, into the trough below, whence it finds its way over the gauge notch into the measuring tank on the left of the apparatus. As both the head of water and the speed of the wheel can be altered independently of one another, this forms a very useful experiment for the student. The records of a test on this wheel with constant head are given.

Test on a Pelton Wheel.

Diameter of bucket wheel $5\frac{3}{4}''$ to centres of buckets.

Diameter of brake wheel 6″ to centre of cord.

Pressure of constant water supply 40 lbs. per sq. inch, giving head $= H = 92\cdot16$ feet.

The theoretical velocity corresponding to this head would be

$$v = \sqrt{2gH} = 76\cdot8 \text{ feet per second.}$$

The actual velocity of the jet was obtained by measuring the water used by the wheel per minute; this amounted to $91\cdot5$ lbs. $= W$.

Water per second $= 0\cdot0244$ cubic feet.

Diam. of tapered nozzle $= 0\cdot25''$.

Area of jet $= 0\cdot000341$ sq. feet.

$$V_1 = \text{Velocity of jet} = \frac{0\cdot0244}{0\cdot000341} = \underline{71\cdot5} \text{ feet per second.}$$

This latter value has been used in working out the results.

$$V_2 = \text{velocity of bucket}$$

$$= \frac{\pi \times 5\frac{3}{4}}{12} \times \text{revolutions of wheel per second,}$$

$$= 1\cdot5 \times \frac{N}{60} \text{ feet per second.}$$

WH = energy supplied per minute $= 91\cdot5 \times 92\cdot16$ ft.-lbs.

$$\text{H.P. supplied in water} = \frac{91\cdot5 \times 92\cdot16}{33{,}000} = \underline{0\cdot278.}$$

$$\text{B.H.P.} = \frac{(P - W)2\pi R \, . \, N}{33{,}000} = \frac{1\cdot57}{33{,}000}(P - W)N$$

$$= 0\cdot000048(P - W)N.$$

Results of Test.

Speed of wheel revolutions per min., N.	Velocity of bucket, V_2 ft. per sec.	Brake loads.		Brake horse-power.	Ratio $\frac{V_2}{V_1}$.	Efficiency $=\frac{\text{B.H.P.}}{\text{H.P. supplied}}\times 100$.
		P lbs.	W lbs.			
650	16·2	2·01	5·75	0·117	0·227	42·1
800	20·0	1·51	5·00	0·134	0·28	48·3
1000	25·0	1·01	4·25	0·155	0·35	55·9
1150	28·74	0·71	3·50	0·154	0·402	55·5
1200	30·0	0·51	3·00	0·143	0·42	51·5
1400	35·0	0·21	2·00	0·120	0·49	43·2
1480	37·0	0·11	1·75	0·116	0·518	41·8

The speed of the wheel was adjusted by varying the brake load, and as the conditions, when altered, settled down at once,

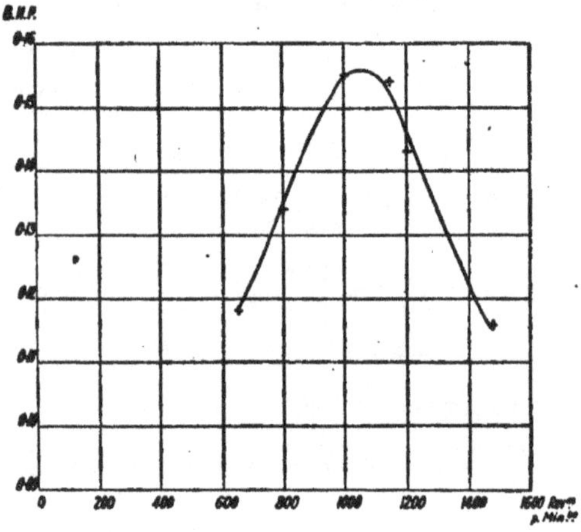

Fig. 345.—Test on a Pelton wheel; curve of B.H.P. and revolutions.

each test lasted only about 3 minutes. Fig. 345 shows a plotted curve of B.H.P. at the different speeds of revolution, and in Fig. 346 the efficiency of the motor for the varying values of $\frac{V_2}{V_1}$ has been plotted. It will be noticed that maximum efficiency is obtained under the given conditions when the speed of the bucket is about 0·38 that of the jet.

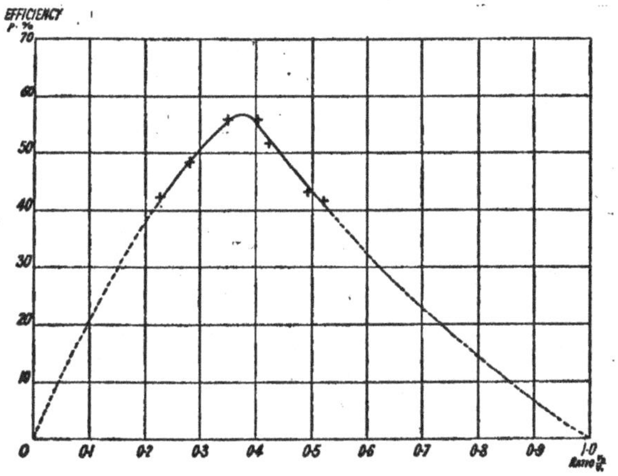

FIG. 346.—Test on a Pelton wheel; curve of efficiency for ratios $\frac{V_2}{V_1}$.

Centrifugal pumps.—In centrifugal pumps, water enters the centre of a wheel furnished with blades. The water enters as nearly as possible with radial velocity only, and the blades are so shaped as to gradually give rotational velocity to the water as it passes outwards. This, of course, simply means that a further store of kinetic energy is added to the water while passing through the wheel. The water escapes at the outer circumference of the wheel into a large circular chamber, called the **whirlpool chamber.** Here its velocity is allowed gradually to diminish, with the effect that its pressure energy increases. It will now consequently be able to overcome the resistance of a considerable head of water and will be able to flow up a pipe into a tank above. Centrifugal pumps deriving their supply of water from a reservoir on a lower level than the pump require a foot valve on the pipe in the reservoir and the supply pipe and pump must be charged with water before starting. When the pump starts, the increase of kinetic energy of the water near the wheel centre causes the pressure energy to diminish there, the pressure consequently falls below atmospheric, and the pressure of the atmosphere on the surface of the water in the reservoir causes it to flow up the pipe into the wheel. At *A* (Fig. 347) therefore, the energy of a pound of water is wholly pressure energy and is that due to atmospheric pressure. At *B*, the wheel centre, the energy is partly potential, due to the elevation *AB*; partly pressure, but less than that due to atmospheric pressure; and partly kinetic. The wheel adds largely to the kinetic energy, and in the whirlpool chamber this partly is changed again into pressure energy. At *C*, the discharge pipe entrance, the energy is largely pressure energy, partly potential due to the elevation *AC* and partly kinetic, this last being due to the velocity in the discharge pipe necessary to produce flow. It should be observed here, that if the discharge pipe is of too great height, the whole of the energy of the water at *C* may be made up of potential and pressure forms only, caused by the head being too great. The absence of kinetic energy means that there will be no velocity of flow and consequently no water will be discharged. Lowering the level of *D* even a small amount may correct this and give a very good efficiency where previously there was none. The potential energy of the water gradually becomes greater as it ascends the discharge pipe and if the pipe is of uniform section, its velocity will remain constant and therefore its kinetic energy also will be constant. The gain of potential energy, is therefore at the expense of pressure energy, which accordingly becomes less. Finally, the water is discharged with potential energy, pressure energy due to atmospheric pressure, and kinetic energy, which last is soon dissipated in the upper tank in surging of the water. The total gain of energy due to the whole arrangement will be

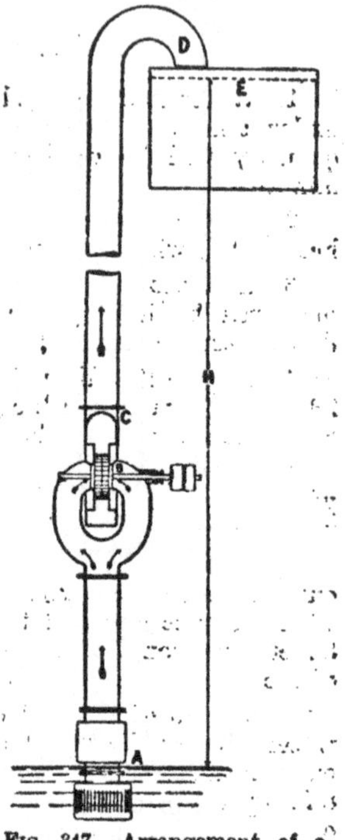

FIG. 347.—Arrangement of a centrifugal pump.

simply the difference between the potential energies of a pound of water in *E* and in *A*, that is, *H* foot-lbs.

EXERCISES ON CHAP. XVIII.

1. Water is discharged through a circular orifice in a thin plate under a head of 60 feet. If the diameter of the orifice is 1″, what quantity will be discharged; give your answer in gallons per minute.

2. Water flows over a V-notch, the angle of which is 90°; if the head is 15″, what quantity passes per minute?

3. A rectangular gauge notch is 2 feet wide. The head of water measured by hook gauge is 5″; calculate the quantity of water flowing per minute. There are two end contractions.

4. A pipe 24″ diam. is 2500 feet long and has a fall of level of 20 feet. What probable quantity of water will flow per hour? State the answer in gallons.

5. 4 cubic feet of water per second enter an overshot wheel whose diameter is 40 feet. Taking an efficiency of 65 per cent., what horse-power can be obtained from the wheel?

6. A wheel has radial blades, and water flows through it from outside to centre. The speed of the wheel at the outer rim is 40 feet per second and the radial velocity of the entering water is 8 feet per second. Find, and show in a diagram, the actual velocity of the entering water, if there is to be no shock.

7. In Question 6, what will be the tangential velocity of the water leaving the wheel blades if the radius there is one half the outer radius? Supposing the radial velocity of the water there is 8 feet per second, show in a diagram what alteration must be made on the blade in order to discharge the water radially?

8. The radial velocity of water in a centrifugal pump wheel is 2 feet per second; the vane makes an angle 35° with the outer circumference; what is the velocity of the water relatively to the wheel? (A graphical method of solution may be used.) (1900.)

9. A horizontal pipe of 12 inches diameter gradually becomes of 3 inches diameter and then becomes of 12 inches diameter again. There is a flow of water of 5 cubic feet per second. Neglecting friction, calculate and state how the pressure alters along the axis of the pipe. (1901.)

10. Ten cubic feet of water per second enters a turbine wheel with a tangential velocity of 50 feet per second; it enters without shock, the velocity of the rim of the wheel being 50 feet per second; the water leaves the centre of the wheel with only a radial velocity; what energy does the water give to the wheel per second? (1901.)

CHAPTER XIX.

MATERIALS.

Iron in its many forms is the metal chiefly used by the engineer and builder. It is reduced from its ores in the *blast furnace*, the resulting product being called **pig iron**. Pig iron contains from 2 to 5 per cent. of *carbon*, which may be simply mechanically mixed with the iron, such iron being called **grey iron**, or the carbon may be in *combination with* or *in solution in* the iron, the iron being then known as **white iron**. Grey iron is used for foundry purposes, as it gives good castings. White iron is brittle and very hard. It fuses at a lower temperature than grey iron, but assumes a pasty condition before becoming quite liquid. White iron is used for wrought iron and steel production.

Cast iron.—Casting into the required forms is performed from a furnace called a *cupola*. The pigs are melted in this furnace, different grades being combined in certain proportions, and the resulting mixture is cast into sand or loam moulds having the desired form. Rapid cooling after the iron is in the mould tends to produce white iron; hence, when castings are required to be very hard they are cast into moulds chilled by circulating water and are known as **chilled castings**. Cast iron is a *crystalline* metal, weak under tension and strong under compression. It lacks ductility and cannot be welded.

Wrought iron is simply iron, as nearly pure as possible, which has been manipulated so as to produce a *fibrous* structure instead of a crystalline. It is produced from pig iron by a preliminary refining process in which many of the impurities are removed. Fusion in a **puddling furnace** follows, the iron being removed from the furnace in a pasty condition to be hammered or squeezed. This process gets rid of slag and consolidates the mass. The iron is then rolled into **puddled bars.** To improve the quality, these bars are cut into short lengths, *bundled*, *heated*, and *hammered* together, the quality of the iron being improved with each repetition of these processes. The iron is then rolled into the market forms of plates and bars. Wrought iron is a ductile, fibrous metal, easily malleable, and can be welded.

Steel is iron as nearly pure as possible with the addition of a small quantity of carbon, the proportion varying according to the purpose for which the steel is intended. *Mild steel*, which closely resembles wrought iron, may contain from 0·15 to 0·25 per cent., and *hard steel* up to 1·5 per cent. of carbon.

The finest qualities of hard steel are produced by the **cementation process,** which consists in baking high quality wrought-iron bars in contact with charcoal for about a fortnight. Carbon from the charcoal enters into the iron during the process, converting it into *blister steel.* The process may be stopped at different stages, giving steel suitable for the manufacture of springs, tools, cutlery, etc., by subsequent reheating, rolling, and hammering. *Cast steel* is produced by remelting blister steel in crucibles. A more uniform metal results than can be obtained by hammering.

In the **Bessemer process** steel suitable for rails, etc., is produced direct from pig iron containing silicon, by burning off the carbon and impurities in a *converter*, a powerful blast being supplied. *Spiegeleisen* (pig iron containing manganese and a large proportion of carbon), or *ferro-manganese*, is added to bring the carbon to the right proportion, and the metal is then cast into *ingots*, being afterwards rolled into plates or bars of the desired form. A "blow" in the Bessemer process occupies about 30 minutes.

In the **Martin process** wrought iron or steel scrap and pig iron are melted together to produce steel; in the **Siemens process** pig iron and ore are used; in the **Siemens-Martin process,** which is a combination of the two, scrap iron, pig iron, and ore are used together. These processes are conducted in a Siemens *regenerative furnace*, and are known as **open-hearth processes.** The steel is cast from the furnace into ingots, which are then cogged, or hammered, to consolidate the metal, and sheared and rolled into plates and bars.

The process of steel production by open-hearth may occupy from 9 to 12 hours, and the whole being under perfect control, steel can be produced to specification. The **mild steel or ingot iron** produced differs from wrought iron only in being more homogeneous, having been produced from a cast ingot instead of from a puddled ball containing slag. It may contain the same proportion of carbon as wrought iron, and has a greater strength and is more ductile. It also forges well.

Hardening and Tempering.—Steel containing 0·5 per cent. and upwards of carbon can be hardened by dipping it when hot into water, or other liquid, so as to cool it suddenly. It is then very hard and brittle and not suitable for many purposes, but can be tempered by carefully reheating it to certain temperatures, when it loses its brittleness and some of its hardness. The temperature at which steel acquires a temper suitable for various purposes is practically ascertained by the *colour* of the skin of oxide which comes over its originally clean surface during reheating. These colours and approximate temperatures are:

COLOURS AND TEMPERATURES DURING TEMPERING.

Temperature.	Colour.	Suitable for
220° C.	Faint yellow.	(Very hard.) Surgical instruments.
230°	Straw.	Razors.
245°	Dark straw.	Wood-working tools, taps, and dies.
255°	Brownish yellow.	Chisels.
265°	Purplish brown.	Axes, planes, chipping chisels.
275°	Purple.	Table knives.
285°	Light blue.	Swords, springs.
295°	Dark blue.	Fine saws, augers.
315°	Blackish blue.	Large saws.

The ordinary workshop method of tempering a turning tool, or chisel, is to heat it to a bright red, then plunge about one inch of the point into water till this portion is cold, withdraw and clean the point with a piece of sandstone or emery; the heat travels down into the point from the hot body of the tool and the colours of the oxide formed on the point are watched. When the proper colour has been attained, the whole tool is dipped again into water and held there till cold.

Malleable castings are produced from the ordinary cast-iron castings by heating them for several days in contact with some substance, such as red haematite, which will remove the carbon.

Case hardening consists in giving a surface of steel to wrought iron articles. This process is effected by heating the articles in contact with some substances, such as charcoal, leather, and ferrocyanide of potassium, which will give carbon to the iron. The operation resembles the cementation process, but is not allowed to go so far, as only a thin layer, $\frac{1}{8}''$ to $\frac{1}{4}''$ thick, of steel is required.

Copper is largely found native, and is also produced by reduction from its ores. It is much used in sheets, bars, and as wire, for fire boxes, sheathing, tubes, boiler stays, nails, electrical conductors, etc. Copper castings are not much used, being poor and expensive. Copper is strong and malleable, is easily wiredrawn, and can be made up into many forms by hammering or brazing. The metal is best forged at a moderate red heat and has its strength improved by the process. It is hardened by working, but may have its ductility restored, and at the same time its strength reduced, by heating and quenching in water. When pure, copper has a high electrical conductivity. It is a good conductor of heat and has a high power of resisting corrosion by air and water.

Tin is not found native, being generally extracted from its ore *tinstone*. Its strength and ductility are low, but it is very malleable and is easily made into foil—tin-foil. It is very brittle when at temperatures near its melting point. When pure tin is bent, it makes a crackling noise. The metal does not tarnish much at ordinary temperatures, and its melting point is low. The alloys of tin with other metals are valuable, and the metal is much used for coating other metals to protect them.

Thin iron plates are coated with tin by first undergoing a process of scouring and pickling in order to thoroughly clean their surfaces and then dipping in a bath of molten tin. The resulting **tin plate** preserves well so long as the tin coat remains unbroken, but rusts very rapidly directly this is damaged, on account of the galvanic action set up. Copper and brass are often tinned.

Zinc is obtained from its ores by a process of distillation. It is brittle at ordinary temperatures, but is malleable and can be rolled into sheets at temperatures from 200° to 300°F. At temperatures about 400°F. it becomes brittle again. Zinc gives good castings suitable for art work. Its principal use in engineering is for coating iron plates in order to protect them, the resulting sheets being known as **galvanised iron.** The protection afforded by zinc is more effective than that by tin.

Lead is reduced from its ores by treatment in a reverberatory furnace connected to a long flue in which the lead particles are condensed. Lead is very soft and heavy. Its tensile strength is low, and it does not give good castings. The metal is easily rolled into sheets and made into pipes. Lead pipes are sometimes lined with tin in order to prevent the formation of poisonous salts under the action of water. Lead is used for a variety of minor purposes in engineering.

Aluminium is a very widely distributed metal, and forms the base of clay. Aluminium resembles zinc in colour and hardness; it is very light, having a specific gravity about one-third that of wrought iron. As the tensile strength of aluminium is also about one-third that of wrought iron, it follows that wrought iron and aluminium, *weight for weight*, have equal tensile strengths. The metal is easily rolled into sheets, and can be wiredrawn, forged, and cast. A skin of oxide forms very rapidly on the surface of the metal, preventing further corrosion, but at the same time making it very difficult to secure satisfactory soldered or brazed joints. Aluminium alloys readily with other metals, and the alloys are very valuable. The strength of the metal is increased by cold hammering.

Copper alloys with zinc are called *brasses*, with tin—*bronzes*.

Brass.—**Ordinary brass** is made of two parts of copper to one part of zinc by weight. A higher proportion of copper gives a better metal. Lead, present in small quantities, gives an alloy easier to machine, but too much lead produces brittleness. **Sterro metal** consists of copper and zinc in about the same proportions as for brass, with the addition of small quantities of tin and iron. The metal is strong, non-corrosive, and non-porous. **Delta metal** consists of brass with small quantities of iron and phosphorus. The metal can be wrought both hot and cold, and can be wire drawn. **Muntz metal** contains 66 per cent. copper, 33 per cent. zinc, and 1 per cent. lead. It can easily be rolled into sheets.

Bronzes.—**Gun-metal** consists of 90 per cent. copper and 10 per cent. tin. This metal is much used by engineers for bearings, cocks, valves, etc. It gives good castings. **Aluminium bronze** contains 90 per cent. copper and 10 per cent. aluminium. The metal can be forged, and is very durable. **Phosphor bronze** consists of any brass or bronze of copper, tin, and zinc, with the addition of phosphorus. The metal is very strong. **Manganese bronze** contains 88 per cent. copper, 10 per cent. tin, and 2 per cent. manganese. Its strength and ductility are about the same as those of mild steel. The metal can be forged well when hot, and may be rolled into sheets.

Building stones.—The commonest stones used for building purposes are granites, sandstones, and limestones. These stones are derived from rocks which have either been first in a state of fusion under the action of heat and have then consolidated, or have been deposited under water and stratified, that is, built up of layers. **Granite** is unstratified, and is composed of quartz, felspar, and mica. The best quality is very strong, hard, and durable, and is much used in engineering work. The stone takes a high polish. **Sandstone** consists of small grains of quartz cemented together. As the quartz is practically indestructible, the strength of the stone depends on the nature of the cementing material, which in the best stone is silica, and in the worst—alumina. **Limestones** vary in compactness from chalk to marble. Portland stone and Bath stone are prized for their durability.

Bricks are made from clay which is dug in the autumn and cleared of gravel, etc., by hand picking, or, if much gravel is present, crushing between rollers is necessary. The clay is left over one or more winters, so that frost may disintegrate it. *Tempering* follows, consisting of working the clay with the spade or by machine. The prepared clay is then moulded into bricks, air dried, and burned either in stacks or kilns.

Lime is made from limestone, or other mineral containing carbonate of lime, by *calcination* in kilns to drive off water and carbonic acid, the resulting material being called *quicklime.* Quicklime, when sprinkled with water, swells and breaks up into powder, and heat is evolved. This is called *slaking.* This powder which results, when made into a paste with water and left, soon hardens, or *sets*, the operation consisting in the absorption of carbonic acid from the air, thereby reconverting the material into carbonate of lime.

Cements are either *natural* or *artificial.* The most important natural cement is **Roman cement**, which is made by calcination from nodules found in London clay. It sets quickly but is not strong. **Portland cement** is artificial, and is made generally from chalk and clay by mixing in suitable proportions, burning in kilns, and then grinding to a fine powder. Portland cement is much used by the engineer and builder. **Sand** is often mixed with cement for building purposes. The strength of Portland cement diminishes as the proportion of sand is increased. The action of the cement is to form a binding material uniting the grains of sand together, consequently the cement and sand should be thoroughly mixed both dry and wet. The strength of Portland cement alone, and also mixed with sand, gradually increases for several months after setting.

Concrete is made from ballast, broken bricks, etc., by mixing with cement or lime, and water. It is much used for foundations.

Timber is obtained from exogenous trees, that is, trees in which the growth takes place by successive additions on the outside of the woody matter already formed. The oak, fir, beech, etc., are examples of these. Endogenous trees, such as palms, are unsuitable for timber, as their structure consists of independent fibres cemented together, thus making the wood too flexible for structural purposes.

If the section of the trunk of a tree suitable for timber be examined, the following facts may be noticed. The pith occupies the centre or nearly the centre ; outside of the pith is seen a broad ring of fully-formed and matured wood called the *heart wood* ; outside of this, again, *sap wood* occurs, and then a thin layer of slimy matter called *cambium* between the wood and the bark, and outside of all the *bark*. The woody matter is made up of more or less distinct rings, one of these being added each year in our climate, hence the name—*annual rings*. Radiating from the centre may be noticed narrow strips called *medullary rays*.

The growth of the tree takes place by the ascent from the roots of water and mineral salts, this ascent being promoted by the medullary rays; these substances on reaching the leaves, and being acted on by the constituents of the atmosphere in the presence of sunlight, take in carbon from the carbonic acid of the atmosphere and then descend to the cambium, which consists of active cells, and transforms the food supplied from the leaves into woody matter to be added to that already existing.

Timber is best felled in winter, as the resulting wood is better than that felled during active growth in the warmer months. It is converted after felling into forms suitable for the market, and seasoned to remove the sap and water which would, if left, destroy it.

Seasoning may be done by *natural means*, viz., stacking the timber in a dry sheltered place and providing for ample ventilation of dry air, or *artificially* by hot air, or otherwise. Timber is liable to crack or develop shakes during seasoning, such as *heart shakes* extending radially from the outer parts, *star shakes* radiating from the centre, and *cup shakes* in which separation occurs along the circumference of the annual rings.

Timber also *shrinks* in seasoning, chiefly circumferentially along the annual rings, not so much radially and very little in the direction of the length. Knowledge of these facts is taken advantage of in the conversion, by cutting up the stem of the tree in such a manner that shrinkage affects the pieces in the least harmful way.

Timber is liable to *dry rot*, in which case the woody structure becomes powdery. This fault arises from want of sufficient ventilation. The growing tree is often injured by attacks of insects and certain plants, and the timber from them also, is liable to insect attacks after the structure is erected.

Timber is preserved by *painting* or *tarring*, but the wood must be thoroughly dry first, or the moisture is simply confined by the coating and the wood will rot. *Creosoting* is the most effective way of preserving timber. It is conducted in a closed vessel, the timber being first subjected to a partial vacuum, to withdraw air and moisture from the pores, and then tarry oil mixed with creosote is forced in hot, under a pressure of about 170 lbs. per square inch. The creosote fills all the pores and the oily matter remains also as a coating to the woody fibres, protecting them from damp. Properly carried out creosoting is very effective.

For our purpose, timber may be divided into hard woods and soft woods.

Hard woods.—*Oak* is one of our strongest and most durable timbers. It does not warp much after it is thoroughly dry and if cut properly along the medullary rays shows a beautiful *silver grain* much prized for ornamental purposes. The acid present in oak corrodes metal fastenings, and oak should therefore be put together with wooden pins called *trenails.* The timber may be used in wet positions.

Beech also corrodes metal fastenings; it stands well if kept constantly dry or constantly wet, but will not preserve its shape with alternate wetting and drying. It can be used for positions under water.

Ash is very tough and flexible, and consequently is useful for structures, etc., subject to shocks.

Elm also will not stand alternate wetting and drying, but if constantly in one or the other condition it can be preserved. It warps greatly, is strong against crushing, and is also strong across the grain.

Mahogany can be preserved well dry and does not warp or shrink much. It takes metal fastenings and glue well.

Teak is a very strong and durable timber for engineer's work. It takes metal fastenings well, is very stiff, and resists insect attacks.

Greenheart is very strong, heavy and hard. It burns freely.

Soft woods.—*Northern pine,* also called red or yellow fir, comes from Scotland, Russia and the Baltic. When of good quality the timber is strong and durable and is much used for carpenter's work.

American yellow pine works very easily and is much used for

Table of Ultimate Strength of Materials.

Material.	Tensile strength, tons per sq. inch.	Compressive strength, tons per sq. inch.	Shearing strength, tons per sq. inch.
Cast iron, ...	5 to 15, average 8	25 to 65	6 to 13
Wrought iron—			
Tested in direction of rolling,	20 to 29	16 to 20	22
Tested across direction of rolling, ...	19 to 24		
Mild steel, ...	27 to 32	...	21 to 25
Cast steel, ...	35 to 70	...	...
Copper, cast, ...	8 to 12	...	...
,, rolled, ...	15	...	...
,, wire (hard drawn), .	28	...	...
Tin, ...	2	...	...
Zinc, cast, ...	1 to 3	...	...
,, rolled, ...	8 to 10	...	...
Lead, ...	1	...	...
Aluminium, cast, ...	5	...	...
,, rolled, . . .	6 to 10	...	...
Brass, ordinary, ...	11	...	...
,, wire, ...	20 to 25.	...	...
Sterro metal, ...	35	...	...
Delta metal, cast, ...	22	...	...
,, forged, . .	34	...	...
,, wiredrawn, .	55	...	...
Muntz metal, ...	22	...	...
Gun metal, ...	15	...	...
Aluminium bronze, . .	40	...	...
Phosphor bronze, annealed, .	25	...	...
,, unannealed,	up to 70	...	...
Manganese bronze, ...	28	...	...
Granite, ...	...	6 to 10	...
Sandstone, ...	...	2 to 5	...
Portland stone, ...	...	2	...
Brick, London stock, . .	...	1	...
,, Staffordshire blue), .	...	2 to 6	...
Pine, ...	5	2½	¼
Oak, ...	7	4½	1
Leather, ...	2	...	...

internal work. It takes glue but not nails well, and is durable in dry climates.

Pitch pine contains much resinous matter, is durable, but difficult to work. It rots soon if not kept dry.

Spruce or white deal is tough, but is much subject to warping and shrinkage, and is inferior to red deal in strength; it also breaks easily under shock.

EXERCISES ON CHAP. XIX.

1. Describe very briefly the differences in composition, properties and uses, of cast iron, wrought iron, mild steel, tool steel, and any three alloys of copper. (1896.)

2. What is the difference between hardening and tempering a piece of steel? Describe the process of hardening and tempering a chisel for cutting wrought iron. In what order do the colours successively appear during the process of tempering? (1897.)

3. Choose some cast iron object, and explain why, and where, it may have initial strains and weaknesses. (1898.)

4. What is a chilled casting? A malleable casting? How is each produced? Describe how a wrought iron body is case-hardened. (1898.)

5. Describe the manufacture of any kind of steel; describe its chemical composition and physical properties. (1901.)

6. How is cement made? What is your notion of what occurs when it sets and gradually hardens? What is the effect of the addition of sand? (1898.)

COURSE OF LABORATORY WORK.

General Instructions.—Two Laboratory Note-books are required; in one rough notes of the experiments should be made, and in the other a fair copy of them in ink should be entered.

Before commencing any experiment, make sure that you understand what its object is; also the construction of the apparatus and instruments employed.

Reasonable care should be exercised in order to avoid damage to apparatus, and to secure fairly accurate results.

In writing up the results, enter the notes in the following order:

(1) The title of the experiment and the date on which it was performed.

(2) Sketches and descriptions of any special apparatus or instruments used.

(3) The object of the experiment.

(4) Dimensions and weights required for working out the results; from these values calculate any constants required.

(5) Log of the experiment, entered in tabular form, together with any remarks necessary.

(6) Work out the results of the experiment and enter them in tabular form.

(7) Plot any curves required.

(8) Work out the equations for the curves where possible.

Notes should not be left in the rough form for several days; it is much better to work out the results and enter them directly after the experiments have been performed.

MEASUREMENTS.

1. Using an ordinary caliper and steel rule, find the dimensions of the pieces of bar metal provided.

2. Find the breadths and thicknesses, or the diameters, of the same pieces of material as in Expt. 1, using a micrometer caliper. First find the zero error, if any, of the instrument and correct your readings accordingly. pp. 3-5.

3. Repeat Expt. 2, using a vernier caliper. pp. 5, 6.

4. Weigh the same pieces of materials; calculate their volumes from the dimensions obtained in Expts. 2 and 3; then calculate the specific gravities of the materials. p. 16.

FORCES. SIMPLE STRUCTURES.

5. **Parallelogram of forces.**—Apparatus as in Figs. 33, 34. Perform the Expt. as directed. p. 24.

6. **Forces acting on a pendulum.**—Apparatus as in Fig. 39. Perform the Expt. as directed. Confirm each reading of P and T graphically by the parallelogram of forces. p. 28.

7. **Forces in a simple roof truss.**—Apparatus as in Fig. 42. Find graphically the forces in all the parts for a given load and confirm the results by experiment. p. 29.

8. **Triangle of forces.**—Apparatus as in Figs. 33, 34. Find E and R for two given forces by applying the triangle of forces. Confirm the result by trial. p. 32.

9. **Resultant of several forces.**—Apparatus as in Fig. 51A. Find E and R for several given forces by repeated applications of the parallelogram of forces. Confirm the result by trial. p. 33.

10. **Polygon of forces.**—Apparatus as in Fig. 51A. Arrange a number of forces in the apparatus and let the ring come to rest. Draw the polygon of forces for them and ascertain if it closes. pp. 34-36.

11. **Forces in a derrick crane.**—Apparatus as in Fig. 52. Perform the Expt. as directed (*a*) when the cord is attached to the top of the jib, (*b*) when the cord passes over a pulley at the top of the jib. pp. 36, 37.

12. **Forces in a wall crane.**—Apparatus as in Fig. 57. Same instructions as for Expt. 11. p. 37.

13. **Forces in sheer legs.**—Apparatus as in Fig. 58. Perform the Expt. as directed. p. 39.

14. **Forces on a carriage on an inclined plane.**—Apparatus as in Fig. 60. Perform the Expt. as directed. p. 40.

MOMENTS. PARALLEL FORCES.

15. **Balance of two forces of equal moment.**—Apparatus as in Figs. 63, 64, 65. Prove experimentally that a given force may be balanced by one of equal moment. pp. 42-44.

16. **Principle of moments.**—Apparatus as in Fig. 66. Perform the Expt. as directed. p. 44.

17. **Reactions of the supports of a beam.**—Apparatus as in Fig. 69. Place given loads on the beam. Calculate the reactions of the supports and confirm by reading the spring balances. pp. 45-47.

18. **Resultant of parallel forces.**—Apparatus as in Fig. 70. Perform the Expts. as directed (*a*) when P and Q have the same sense, (*b*) when P and Q are of opposite sense. pp. 47, 48.

19. **Couples.**—Apparatus as in Fig. 72. Perform the Expts. as directed. pp. 49, 50.

20. **Forces in the parts of an engine.**—Apparatus as in Fig. 77. Take $P=10$ lbs. Find by experiment Q, T, S, and V for crank angles 45°, 90°, 135°; confirm the results by applications of the parallelogram of forces. Draw a turning moment diagram for the crank moving from 0° to 180°. pp. 50-52.

21. **Centre of gravity.**—Apparatus as in Figs. 79, 80. Find experimentally the centres of gravity of the thin plates provided. pp. 55, 56.

22. **Tensions in a hanging cord.**—Apparatus as in Fig. 92. Perform the Expt. as directed. p. 60.

23. **Tensions in a stretched chain.**—Apparatus as in Fig. 96. Measure the dip and span of the chain; calculate from these and the weight of the chain H, T_v, T_h, and T. Confirm the results by experiment. pp. 62, 63.

STRENGTH AND ELASTICITY OF MATERIALS.

24. **Extensions of pulled rubber.**—Apparatus as in Fig. 110. Gradually increase the load, noting the extension of the portion CD for each increment; then gradually diminish the load, again noting the changes of length of CD. Plot loads and extensions for both sets of readings. p. 70.

25. **Extensions of pulled wires.**—Apparatus as in Fig. 111. Use wires of steel, iron, copper, brass, etc. Same instructions as for Expt. 24. p. 70.

26. **Wires loaded to breaking under tension.**—Apparatus as in Fig. 111. Perform the Expt. as directed. pp. 77-79.

27. **Stiffness of beams.**—Apparatus as in Fig. 150. Use pieces of tool steel, each 3′ 3″ long, sections $1'' \times \frac{1}{4}''$, $\frac{3}{4}'' \times \frac{1}{4}''$, $\frac{1}{2}'' \times \frac{1}{4}''$, $\frac{1}{2}'' \times \frac{1}{2}''$, $1'' \times 1''$, respectively, arranging them (*a*) as beams supported at

two places, (*b*) as cantilevers. Note the deflections produced by gradually increasing and diminishing the loads. From the results, verify the laws stated for proportional stiffness (p. 99). Plot deflections and loads for each case. From the results work out the value of Young's modulus for each sample. pp. 103-105.

28. **Strength of timber test bars.**—Apparatus as in Fig. 150. Perform the Expt. as directed. Calculate the value of the modulus of Transverse Rupture (p. 102) for each sample. p. 106.

29. **Stiffness of wires under torsion.**—Apparatus as in Fig. 162. Use wires of the same material and having different diameters. Vary the length under test in each case. Plot angles of twist and torque. From the results verify the laws of proportional stiffness. pp. 116-118.

30. **Elastic extensions of a spring.**—Apparatus as in Fig. 170. Perform the Expt. as directed. Plot extensions and loads. From the results verify the proportional laws. p. 120.

LAWS OF FRICTION.

31. **Friction of a slider.**—Apparatus as in Figs. 183, 184. Perform the Expt. as directed. pp. 138-140.

32. **Slider on an inclined plane.**—Apparatus as in Fig. 186. Perform the Expt. as directed. pp. 140, 141.

33. **Effect of extent of surfaces in contact.**—Apparatus as in Figs. 184, 188. Perform the Expt. in the same manner as for Expt. 32. p. 142.

34. **Friction of a cord coiled on a drum.**—Apparatus as in Fig. 191. Determine the ratio of the pulls for angles of lap differing by 90°. pp. 144-146.

35. **Rolling friction**—Apparatus as in Fig. 184 with the addition of the small carriage shown in Fig. 60. Determine, for different loads, the resistances offered to rolling on roads of cast iron, teak, and rubber. Plot loads and resistances. pp. 146-148.

VELOCITY. ACCELERATION. MECHANISM.

36. **The law,** $P=\frac{ma}{g}$.—Apparatus as in Fig. 202. Perform the Expt. as directed. p. 160.

37. **Crank and connecting-rod.**—Apparatus as in Fig. 235. Determine, from the model, piston positions corresponding to crank angles differing by 30°. Plot these in a diagram. p. 183.

38. **Infinite connecting-rod.**—Apparatus as in Fig. 238. Same instructions as for Expt. 38. p. 185.

39. **Oscillating engine.**—Apparatus as in Fig. 239. Same instructions as for Expt. 38. p. 186.

MECHANICAL ADVANTAGE. EFFICIENCY OF MACHINES.

In all the following Expts., follow the procedure given on pp. 196-202.

40. **Simple pulley block.**—Apparatus as in Fig. 252. p. 195.

41. **Pulley blocks.**—Apparatus as in Fig. 253. p. 197.

42. **Rope blocks.**—Apparatus as in Fig. 257. p. 202.

43. **Weston's blocks.**—Apparatus as in Fig. 258. p. 202.

44. **Wheel and differential axle.**—Apparatus as in Fig. 261. p. 205.

45. **Helical blocks.**—Apparatus as in Fig. 263. p. 206.

46. **The crab.**—Apparatus as in Fig. 264. pp. 207-208.

47. **Screw jack.**—Apparatus as in Fig. 266. p. 208.

48. **Hydraulic jack.**—Apparatus as in Fig. 317. p. 264.

MISCELLANEOUS.

49. **Energy of a flywheel.**—Apparatus as in Fig. 277. Perform the Expt. as directed. pp. 223-226.

50. **Simple pendulum.**—Perform the Expt. as directed. p. 243.

51. **Vibrations of springs.**—Perform the Expt. as directed. p. 243.

52. **Specific gravity.**—Find the specific gravities of the pieces of metal supplied. p. 254.

53. **Pressure of the atmosphere.**—Apparatus as in Fig. 304. Perform the Expt. as directed. Compare the result with the reading of a standard barometer and account for any difference noted. p. 255.

HYDRAULICS.

54. **Discharge from an orifice.**—Apparatus as in Fig. 321. Compare the actual flow with the result obtained by calculation. pp. 268-273.

55. **Flow-over gauge notches.**—Apparatus as in Figs. 321, 323. Using (*a*) the V-notch, (*b*) the rectangular notch, compare the actual flow with the result obtained by calculation. pp. 274-276.

56. **Pipe of varying section.**—Apparatus as in Fig. 332. Perform the Expt. as directed. p. 278.

MATHEMATICAL TABLES,

ANSWERS, AND

INDEX.

LOGARITHMS.

	0	1	2	3	4	5	6	7	8	9	1	2	3	4	5	6	7	8	9
10	0000	0043	0086	0128	0170	0212	0253	0294	0334	0374	4	8	12	17	21	25	29	33	37
11	0414	0453	0492	0531	0569	0607	0645	0682	0719	0755	4	8	11	15	19	23	26	30	34
12	0792	0828	0864	0899	0934	0969	1004	1038	1072	1106	3	7	10	14	17	21	24	28	31
13	1139	1173	1206	1239	1271	1303	1335	1367	1399	1430	3	6	10	13	16	19	23	26	29
14	1461	1492	1523	1553	1584	1614	1644	1673	1703	1732	3	6	9	12	15	18	21	24	27
15	1761	1790	1818	1847	1875	1903	1931	1959	1987	2014	3	6	8	11	14	17	20	22	25
16	2041	2068	2095	2122	2148	2175	2201	2227	2253	2279	3	5	8	11	13	16	18	21	24
17	2304	2330	2355	2380	2405	2430	2455	2480	2504	2529	2	5	7	10	12	15	17	20	22
18	2553	2577	2601	2625	2648	2672	2695	2718	2742	2765	2	5	7	9	12	14	16	19	21
19	2788	2810	2833	2856	2878	2900	2923	2945	2967	2989	2	4	7	9	11	13	16	18	20
20	3010	3032	3054	3075	3096	3118	3139	3160	3181	3201	2	4	6	8	11	13	15	17	19
21	3222	3243	3263	3284	3304	3324	3345	3365	3385	3404	2	4	6	8	10	12	14	16	18
22	3424	3444	3464	3483	3502	3522	3541	3560	3579	3598	2	4	6	8	10	12	14	15	17
23	3617	3636	3655	3674	3692	3711	3729	3747	3766	3784	2	4	6	7	9	11	13	15	17
24	3802	3820	3838	3856	3874	3892	3909	3927	3945	3962	2	4	5	7	9	11	12	14	16
25	3979	3997	4014	4031	4048	4065	4082	4099	4116	4133	2	3	5	7	9	10	12	14	15
26	4150	4166	4183	4200	4216	4232	4249	4265	4281	4298	2	3	5	7	8	10	11	13	15
27	4314	4330	4346	4362	4378	4393	4409	4425	4440	4456	2	3	5	6	8	9	11	13	14
28	4472	4487	4502	4518	4533	4548	4564	4579	4594	4609	2	3	5	6	8	9	11	12	14
29	4624	4639	4654	4669	4683	4698	4713	4728	4742	4757	1	3	4	6	7	9	10	12	13
30	4771	4786	4800	4814	4829	4843	4857	4871	4886	4900	1	3	4	6	7	9	10	11	13
31	4914	4928	4942	4955	4969	4983	4997	5011	5024	5038	1	3	4	6	7	8	10	11	12
32	5051	5065	5079	5092	5105	5119	5132	5145	5159	5172	1	3	4	5	7	8	9	11	12
33	5185	5198	5211	5224	5237	5250	5263	5276	5289	5302	1	3	4	5	6	8	9	10	12
34	5315	5328	5340	5353	5366	5378	5391	5403	5416	5428	1	3	4	5	6	8	9	10	11
35	5441	5453	5465	5478	5490	5502	5514	5527	5539	5551	1	2	4	5	6	7	9	10	11
36	5563	5575	5587	5599	5611	5623	5635	5647	5658	5670	1	2	4	5	6	7	8	10	11
37	5682	5694	5705	5717	5729	5740	5752	5763	5775	5786	1	2	3	5	6	7	8	9	10
38	5798	5809	5821	5832	5843	5855	5866	5877	5888	5899	1	2	3	5	6	7	8	9	10
39	5911	5922	5933	5944	5955	5966	5977	5988	5999	6010	1	2	3	4	5	7	8	9	10
40	6021	6031	6042	6053	6064	6075	6085	6096	6107	6117	1	2	3	4	5	6	8	9	10
41	6128	6138	6149	6160	6170	6180	6191	6201	6212	6222	1	2	3	4	5	6	7	8	9
42	6232	6243	6253	6263	6274	6284	6294	6304	6314	6325	1	2	3	4	5	6	7	8	9
43	6335	6345	6355	6365	6375	6385	6395	6405	6415	6425	1	2	3	4	5	6	7	8	9
44	6435	6444	6454	6464	6474	6484	6493	6503	6513	6522	1	2	3	4	5	6	7	8	9
45	6532	6542	6551	6561	6571	6580	6590	6599	6609	6618	1	2	3	4	5	6	7	8	9
46	6628	6637	6646	6656	6665	6675	6684	6693	6702	6712	1	2	3	4	5	6	7	7	8
47	6721	6730	6739	6749	6758	6767	6776	6785	6794	6803	1	2	3	4	5	5	6	7	8
48	6812	6821	6830	6839	6848	6857	6866	6875	6884	6893	1	2	3	4	4	5	6	7	8
49	6902	6911	6920	6928	6937	6946	6955	6964	6972	6981	1	2	3	4	4	5	6	7	8
50	6990	6998	7007	7016	7024	7033	7042	7050	7059	7067	1	2	3	3	4	5	6	7	8
51	7076	7084	7093	7101	7110	7118	7126	7135	7143	7152	1	2	3	3	4	5	6	7	8
52	7160	7168	7177	7185	7193	7202	7210	7218	7226	7235	1	2	2	3	4	5	6	7	7
53	7243	7251	7259	7267	7275	7284	7292	7300	7308	7316	1	2	2	3	4	5	6	6	7
54	7324	7332	7340	7348	7356	7364	7372	7380	7388	7396	1	2	2	3	4	5	6	6	7

LOGARITHMS.

	0	1	2	3	4	5	6	7	8	9	1	2	3	4	5	6	7	8	9
55	7404	7412	7419	7427	7435	7443	7451	7459	7466	7474	1	2	2	3	4	5	5	6	7
56	7482	7490	7497	7505	7513	7520	7528	7536	7543	7551	1	2	2	3	4	5	5	6	7
57	7559	7566	7574	7582	7589	7597	7604	7612	7619	7627	1	2	2	3	4	5	5	6	7
58	7634	7642	7649	7657	7664	7672	7679	7686	7694	7701	1	1	2	3	4	4	5	6	7
59	7709	7716	7723	7731	7738	7745	7752	7760	7767	7774	1	1	2	3	4	4	5	6	7
60	7782	7789	7796	7803	7810	7818	7825	7832	7839	7846	1	1	2	3	4	4	5	6	6
61	7853	7860	7868	7875	7882	7889	7896	7903	7910	7917	1	1	2	3	4	4	5	6	6
62	7924	7931	7938	7945	7952	7959	7966	7973	7980	7987	1	1	2	3	3	4	5	6	6
63	7993	8000	8007	8014	8021	8028	8035	8041	8048	8055	1	1	2	3	3	4	5	5	6
64	8062	8069	8075	8082	8089	8096	8102	8109	8116	8122	1	1	2	3	3	4	5	5	6
65	8129	8136	8142	8149	8156	8162	8169	8176	8182	8189	1	1	2	3	3	4	5	5	6
66	8195	8202	8209	8215	8222	8228	8235	8241	8248	8254	1	1	2	3	3	4	5	5	6
67	8261	8267	8274	8280	8287	8293	8299	8306	8312	8319	1	1	2	3	3	4	5	5	6
68	8325	8331	8338	8344	8351	8357	8363	8370	8376	8382	1	1	2	3	3	4	4	5	6
69	8388	8395	8401	8407	8414	8420	8426	8432	8439	8445	1	1	2	2	3	4	4	5	6
70	8451	8457	8463	8470	8476	8482	8488	8494	8500	8506	1	1	2	2	3	4	4	5	6
71	8513	8519	8525	8531	8537	8543	8459	8555	8561	8567	1	1	2	2	3	4	4	5	5
72	8573	8579	8585	8591	8597	8603	8609	8615	8621	8627	1	1	2	2	3	4	4	5	5
73	8633	8639	8645	8651	8657	8663	8669	8675	8681	8686	1	1	2	2	3	4	4	5	5
74	8692	8698	8704	8710	8716	8722	8727	8733	8739	8745	1	1	2	2	3	4	4	5	5
75	8751	8756	8762	8768	8774	8779	8785	8791	8797	8802	1	1	2	2	3	3	4	5	5
76	8808	8814	8820	8825	8831	8837	8842	8848	8854	8859	1	1	2	2	3	3	4	5	5
77	8865	8871	8876	8882	8887	8893	8899	8904	8910	8915	1	1	2	2	3	3	4	4	5
78	8921	8927	8932	8938	8943	8949	8954	8960	8965	8971	1	1	2	2	3	3	4	4	5
79	8976	8982	8987	8993	8998	9004	9009	9015	9020	9025	1	1	2	2	3	3	4	4	5
80	9031	9036	9042	9047	9053	9058	9063	9069	9074	9079	1	1	2	2	3	3	4	4	5
81	9085	9090	9096	9101	9106	9112	9117	9122	9128	9133	1	1	2	2	3	3	4	4	5
82	9138	9143	9149	9154	9159	9165	9170	9175	9180	9186	1	1	2	2	3	3	4	4	5
83	9191	9196	9201	9206	9212	9217	9222	9227	9232	9238	1	1	2	2	3	3	4	4	5
84	9243	9248	9253	9258	9263	9269	9274	7279	9284	9289	1	1	2	2	3	3	4	4	5
85	9294	9299	9304	9309	9315	9320	9325	9330	9335	9340	1	1	2	2	3	3	4	4	5
86	9345	9350	9355	9360	9365	9370	9375	9380	9385	9390	1	1	2	2	3	3	4	4	5
87	9395	9400	9405	9410	9415	9420	9425	9430	9435	9440	0	1	1	2	2	3	3	4	4
88	9445	9450	9455	9460	9465	9469	9474	9479	9484	9489	0	1	1	2	2	3	3	4	4
89	9494	9499	9504	9509	9513	9518	9523	9528	9533	9538	0	1	1	2	2	3	3	4	4
90	9542	9547	9552	9557	9562	9566	9571	9576	9581	9586	0	1	1	2	2	3	3	4	4
91	9590	9595	9600	9605	9609	9614	9619	9624	9628	9633	0	1	1	2	2	3	3	4	4
92	9638	9643	9647	9652	9657	9661	9666	9671	9675	9680	0	1	1	2	2	3	3	4	4
93	9685	9689	9694	9699	9703	9708	9713	9717	9722	9727	0	1	1	2	2	3	3	4	4
94	9731	9736	9741	9745	9750	9754	9759	9763	9768	9773	0	1	1	2	2	3	3	4	4
95	9777	9782	9786	9791	9795	9800	9805	9809	9814	9818	0	1	1	2	2	3	3	4	4
96	9823	9827	9832	9836	9841	9845	9850	9854	9859	9863	0	1	1	2	2	3	3	4	4
97	9868	9872	9877	9881	9886	9890	9894	9899	9903	9908	0	1	1	2	2	3	3	4	4
98	9912	9917	9921	9926	9930	9934	9939	9943	9948	9952	0	1	1	2	2	3	3	4	4
99	9956	9961	9965	9969	9974	9978	9983	9987	9991	9996	0	1	1	2	2	3	3	3	4

ANTILOGARITHMS.

	0	1	2	3	4	5	6	7	8	9	1	2	3	4	5	6	7	8	9
·00	1000	1002	1005	1007	1009	1012	1014	1016	1019	1021	0	0	1	1	1	1	2	2	2
·01	1023	1026	1028	1030	1033	1035	1038	1040	1042	1045	0	0	1	1	1	1	2	2	2
·02	1047	1050	1052	1054	1057	1059	1062	1064	1067	1069	0	0	1	1	1	1	2	2	2
·03	1072	1074	1076	1079	1081	1084	1086	1089	1091	1094	0	0	1	1	1	1	2	2	2
·04	1096	1099	1102	1104	1107	1109	1112	1114	1117	1119	0	1	1	1	1	2	2	2	2
·05	1122	1125	1127	1130	1132	1135	1138	1140	1143	1146	0	1	1	1	1	2	2	2	2
·06	1148	1151	1153	1156	1159	1161	1164	1167	1169	1172	0	1	1	1	1	2	2	2	2
·07	1175	1178	1180	1183	1186	1189	1191	1194	1197	1199	0	1	1	1	1	2	2	2	2
·08	1202	1205	1208	1211	1213	1216	1219	1222	1225	1227	0	1	1	1	1	2	2	2	3
·09	1230	1233	1236	1239	1242	1245	1247	1250	1253	1256	0	1	1	1	1	2	2	2	3
·10	1259	1262	1265	1268	1271	1274	1276	1279	1282	1285	0	1	1	1	1	2	2	2	3
·11	1288	1291	1294	1297	1300	1303	1306	1309	1312	1315	0	1	1	1	2	2	2	2	3
·12	1318	1321	1324	1327	1330	1334	1337	1340	1343	1346	0	1	1	1	2	2	2	2	3
·13	1349	1352	1355	1358	1361	1365	1368	1371	1374	1377	0	1	1	1	2	2	2	3	3
·14	1380	1384	1387	1390	1393	1396	1400	1403	1406	1409	0	1	1	1	2	2	2	3	3
·15	1413	1416	1419	1422	1426	1429	1432	1435	1439	1442	0	1	1	1	2	2	2	3	3
·16	1445	1449	1452	1455	1459	1462	1466	1469	1472	1476	0	1	1	1	2	2	2	3	3
·17	1479	1483	1486	1489	1493	1496	1500	1503	1507	1510	0	1	1	1	2	2	2	3	3
·18	1514	1517	1521	1524	1528	1531	1535	1538	1542	1545	0	1	1	1	2	2	2	3	3
·19	1549	1552	1556	1560	1563	1567	1570	1574	1578	1581	0	1	1	1	2	2	3	3	3
·20	1585	1589	1592	1596	1600	1603	1607	1611	1614	1618	0	1	1	1	2	2	3	3	3
·21	1622	1626	1629	1633	1637	1641	1644	1648	1652	1656	0	1	1	2	2	2	3	3	3
·22	1660	1663	1667	1671	1675	1679	1683	1687	1690	1694	0	1	1	2	2	2	3	3	3
·23	1698	1702	1706	1710	1714	1718	1722	1726	1730	1734	0	1	1	2	2	2	3	3	4
·24	1738	1742	1746	1750	1754	1758	1762	1766	1770	1774	0	1	1	2	2	2	3	3	4
·25	1778	1782	1786	1791	1795	1799	1803	1807	1811	1816	0	1	1	2	2	2	3	3	4
·26	1820	1824	1828	1832	1837	1841	1845	1849	1854	1858	0	1	1	2	2	3	3	3	4
·27	1862	1866	1871	1875	1879	1884	1888	1892	1897	1901	0	1	1	2	2	3	3	3	4
·28	1905	1910	1914	1919	1923	1928	1932	1936	1941	1945	0	1	1	2	2	3	3	4	4
·29	1950	1954	1959	1963	1968	1972	1977	1982	1986	1991	0	1	1	2	2	3	3	4	4
·30	1995	2000	2004	2009	2014	2018	2023	2028	2032	2037	0	1	1	2	2	3	3	4	4
·31	2042	2046	2051	2056	2061	2065	2070	2075	2080	2084	0	1	1	2	2	3	3	4	4
·32	2089	2094	2099	2104	2109	2113	2118	2123	2128	2133	0	1	1	2	2	3	3	4	4
·33	2138	2143	2148	2153	2158	2163	2168	2173	2178	2183	0	1	1	2	2	3	3	4	4
·34	2188	2193	2198	2203	2208	2213	2218	2223	2228	2234	1	1	2	2	3	3	4	4	5
·35	2239	2244	2249	2254	2259	2265	2270	2275	2280	2286	1	1	2	2	3	3	4	4	5
·36	2291	2296	2301	2307	2312	2317	2323	2328	2333	2339	1	1	2	2	3	3	4	4	5
·37	2344	2350	2355	2360	2366	2371	2377	2382	2388	2393	1	1	2	2	3	3	4	4	5
·38	2399	2404	2410	2415	2421	2427	2432	2438	2443	2449	1	1	2	2	3	3	4	4	5
·39	2455	2460	2466	2472	2477	2483	2489	2495	2500	2506	1	1	2	2	3	3	4	5	5
·40	2512	2518	2523	2529	2535	2541	2547	2553	2559	2564	1	1	2	2	3	4	4	5	5
·41	2570	2576	2582	2588	2594	2600	2606	2612	2618	2624	1	1	2	2	3	4	4	5	5
·42	2630	2636	2642	2649	2655	2661	2667	2673	2679	2685	1	1	2	2	3	4	4	5	6
·43	2692	2698	2704	2710	2716	2723	2729	2735	2742	2748	1	1	2	3	3	4	4	5	6
·44	2754	2761	2767	2773	2780	2786	2793	2799	2805	2812	1	1	2	3	3	4	4	5	6
·45	2818	2825	2831	2838	2844	2851	2858	2864	2871	2877	1	1	2	3	3	4	5	5	6
·46	2884	2891	2897	2904	2911	2917	2924	2931	2938	2944	1	1	2	3	3	4	5	5	6
·47	2951	2958	2965	2972	2979	2985	2992	2999	3006	3013	1	1	2	3	3	4	5	5	6
·48	3020	3027	3034	3041	3048	3055	3062	3069	3076	3083	1	1	2	3	4	4	5	6	6
·49	3090	3097	3105	3112	3119	3126	3133	3141	3148	3155	1	1	2	3	4	4	5	6	6

ANTILOGARITHMS.

	0	1	2	3	4	5	6	7	8	9	1	2	3	4	5	6	7	8	9
·50	3162	3170	3177	3184	3192	3199	3206	3214	3221	3228	1	1	2	3	4	4	5	6	7
·51	3236	3243	3251	3258	3266	3273	3281	3289	3296	3304	1	2	2	3	4	5	5	6	7
·52	3311	3319	3327	3334	3342	3350	3357	3365	3373	3381	1	2	2	3	4	5	5	6	7
·53	3388	3396	3404	3412	3420	3428	3436	3443	3451	3459	1	2	2	3	4	5	6	6	7
·54	3467	3475	3483	3491	3499	3508	3516	3524	3532	3540	1	2	2	3	4	5	6	6	7
·55	3548	3556	3565	3573	3581	3589	3597	3606	3614	3622	1	2	2	3	4	5	6	7	7
·56	3631	3639	3648	3656	3664	3673	3681	3690	3698	3707	1	2	3	3	4	5	6	7	8
·57	3715	3724	3733	3741	3750	3758	3767	3776	3784	3793	1	2	3	3	4	5	6	7	8
·58	3802	3811	3819	3828	3837	3846	3855	3864	3873	3882	1	2	3	4	4	5	6	7	8
·59	3890	3899	3908	3917	3926	3936	3945	3954	3963	3972	1	2	3	4	5	5	6	7	8
·60	3981	3990	3999	4009	4018	4027	4036	4046	4055	4064	1	2	3	4	5	6	6	7	8
·61	4074	4083	4093	4102	4111	4121	4130	4140	4150	4159	1	2	3	4	5	6	7	8	9
·62	4169	4178	4188	4198	4207	4217	4227	4236	4246	4256	1	2	3	4	5	6	7	8	9
·63	4266	4276	4285	4295	4305	4315	4325	4335	4345	4355	1	2	3	4	5	6	7	8	9
·64	4365	4375	4385	4395	4406	4416	4426	4436	4446	4457	1	2	3	4	5	6	7	8	9
·65	4467	4477	4487	4498	4508	4519	4529	4539	4550	4560	1	2	3	4	5	6	7	8	9
·66	4571	4581	4592	4603	4613	4624	4634	4645	4656	4667	1	2	3	4	5	6	7	9	10
·67	4677	4688	4699	4710	4721	4732	4742	4753	4764	4775	1	2	3	4	5	7	8	9	10
·68	4786	4797	4808	4819	4831	4842	4853	4864	4875	4887	1	2	3	4	6	7	8	9	10
·69	4898	4909	4920	4932	4943	4955	4966	4977	4989	5000	1	2	3	5	6	7	8	9	10
·70	5012	5023	5035	5047	5058	5070	5082	5093	5105	5117	1	2	4	5	6	7	8	9	11
·71	5129	5140	5152	5164	5176	5188	5200	5212	5224	5236	1	2	4	5	6	7	8	10	11
·72	5248	5260	5272	5284	5297	5309	5321	5333	5346	5358	1	2	4	5	6	7	9	10	11
·73	5370	5383	5395	5408	5420	5433	5445	5458	5470	5483	1	3	4	5	6	8	9	10	11
·74	5495	5508	5521	5534	5546	5559	5572	5585	5598	5610	1	3	4	5	6	8	9	10	12
·75	5623	5636	5649	5662	5675	5689	5702	5715	5728	5741	1	3	4	5	7	8	9	10	12
·76	5754	5768	5781	5794	5808	5821	5834	5848	5861	5875	1	3	4	5	7	8	9	11	12
·77	5888	5902	5916	5929	5943	5957	5970	5984	5998	6012	1	3	4	5	7	8	10	11	12
·78	6026	6039	6053	6067	6081	6095	6109	6124	6138	6152	1	3	4	6	7	8	10	11	13
·79	6166	6180	6194	6209	6223	6237	6252	6266	6281	6295	1	3	4	6	7	9	10	11	13
·80	6310	6324	6339	6353	6368	6383	6397	6412	6427	6442	1	3	4	6	7	9	10	12	13
·81	6457	6471	6486	6501	6516	6531	6546	6561	6577	6592	2	3	5	6	8	9	11	12	14
·82	6607	6622	6637	6653	6668	6683	6699	6714	6730	6745	2	3	5	6	8	9	11	12	14
·83	6761	6776	6792	6808	6823	6839	6855	6871	6887	6902	2	3	5	6	8	9	11	13	14
·84	6918	6934	6950	6966	6982	6998	7015	7031	7047	7063	2	3	5	6	8	10	11	13	15
·85	7079	7096	7112	7129	7145	7161	7178	7194	7211	7228	2	3	5	7	8	10	12	13	15
·86	7244	7261	7278	7295	7311	7328	7345	7862	7879	7396	2	3	5	7	8	10	12	13	15
·87	7413	7430	7447	7464	7482	7499	7516	7534	7551	7568	2	3	5	7	9	10	12	14	16
·88	7586	7603	7621	7638	7656	7674	7691	7709	7727	7745	2	4	5	7	9	11	12	14	16
·89	7762	7780	7798	7816	7834	7852	7870	7889	7907	7925	2	4	5	7	9	11	13	14	16
·90	7943	7962	7980	7998	8017	8035	8054	8072	8091	8110	2	4	6	7	9	11	13	15	17
·91	8128	8147	8166	8185	8204	8222	8241	8260	8279	8299	2	4	6	8	9	11	13	15	17
·92	8318	8337	8356	8375	8395	8414	8433	8453	8472	8492	2	4	6	8	10	12	14	15	17
·93	8511	8531	8551	8570	8590	8610	8630	8650	8670	8690	2	4	6	8	10	12	14	16	18
·94	8710	8730	8750	8770	8790	8810	8831	8851	8872	8892	2	4	6	8	10	12	14	16	18
·95	8913	8933	8954	8974	8995	9016	9036	9057	9078	9099	2	4	6	8	10	12	15	17	19
·96	9120	9141	9162	9183	9204	9226	9247	9268	9290	9311	2	4	6	8	11	13	15	17	19
·97	9333	9354	9376	9397	9419	9441	9462	9484	9506	9528	2	4	7	9	11	13	15	17	20
·98	9550	9572	9594	9616	9638	9661	9683	9705	9727	9750	2	4	7	9	11	13	16	18	20
·99	9772	9795	9817	9840	9863	9886	9908	9931	9954	9977	2	5	7	9	11	14	16	18	20

TRIGONOMETRICAL TABLE.

Angle.	Radians.	Sine.	Tangent.	Cotangent.	Cosine.		
0°	0	0	0	∞	1	1·5708	90°
1	·0175	·0175	·0175	57·2900	·9998	1·5533	89
2	·0349	·0349	·0349	28·6363	·9994	1·5359	88
3	·0524	·0523	·0524	19·0811	·9986	1·5184	87
4	0698	·0698	·0699	14·3006	·9976	1·5010	86
5	·0873	·0872	·0875	11·4301	·9962	1·4835	85
6	·1047	·1045	·1051	9·5144	·9945	1·4661	84
7	·1222	·1219	·1228	8·1443	·9925	1·4486	83
8	·1396	·1392	·1405	7·1154	·9903	1·4312	82
9	·1571	·1564	·1584	6·3138	·9877	1·4137	81
10	·1745	·1736	·1763	5·6713	·9848	1·3963	80
11	·1920	·1908	·1944	5·1446	·9816	1·3788	79
12	·2094	·2079	·2126	4·7046	·9781	1·3614	78
13	·2269	·2250	·2309	4·3315	·9744	1·3439	77
14	·2443	·2419	·2493	4·0108	·9703	1·3265	76
15	·2618	·2588	·2679	3·7321	·9659	1·3090	75
16	·2793	·2756	·2867	3·4874	·9613	1·2915	74
17	·2967	·2924	·3057	3·2709	·9563	1·2741	73
18	·3142	·3090	·3249	3·0777	·9511	1·2566	72
19	·3316	·3256	·3443	2·9042	·9455	1·2392	71
20	·3491	·3420	·3640	2·7475	·9397	1·2217	70
21	·3665	·3584	·3839	2·6051	·9336	1·2043	69
22	·3840	·3746	·4040	2·4751	·9272	1·1868	68
23	·4014	·3907	·4245	2·3559	·9205	1·1694	67
24	·4189	·4067	·4452	2·2460	·9135	1·1519	66
25	·4363	·4226	·4663	2·1445	·9063	1·1345	65
26	·4538	·4384	·4877	2·0503	·8988	1·1170	64
27	·4712	·4540	·5095	1·9626	·8910	1·0996	63
28	·4887	·4695	·5317	1·8807	·8830	1·0821	62
29	·5061	·4848	·5543	1·8040	·8746	1·0647	61
30	·5236	·5000	·5774	1·7321	·8660	1·0472	60
31	·5411	·5150	·6009	1·6643	·8572	1·0297	59
32	·5585	·5299	·6249	1·6003	·8480	1·0123	58
33	·5760	·5446	·6494	1·5399	·8387	·9948	57
34	·5934	·5592	·6745	1·4826	·8290	·9774	56
35	·6109	·5736	·7002	1·4281	·8192	·9599	55
36	·6283	·5878	·7265	1·3764	·8090	·9425	54
37	·6458	·6018	·7536	1·3270	·7986	·9250	53
38	·6632	·6157	·7813	1·2799	·7880	·9076	52
39	·6807	·6293	·8098	1·2349	·7771	·8901	51
40	·6981	·6428	·8391	1·1918	·7660	·8727	50
41	·7156	·6561	·8693	1·1504	·7547	·8552	49
42	·7330	·6691	·9004	1·1106	·7431	·8378	48
43	·7505	·6820	·9325	1·0724	·7314	·8203	47
44	·7679	·6947	·9657	1·0355	·7193	·8029	46
45	·7854	·7071	1·0000	1·0000	·7071	·7854	45
		Cosine.	Cotangent.	Tangent.	Sine.	Radians.	Angle.

ANSWERS.

Chapter I., p. 11.

1. 2·908 metres. 2. 9 feet 7·74 inches. 3. 5·129 kilometres.
4. 2·114 inches. 5. 12·58 square inches. 6. 53 square cms.
7. 7·3 square inches. 8. (*a*) 44 cms.; (*b*) 154 square cms.
9. 381·9 cubic inches. 10. 5·44 square inches. 11. 14·5 square inches.

Chapter II., p. 19.

1. 6·72 lbs. 2. 2·21 lbs. 3. 277 lbs. 4. 17·26. lbs.
5. 82·9. lbs. 6. 17·49 lbs. 7. 3·27 lbs. 8. 3·05 lbs.
9. 14·4 lbs. 10. 12·4 tons. 11. 496 lbs. 12. 8·7″.

Chapter III., p. 30.

2. 5·7 lbs. 3. 5 lbs. 4. 29 lbs.
5. 21 lbs., acting towards the left.
6. (*a*) 11·7 lbs.; (*b*) 8·7 lbs.; (*c*) 14 lbs.
7. 19·5 lbs. 8. 14·5 lbs. 10. 8·96 lbs.; 7·32 lbs.
11. 9·65 cwts. 13. 108·6 lbs.; 51 lbs. 14. 160 lbs.; 81 lbs.
15. 10 lbs.; 17·32 lbs.

Chapter IV., p. 40.

1. Push in AB = 0·16 ton; push in AC = 0·89 ton.
2. Push in jib = 7·12 lbs.; pull in tie = 5·75 lbs.
3. Push in jib = 5·28 tons; pull in tie = 4·36 tons.
4. Push in jib = 7·78 tons; pull in tie = 4·86 tons.
5. Pull in AC = 0·16 ton; push in BC = 0·83 ton.
6. Push in AC = 1·16 tons; pull in BC = 0·83 ton.

7. Pull in AB =3·0 tons; pull in BC =2·425 tons;
push in CD =0·97 ton; push in BD =2·55 tons.
8. Push in each leg=21 tons; pull in back leg=23 tons.
9. Pull in AP =2000 lbs; push in BP =3464 lbs.
11. (*a*) 4·66 lbs.; (*b*) 4·23 lbs.; (*c*) 4·88 lbs.

Chapter V., p. 53.

1. At 25″ from C, on the other side of the pivot from W.
2. 69·2 lbs. 3. P=64·1 lbs. 4. At 2·25 feet from the 3 lb. load.
6. 0·833 ton; 0·417 ton. 7. 3·55 tons; 2·45 tons.
8. 1·33 tons at A; 4·66 tons at C.
10. (*a*) 2400 lbs.; (*b*) 5376 lbs. 11. 41·8″.

Chapter VI., p. 65.

1. 345 lbs.; 245 lbs. 2. 5·417 tons; 4·583 tons.
3. 200 lbs.; 100 lbs. 4. 663 lbs. 5. 67 lbs.
6. 1600 lb.-feet; 707 lb.-feet; wall will fall.
8. 536 lbs. 9. 13·25 tons; 9·75 tons.
10. On the shorter portion of the beam, 2·24 ft. from the pivot.
11. 1·77 ft. from the heavier end of the ladder.
12. Pull at middle=25 lbs.; pull at each end=26·9 lbs.
13. 177·5 lbs.; 216·9 lbs. 14. 25·3 lbs.; tension increases.

Chapter VII., p. 86.

1. 2·8 tons per sq. inch. 2. 12·5 tons.
3. Strain=0·00083; E=30,000,000 lbs. per sq. inch.
4. 0·504″. 5. 0·0556″. 6. 0·076 lb.
7. 9062 lbs. 8. 29·46 tons; 66·6 tons per sq. inch.
9. 9·82 tons; 199·8 tons per sq. inch.
10. E=15,700,000 lbs. per sq. inch. 11. 0·747 sq. inch.
13. 3600 lbs. per sq. inch; 0·000125; 28,800,000 lbs. per sq. inch.

Chapter VIII., p. 107.

1. (*a*) M=0; S=300 lbs. (*b*) M=1500 lb.-ft.; S=300 lbs.
2. (*a*) M=3000 lb.-ft.; S=500 lbs. (*b*) M=1500 lb.-ft.; S=500 lbs.
3. Bending moments,—at middle, 15 ton-feet; at each 1 ton load, 10 ton-feet. Shearing force=1 ton.

4.

Place.	Bending Moment.	Shearing Force.
At wall	1600 lb.-ft.	400 lbs.
2 ft. from wall	900 ,,	300 ,,
4 ft. ,,	400 ,,	200 ,,
6 ft. ,,	100 ,,	100 ,,
8 ft. ,,	0	0

5. At middle—$M = 3600$ lb.-feet; $S = 0$.
At 3 feet from each end, $M = 2500$ lb.-feet; $S = 600$ lbs.
6. 1·375 tons. 7. 2·875 tons. 8. 18·16 tons; 4·83 tons.
9. 122·4 ton-feet; 5·1 sq. inches.
10. Tensile stress = 1·66 tons per sq. inch; compressive stress = 8·33 tons per sq. inch.
11. 5·21 cwts. 12. 1·8 tons; 3·6 tons. 13. 4·77 tons per sq. in.
14. 30,190,000 lbs. per sq. inch.

Chapter IX., p. 121.

1. 24·75 tons. 2. $\frac{13}{16}''$; $3\frac{1}{8}''$; 74 per cent. 3. (*a*) 0·5″; (*b*) $\frac{11}{16}''$.
4. 1″. 5. 1·42 : 1. 6. 7636 lbs. per sq. inch. 7. 3·126″.
8. 785·7 ton-inches; 29·1 tons. 9. $M : T = 1 : 2$.
11. (*a*) 15710 lb.-inches; (*b*) 0·426″. 12. 2·28″.

Chapter X., p. 133.

1. 600 ft.-tons. 2. 135,000 ft.-lbs. 3. 697,000 ft.-lbs.
4. 375 ft.-tons. 5. 3400 ft.-lbs. 6. (*a*) 140 ft.-lbs.; (*b*) 220 ft.-lbs.
7. 2,250,000 ft.-lbs. 8. 324 ft.-lbs. 9. 24,550,000 ft.-lbs.
10. 7500 ft.-lbs. 11. 463,200 ft.-lbs. 12. 15,440,000 ft.-lbs.
13. 0·64 H.P. 14. 318 H.P. 15. 101 H.P.
16. 19·8 per cent. 17. 160 H.P. 18. 0·215 inch-ton.
19. 3,780,000 ft.-lbs. 20. 0·091 H.P.; 0·68 H.P. 21. 288,000 ft.-lbs.
22. 247,000 ft.-lbs.; 112,700 ft.-lbs.; 134,300 ft.-lbs.

Chapter XI., p. 151.

1. 0·266. 2. 0·433. 3. 0·288. 4. 147 lbs.
5. 1530 lbs.; 1·85 H.P. 6. 3·28 feet. 7. 416 lbs.; 166 lbs.
8. 1800 lbs.; 144 H.P. 9. 0·8 H.P.; 34·2 B.T.U. 10. 0·179.
11. 423·4 ft.-lbs. 13. 192 H.P.; 480 H.P.

Chapter XII., p. 164.

1. 440 feet.
2. 32·7 miles per hour.
3. 17·4 feet per sec.
4. 47 miles per hour.
5. 0·084 feet per sec. per sec.
6. 12 feet.
7. 1·703 tons.
8. 400 feet per min.
9. 1863 lbs.
10. 1500 ft.-lbs.
11. 49·7 ft.-lbs.
12. 265,600 ft.-tons.
14. 10 feet per sec.
15. 43 feet per sec.
16. Heights fallen—9·82 feet; 0·9676 feet; 0·0966 feet.
 Average velocities—98·21 ft. per sec.; 96·76 ft. per sec.; 96·616 ft. per sec.
17. 1770 ft.-lbs.

Chapter XIII., p. 192.

1. (*a*) 212; 318. (*b*) 207·8; 305·5 revs. per min.
2. 14·5 H.P.
5. 25; 50.
7. 14·66 feet per min.
8. *B*, 3 clockwise; *C*, $\frac{1}{3}$ anticlockwise; *D*, $3\frac{2}{3}$ clockwise.
9. 80 on leading screw; 45 on pin, gearing with the 80 wheel; 60 on same pin, gearing with the 20 wheel.
10. (*a*) 628·6 feet per min.; (*b*) 400 feet per min.;

(*c*)

Crank angle	0°	30°	60°	90°	120°	150°	180°
Piston speed, feet per min.	0	392	624	628·6	464	236	0

12. 2·25″.
13. 19·1″; 22 revs.
16. 2514 feet per min.; 262 lbs.
17. 10″.
18. $\frac{1}{4}$″.

Chapter XIV., p. 209.

1. 46·6 lbs.
2. 15.
3. 238 lbs.
4. 48.
5. 940 degrees.
6. 71·6 lbs.; 301·6 lbs.

Chapter XV., p. 231.

1. 2·86.
2. 5·62.
3. 240.
4. 86·4 per cent.; 29040 ft.-lbs.
5. 896,100 ft.-lbs.
6. 84·7 revs. per min.
7. 2,740 lbs.
8. 5 tons weight.
9. 62,100 lbs. weight.
10. 3,882 lbs. weight.
11. 5·24 tons weight.
12. 100, 320 lb. ft.-sec.; 1,557 lbs weight; 8·8 feet.
13. 126,200 ft.-lbs.
14. 12·9 lbs. weight.
15. 150,000 ft.-lbs.
16. 1·03 lbs. weight.
17. 11,880 ft.-lbs.
18. (*a*) 42,520 ft.-lbs.; (*b*) 75 revs.; (*c*) 567 ft.-lbs.

Chapter XVI., p. 245.

1. 1,325 lbs. **2.** 15 lbs. **3.** 234·4 lb.-feet.
4. 428 ft. per sec. **5.** 0·8 foot; 0·6 inch.

6.

Angle.	0° 360°	330° 30°	60° 300°	90° 270°	120° 240°	150° 210°	180°
Acceleration, ft. sec. sec.	+200	+173·2	+100	0	−100	−173·2	−200

7. 39·12 inches. **8.** 1·7 tons.

Chapter XVII., p. 266.

1. 30,720 lbs.; 9,600 lbs.; 6,400 lbs. **2.** 22,500 lbs.; 45,000 lb.-feet.
3. 3·25 feet. **4.** 0·316 cubic foot. **5.** 33·9 feet.
6. 44,000 ft.-lbs.; 2·2 H.P. **7.** 6·19 ft.; 9,000 lbs.; 8,400 lbs.; 4,000 lbs.
8. 55,000 lbs.; 440,000 ft. lbs.; 4·36 cubic feet. **9.** 166·6 lbs.
11. 31,250 lbs. **12.** 108,000 ft.-lbs.; 17,280 ft.-lbs.; 960,000 ft.-lbs.
13. 470; 425; 90·4 per cent. **14.** 0·033 H.P.
15. 201,600 ft.-lbs.; 107,520 ft.-lbs.
16. 0·41 ton per sq. inch; 1,950 ft.-tons. **17.** 1,017 lbs.
18. 3·21 H.P.; 67 per cent.

Chapter XVIII., p. 290.

1. 78·8 gallons per min. **2.** 276 cubic feet per min.
3. 102 cubic feet per min. **4.** 349,000 gallons per hour.
5. 11·8 H.P. **6.** 40·7 feet per sec. **7.** 20 feet per sec.
8. 3·48 feet per sec. **9.** The pressure in the 12″ part exceeds that in the 3″ part by 70 lbs. per sq. inch. **10.** 48,830 ft.-lbs.

INDEX.

GLASGOW: PRINTED AT THE UNIVERSITY PRESS BY ROBERT MACLEHOSE AND CO.

XII. GEOLOGY.

Geology for Beginners. By W. W. WATTS, M.A., F.G.S. Adapted to the Elementary Stage of the South Kensington Syllabus. 2*s*. 6*d*.

XIV. HUMAN PHYSIOLOGY.

Physiology for Beginners. By Sir MICHAEL FOSTER and Dr. L. E. SHORE. Adapted to the Elementary Stage of the South Kensington Syllabus. 2*s*. 6*d*.

Lessons in Elementary Physiology. By the Right Hon. T. H. HUXLEY, F.R.S. 4*s*. 6*d*. Questions, 1*s*. 6*d*.

XVI. ZOOLOGY.

A Manual of Zoology. By the late T. JEFFERY PARKER, D.Sc., F.R.S., and W. A. HASWELL, M.A., D.Sc., F.R.S. Illustrated. 10*s*. 6*d*.

XVII. BOTANY.

Botany for Beginners. Adapted to the Elementary Stage of the South Kensington Syllabus. By ERNEST EVANS, Burnley Technical School. 2*s*. 6*d*.

XVIII. PRINCIPLES OF MINING.

Coal Mining for Beginners. By MORGAN W. DAVIES, M.I.M.E. [*In the press.*

XIX. METALLURGY.

A Text-Book of Elementary Metallurgy. By A. H. HIORNS, Principal of the School of Metallurgy, Birmingham and Midland Institute. 3*s*. Questions, 1*s*.

XXII. STEAM.

Steam, Gas, and Oil Engines. By JOHN PERRY, F.R.S. 7*s*. 6*d*. net.

XXIII. PHYSIOGRAPHY.

Experimental Science (Section I. Physiography). By Prof. R. A. GREGORY and A. T. SIMMONS, B.Sc. 2*s*. 6*d*.

Physiography for Beginners. By A. T. SIMMONS, B.Sc. 2*s*. 6*d*. Adapted to the Elementary Stage of the South Kensington Syllabus.

Physiography for Advanced Students. By A. T. SIMMONS, B.Sc. 4*s*. 6*d*.

Elementary Lessons in Astronomy. By Sir NORMAN LOCKYER. New Edition. 5*s*. 6*d*. This book contains all the Astronomy required for the Advanced and Honours.

XXIV. THE PRINCIPLES OF AGRICULTURE.

Elementary Lessons in the Science of Agricultural Practice. By H. TANNER, F.C.S. 3*s*. 6*d*.

XXV. HYGIENE.

Hygiene for Beginners. By E. S. REYNOLDS, M.D. Adapted to the Elementary Stage of the South Kensington Syllabus. 2*s*. 6*d*.

Experimental Hygiene (Section I.). By A. T. SIMMONS, B.Sc., and E. STENHOUSE, B.Sc. 2*s*. 6*d*.

Hygiene for Students. By E. F. WILLOUGHBY, M.B. Adapted to the Advanced Stage of the South Kensington Syllabus. Being a New and Revised Edition. 4*s*. 6*d*.

MACMILLAN AND CO., LTD., LONDON

Spotted something that isn't quite right?

Let us know and we'll fix it for future reprints. And as a thank you for your diligence, we'll send you a free Sothis book of your choice.

publisher@sothis.press

This page intentionally left blank

Made in the USA
Monee, IL
08 November 2026

67812149R00204